Applied Mathematics Toolkit

Modern scientific and engineering systems are increasingly defined by nonlinearity, high dimensionality, partial observability, and uncertainty. Addressing such complexity requires more than isolated techniques. It calls for an integrated applied mathematics perspective that connects modeling, data analysis, and computation within a coherent framework. This book develops such a unified toolkit. Rather than treating mathematical methods as separate disciplines, it emphasizes their complementary roles in understanding, analyzing, and predicting complex systems, with a focus on intuition, practical relevance, and transferability across scientific and engineering fields.

- Presents a broad collection of essential applied mathematics tools, organized to emphasize their interplay rather than treating them as isolated topics.
- Each method is introduced through simple yet illuminating examples that make advanced concepts accessible without sacrificing rigor.
- For every technique, the presentation highlights the underlying motivation, key assumptions, strengths, limitations, and the contexts in which it is most effective.
- Theory, data, and algorithms are consistently integrated to demonstrate how they work together in practical applications.
- Is accompanied by short "one concept-one example" videos, each 5–10 minutes long, focusing on a single idea and using deliberately simple examples to support initial exposure, intuition building, and efficient review.
- Example codes included to reinforce the concepts and facilitate hands-on learning.

This book is intended for advanced undergraduate students, graduate students, and researchers in applied mathematics, science, and engineering who seek practical mastery of modern mathematical tools. It is well suited to topic courses and self-study, particularly for readers working at the intersection of modeling, data, and computation. By combining conceptual clarity with hands-on examples and an integrated methodology, the book provides a foundation for tackling real-world problems in a wide range of interdisciplinary settings.

Applied Mathematics Toolkit

Modeling, Data, and Algorithms for Scientists and Engineers

Nan Chen and Charlotte Moser

CRC Press
Taylor & Francis Group
Boca Raton London New York

CRC Press is an imprint of the
Taylor & Francis Group, an **informa** business
A CHAPMAN & HALL BOOK

Designed cover image: Shutterstock

First edition published 2027
by CRC Press
2385 NW Executive Center Drive, Suite 320, Boca Raton FL 33431

and by CRC Press
4 Park Square, Milton Park, Abingdon, Oxon, OX14 4RN

CRC Press is an imprint of Taylor & Francis Group, LLC

ISBN: 978-1-041-23826-3 (hbk)
ISBN: 978-1-041-23824-9 (pbk)
ISBN: 978-1-003-73845-9 (ebk)

DOI: 10.1201/9781003738459

Typeset in CMR10 font
by KnowledgeWorks Global Ltd.

Publisher's note: This book has been prepared from camera-ready copy provided by the authors.

Contents

Preface

Over the past several decades, scientific and engineering disciplines have undergone a profound transformation. The rapid growth of computational power, the explosion of data availability, and the increasing ambition of predictive modeling have fundamentally reshaped how we study complex systems. Yet alongside these advances, fundamental challenges persist. Many complex systems are nonlinear, multiscale, partially observed, and subject to inherent uncertainty. Addressing these challenges requires not only accurate models and efficient algorithms but also coherent and integrative mathematical thinking.

This book grew out of years of teaching, research, and collaboration across applied mathematics, geophysical sciences, engineering, and data science. In numerous conversations with students and colleagues, a recurring challenge in teaching applied mathematics became evident: many powerful ideas are first introduced through examples that are already too complex for initial understanding. The central motivation of this book is to present these areas not as disconnected techniques but as components of an integrated applied mathematics toolkit. Modeling, inference, uncertainty quantification, and computation are deeply intertwined. When these tools are combined thoughtfully and systematically, they reinforce one another and lead to more robust and interpretable solutions.

The book is structured around five foundational toolkits (statistical methods, machine learning, dynamical systems, applied analysis, and stochastic processes) followed by chapters that demonstrate how these foundations converge in data assimilation, information-theoretic uncertainty quantification, causal inference, and reduced-order modeling. The aim is to provide both a coherent conceptual framework and practical mathematical tools that readers can adapt to a wide range of scientific and engineering problems.

Throughout the book, carefully designed examples play a central role in developing intuition and clarity. Rather than introducing concepts through overly complicated formulations, many ideas are first illustrated using deliberately simple settings. This progression, from accessible examples to more general frameworks, helps reveal the underlying structure shared across seemingly different methodologies.

The book is largely self-contained and can be followed with a modest mathematical background. It is intended for advanced undergraduate students, graduate students, postdoctoral researchers, and early-career researchers in applied mathematics, physics, engineering, and related disciplines. It may serve either as a textbook for undergraduate or graduate courses or as a reference for researchers seeking a coherent mathematical perspective on modern modeling and inference.

To further support learning, MATLAB® codes are provided to validate concepts and reproduce results from the illustrative examples in the book. In parallel, a short video series built around the simple principle "one concept · one example" accompanies the toolkit chapters (Chapters 2–6). Each video is 5–10 minutes long, focuses on a single idea, and uses deliberately simple examples to support initial exposure, intuition building, and efficient review. The corresponding QR codes for accessing these videos are provided below.

Madison, Wisconsin,
2026 *Nan Chen and Charlotte Moser*

Acknowledgment

We would like to thank Dr. Jeffrey Covington, Mr. Zhang Jiang, and Mr. Marios Andreou for their technical support in developing several examples presented in this book. We are also grateful to Dr. Stephen Wiggins, Dr. Reza Malek-Madani, and Dr. Robert Martin for many insightful discussions over the past few years. These interactions played an important role in shaping our approach of presenting complex applied mathematical concepts through simple and illustrative examples. This work, along with its associated teaching and research components, was supported by the Office of Naval Research (N00014-24-1-2244) and the Army Research Office (W911NF-23-1-0118).

We also thank Ms. Sherry Thomas, Senior Editorial Assistant at Taylor & Francis-CRC Press, and Dr. Daina Habdankaite, Mathematics Editor at Chapman & Hall/CRC Press, Taylor & Francis Group, for their guidance and support throughout the preparation and publication of this book.

Authors

Nan Chen is Associate Professor in the Department of Mathematics at the University of Wisconsin-Madison and is affiliated with the Institute for Foundations of Data Science. He received his PhD from the Courant Institute of Mathematical Sciences at New York University in Applied Mathematics and Atmospheric and Ocean Sciences, followed by two years as a Postdoctoral Research Associate at Courant. Dr. Chen's research spans applied mathematics, atmospheric and ocean sciences, materials science, and data science. He has received several major honors, including the Kurt O. Friedrichs Prize, the Silver Medal of the New World Mathematics Awards, the Office of Naval Research Young Investigator Award, and multiple teaching awards. He has published over 100 papers in leading journals, including *Proceedings of the National Academy of Sciences*, *Nature Geoscience*, *Nature Communications*, and *Notices of the American Mathematical Society*. Dr. Chen has delivered more than 100 invited presentations worldwide, and his work has been featured in *SIAM News*, *SIAM DSWeb*, *Eos*, and many other media outlets. He has organized nearly 30 workshops and mini-symposia at major conferences. Dr. Chen has mentored postdoctoral researchers, PhD students, and undergraduates, many of whom have received major research and presentation awards. He is committed to supporting early-career scientists and promoting interdisciplinary and international collaboration at the interface of applied mathematics, engineering, and geoscience. He currently serves as Secretary of SIAM Mathematics of Planet Earth and Chair of the AGU session "Applied Math Perspectives on Modeling, Analyzing, and Predicting Complex Geophysical Systems", playing an active role in shaping interdisciplinary efforts at the interface of applied mathematics and other disciplines.

Charlotte Moser is a PhD researcher in Mathematics at the University of Wisconsin-Madison, where she works as a research assistant at the interface of applied mathematics and environmental science. Her research focuses on mathematical modeling, uncertainty quantification, data assimilation, machine learning, Bayesian inference, and the analysis of complex dynamical systems, with applications to atmospheric and oceanic phenomena. She received her B.S. in Applied Mathematics, with minors in Environmental Science and Psychology. Ms. Moser has received multiple honors, including the DANOC Best Student Presentation Award, the Henry Schaerf Mathematics Graduate Award, and the AGU Outstanding Student Presentation Award, as well as the Dean's Distinguished Fellowship, the Mathematics Honors Program distinction, the IBA Biomathematics Education with Applications and Methods Grant, and the JD and Marcia Wine Mathematics Award. She has presented her research at leading national and international conferences and is emerging as a young researcher at the intersection of applied mathematics and environmental science. In addition to her research, she is actively engaged in scientific communication and education. She develops accessible content on applied mathematics and Earth science through online platforms, aiming to broaden participation and foster interdisciplinary learning. She is also committed to broadening participation in STEM and fostering collaborative scientific communities. Her work is driven by a strong interest in connecting mathematical theory with real-world environmental challenges.

1

Introduction

1.1 Challenges in modern modeling and inference: The need for integrated mathematical toolkits

Modern scientific and engineering systems are increasingly characterized by complexity, nonlinearity, partial observability, and uncertainty. Examples range from atmospheric and oceanic systems to engineered devices, biological networks, and socio-technical systems. In such settings, the governing dynamics are often nonlinear and multiscale, observations are incomplete and noisy, and model error is unavoidable. These features fundamentally limit the effectiveness of traditional modeling and inference approaches.

Owing to the intrinsic complexity of these problems, no single mathematical discipline is sufficient to address the challenges outlined above. Standard dynamical systems theory provides powerful insight into qualitative behavior and long-term dynamics, but often neglects uncertainty. Probability and statistics offer principled tools for quantifying uncertainty, yet are frequently disconnected from the underlying dynamical structure. Machine learning excels at extracting patterns from data, but when applied in isolation, it can suffer from limited interpretability, robustness, and generalization. Standard analysis tools provide useful theoretic guidance but rely critically on appropriate statistical and dynamical assumptions. As a result, effective solutions require integrating tools across multiple areas of applied mathematics.

This book, therefore, emphasizes a coherent applied mathematics framework in which modeling, data analysis, and computation are not treated as separate disciplines, but as complementary components of a unified toolkit. By combining ideas from dynamical systems, statistics, stochastic processes, information theory, machine learning, and numerical analysis, it becomes possible to advance uncertainty quantification, optimal inference, and structure-aware algorithm design.

Several examples illustrate the benefits of such integration. Stochastic modeling naturally combines statistical tools with dynamical systems to describe the time evolution of probability distributions, represent unresolved processes, and effectively quantify the uncertainty. Physics-informed machine learning incorporates physical structure or prior knowledge into data-driven models, improving interpretability and robustness. Data assimilation integrates dynamical models with observational data and often outperforms purely data-driven methods, particularly in regimes with limited or noisy observations.

More broadly, applied mathematics spans a wide range of areas, from classical disciplines such as dynamical systems and analysis to modern tools in statistics, stochastic modeling, uncertainty quantification, and machine learning. Integrating these tools not only expands the scope of problems that can be addressed but also allows the development of new methodologies. For example, improving computational efficiency does not necessarily mean the development of higher-order numerical schemes. Designing appropriate reduced-order or approximate models with desirable mathematical structures can significantly facilitate the development of efficient computational algorithms. Similarly, identifying dominant patterns

DOI: 10.1201/9781003738459-1

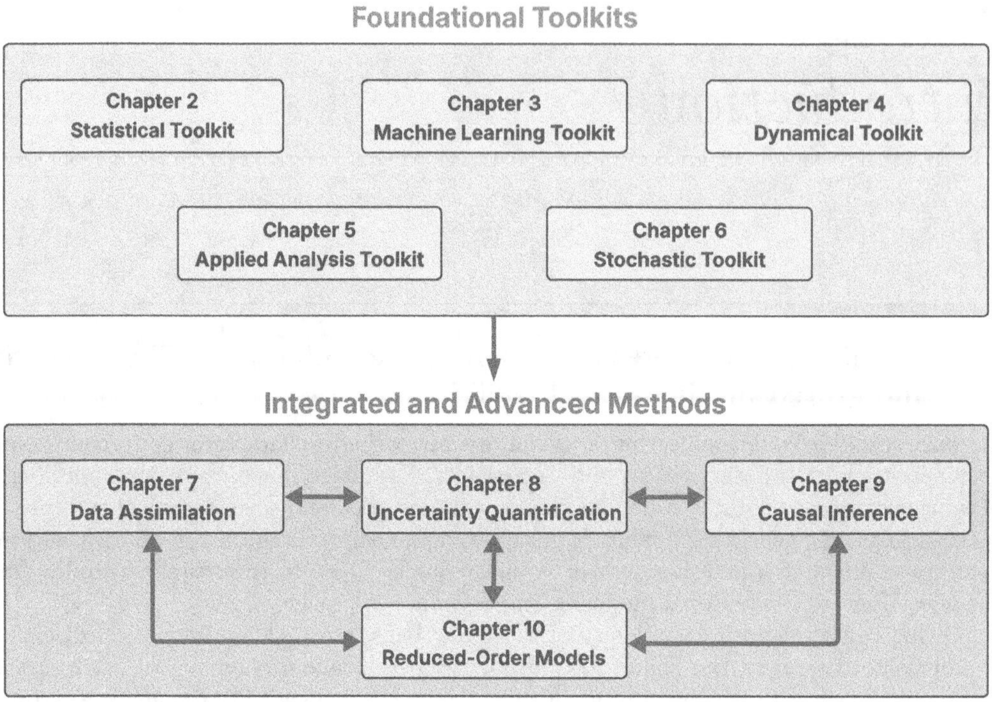

FIGURE 1.1: Overview of the organization of the book.

through dimensionality reduction can simplify model construction and enhance understanding of complex system behavior.

1.2 Organization of the book

The book is organized into a sequence of chapters that introduce the essential toolkits of applied mathematics and progressively build toward integrated and advanced methods. See Figure 1.1 for an overview.

Chapters 2–6 present five foundational toolkits, including statistical methods, machine learning techniques, dynamical systems and differential equations, applied analysis, and stochastic processes. These chapters provide the mathematical building blocks needed to address a wide range of problems and to develop advanced techniques in science and engineering.

Chapters 7–10 focus on inference, uncertainty quantification, and reduced-order modeling. Data assimilation is introduced as a systematic framework for integrating models and data, followed by information-theoretic approaches to uncertainty quantification and experimental design, as well as causal inference methods for identifying directed relationships in complex systems. Reduced-order modeling has broad applications in science and engineering. These chapters emphasize effective strategies that preserve essential dynamical and statistical features while enabling efficient forecasting, data assimilation, and uncertainty quantification. These topics are not independent. Instead, they strongly rely on one another and on the foundational tools developed in earlier chapters.

1.3 Audience and how to use this book

This book is designed for STEM students, postdoctoral researchers, and early-career researchers, particularly those working in applied science and engineering fields such as atmospheric and ocean sciences, geophysics, fluid mechanics, materials science, neuroscience, and related disciplines.

The book is suitable for advanced undergraduate and graduate-level topic courses in STEM disciplines. It is particularly appropriate for advanced undergraduates in applied mathematics, physics, or engineering programs, as well as graduate students across science and engineering fields who require practical mathematical tools. Readers are assumed to have a working knowledge of calculus, linear algebra, and basic probability. Familiarity with numerical computation is helpful but not required. By placing greater emphasis on Chapters 2–6, the book can be used as a textbook for foundational courses. The material in Chapters 7–10 is well suited for use as a reference and for specialized courses in applied mathematical methods, computational modeling, or data-driven approaches in science and engineering.

The chapters are largely modular. Readers interested primarily in inference may focus on the data assimilation, information theory, and causal inference chapters, while those interested in modeling and computation may emphasize the dynamical systems, stochastic processes, and reduced-order modeling chapters. Together, the chapters form a flexible toolkit that can be adapted to a wide range of scientific and engineering applications.

Throughout the book, simple examples are utilized to facilitate the understanding of concepts and methods and help build intuitions.

2

Statistical Toolkit

2.1 Foundations of probability

Throughout this book, vectors are often written in bold font while scalars are in plain or italic font.

2.1.1 Random variables and probability measure

A random variable $\mathbf{X}$ is a fundamental concept in probability and statistics that assigns numerical values to the outcomes of a random experiment [210, 158]. It is a function that maps outcomes from a sample space Ω to a measurable space E. In many cases, $\mathbf{X}$ is real-valued, i.e., $E = \mathbb{R}$, the one-dimensional real space. When the image of $\mathbf{X}$, namely the set of all possible numerical values that $\mathbf{X}$ can take, is finite or countably infinite, the random variable is called a discrete random variable, and its distribution is a discrete probability distribution. In such a case, it can be described by a probability mass function (PMS) that assigns a probability to each value in the image of $\mathbf{X}$. On the other hand, if the image is uncountably infinite (usually an interval or the entire real line), then $\mathbf{X}$ is a continuous random variable, and if absolutely continuous, its distribution is described by a probability density function (PDF), where probabilities are defined over intervals and individual points have zero probability.

Following standard probabilistic notation, the uppercase $\mathbf{X}$ denotes the random variable itself (a function mapping outcomes to values), while the lowercase $\mathbf{x}$ represents a specific realization or observed value of $\mathbf{X}$.

Example 2.1 (Discrete and continuous random variables). Example of a discrete random variable: Number of heads in two coin tosses. Possible values are $0, 1, 2$. Example of a continuous random variable: Height of a randomly selected adult. Possible values are any real number in a plausible range (e.g., 150 cm to 200 cm). □

Since many natural phenomena are modeled by continuous random variables, understanding their associated PDF is essential. Let $\mathbf{X} = (X_1, \ldots, X_n)^{\mathrm{T}}$ be an n-dimensional continuous random vector with joint PDF $p(\mathbf{x})$, where $\mathbf{x} = (x_1, \ldots, x_n)^{\mathrm{T}}$ and $\cdot^{\mathrm{T}}$ denotes the transpose. The PDF satisfies the following fundamental properties:

1. Non-negativity:
$$p(\mathbf{x}) \geq 0 \quad \text{for all } \mathbf{x} \in \mathbb{R}^n. \tag{2.1}$$

2. Normalization:
$$\int_{\mathbb{R}^n} p(\mathbf{x})\,\mathrm{d}\mathbf{x} = 1. \tag{2.2}$$

DOI: 10.1201/9781003738459-2

3. Probability assignment: For any hyperrectangle $[\mathbf{a}, \mathbf{b}] = \prod_{i=1}^{n} [a_i, b_i]$,

$$P(\mathbf{a} \leq \mathbf{X} \leq \mathbf{b}) = \int_{a_1}^{b_1} \cdots \int_{a_n}^{b_n} p(\mathbf{x}) \, \mathrm{d}\mathbf{x}. \tag{2.3}$$

Here, $P(\cdot)$ denotes the probability measure induced by $\mathbf{X}$, and $p(\mathbf{x})$ is the joint PDF of components $X_1, \ldots, X_n$.

2.1.2 Marginal and conditional distributions

Beyond the joint distribution, two key concepts arise: the marginal distribution (which simplifies a joint distribution to a subset of variables) and the conditional distribution (which describes dependencies between variables).

The marginal distribution of a subset of variables is obtained by integrating out the remaining variables from the joint PDF. For $\mathbf{X}$ and $\mathbf{Y}$ with joint PDF $p(\mathbf{x}, \mathbf{y})$, the marginal PDF of $\mathbf{X}$ is:

$$p(\mathbf{x}) = \int_{\Omega_{\mathbf{y}}} p(\mathbf{x}, \mathbf{y}) \, \mathrm{d}\mathbf{y}, \tag{2.4}$$

where $\Omega_{\mathbf{y}}$ is the support of $\mathbf{Y}$.

The conditional distribution describes the behavior of $\mathbf{Y}$ given $\mathbf{X} = \mathbf{x}$. Its PDF is:

$$p(\mathbf{y} \mid \mathbf{x}) = \frac{p(\mathbf{x}, \mathbf{y})}{p(\mathbf{x})}, \qquad \text{provided } p(\mathbf{x}) > 0. \tag{2.5}$$

Analogous definitions apply to discrete cases by replacing integrals with sums.

In practice, marginal distributions are commonly used to study selected features in high-dimensional systems. Conditional distributions widely appear as building blocks in Bayesian inference, data assimilation, and machine learning. Different distributions play crucial roles in uncertainty quantification, model reduction, and the statistical analysis of complex systems.

2.1.3 Gaussian distribution

The Gaussian distribution, also known as the normal distribution, is a continuous probability distribution characterized by its symmetric, bell-shaped curve. It is one of the most widely used distributions in various disciplines due to its mathematical tractability and frequent occurrence in natural phenomena.

A random vector $\mathbf{X} = (X_1, \ldots, X_n)^{\mathtt{T}} \in \mathbb{R}^n$ follows a Gaussian distribution with mean $\boldsymbol{\mu}$ (measuring central tendency) and covariance $\boldsymbol{\Sigma}$ (measuring spread), denoted by

$$\mathbf{X} \sim \mathcal{N}(\boldsymbol{\mu}, \boldsymbol{\Sigma}), \tag{2.6}$$

if its PDF is:

$$p(\mathbf{x}) = \frac{1}{\sqrt{(2\pi)^n |\boldsymbol{\Sigma}|}} e^{-\frac{1}{2}(\mathbf{x}-\boldsymbol{\mu})^{\mathtt{T}} \boldsymbol{\Sigma}^{-1} (\mathbf{x}-\boldsymbol{\mu})}. \tag{2.7}$$

Here, the mean is given by a vector $\boldsymbol{\mu} = (\mu_1, \ldots, \mu_n)^{\mathtt{T}}$, and the covariance is given by a $n \times n$ matrix,

$$\boldsymbol{\Sigma} \equiv \mathbb{E}\left[(\mathbf{X} - \boldsymbol{\mu})(\mathbf{X} - \boldsymbol{\mu})^{\mathtt{T}}\right], \tag{2.8}$$

where the (i, j)-th entry in this matrix is $\mathrm{Cov}(X_i, X_j) = \mathbb{E}[(X_i - \mu_i)(X_j - \mu_j)]$, measuring the statistical dependence of the components X_i andX_j. The covariance matrix is symmetric and positive semi-definite. The Gaussian distribution is uniquely determined by $\boldsymbol{\mu}$ and $\boldsymbol{\Sigma}$. The standard Gaussian has $\boldsymbol{\mu} = \mathbf{0}$ and $\boldsymbol{\Sigma} = \mathbf{I}$ (identity matrix).

One key property of Gaussian distributions is that both marginal and conditional distributions of a Gaussian distribution are themselves Gaussian.

For a Gaussian random vector $\mathbf{X} = (X_1, \ldots, X_n)^{\mathrm{T}}$, the marginal distribution of any subset of variables is obtained by restricting the mean and covariance to the corresponding components.

Example 2.2 (Marginal distribution of a three-dimensional Gaussian). The marginal density over a subset of components is obtained by simply *dropping the remaining components.* For example, given a three-dimensional Gaussian random variable, if $\mathbf{X} = (X_1, X_2, X_3)^{\mathrm{T}} \sim \mathcal{N}(\boldsymbol{\mu}, \boldsymbol{\Sigma})$ with

$$\boldsymbol{\mu} = \begin{pmatrix} \mu_1 \\ \mu_2 \\ \mu_3 \end{pmatrix}, \qquad \boldsymbol{\Sigma} = \begin{pmatrix} \Sigma_{11} & \Sigma_{12} & \Sigma_{13} \\ \Sigma_{21} & \Sigma_{22} & \Sigma_{23} \\ \Sigma_{31} & \Sigma_{32} & \Sigma_{33} \end{pmatrix},$$

then the marginal distribution of $(X_1, X_3)^{\mathrm{T}}$ is

$$\tilde{\boldsymbol{\mu}} = \begin{pmatrix} \mu_1 \\ \mu_3 \end{pmatrix}, \qquad \tilde{\boldsymbol{\Sigma}} = \begin{pmatrix} \Sigma_{11} & \Sigma_{13} \\ \Sigma_{31} & \Sigma_{33} \end{pmatrix}.$$

□

For a partitioned Gaussian vector $\mathbf{X} = (\mathbf{X_1}, \mathbf{X_2})^{\mathrm{T}}$ with

$$\boldsymbol{\mu} = \begin{pmatrix} \boldsymbol{\mu_1} \\ \boldsymbol{\mu_2} \end{pmatrix}, \qquad \boldsymbol{\Sigma} = \begin{pmatrix} \boldsymbol{\Sigma_{11}} & \boldsymbol{\Sigma_{12}} \\ \boldsymbol{\Sigma_{21}} & \boldsymbol{\Sigma_{22}} \end{pmatrix},$$

the conditional distribution of $\mathbf{X_1}$ given $\mathbf{X_2} = \mathbf{x_2}$ is Gaussian

$$p(\mathbf{x_1}|\mathbf{x_2}) \sim \mathcal{N}(\overline{\boldsymbol{\mu}}, \overline{\boldsymbol{\Sigma}}) \tag{2.9}$$

where

$$\overline{\boldsymbol{\mu}} = \boldsymbol{\mu_1} + \boldsymbol{\Sigma_{12}}\boldsymbol{\Sigma_{22}^{-1}}(\mathbf{x_2} - \boldsymbol{\mu_2}) \qquad \text{and} \qquad \overline{\boldsymbol{\Sigma}} = \boldsymbol{\Sigma_{11}} - \boldsymbol{\Sigma_{12}}\boldsymbol{\Sigma_{22}^{-1}}\boldsymbol{\Sigma_{21}}. \tag{2.10}$$

One immediate conclusion drawn from (2.10) is that, for a Gaussian distribution, the conditional covariance $\overline{\boldsymbol{\Sigma}}$ is always "no bigger" than the marginal one $\boldsymbol{\Sigma_{11}}$, since the matrix $\boldsymbol{\Sigma_{12}}\boldsymbol{\Sigma_{22}^{-1}}\boldsymbol{\Sigma_{21}}$ is semi-positive definite. When both $\mathbf{x_1}$ and $\mathbf{x_2}$ are scalar variables, the above two covariance matrices degenerate to the variances, and the conditional variance is expected to be equal to or less than the marginal variance. In the multi-dimensional case, the comparison can be made by computing the areas of the hyperellipses associated with the covariance matrices.

Example 2.3 (Numerical validation of the marginal and conditional distributions of Gaussian). Panel (a) of Figure 2.1 shows the scatter plot of a bivariate Gaussian distribution $\mathcal{N}(\boldsymbol{\mu}, \boldsymbol{\Sigma})$ with variables (x_1, x_2), where

$$\boldsymbol{\mu} = \begin{pmatrix} 1 \\ 2 \end{pmatrix} \qquad \text{and} \qquad \boldsymbol{\Sigma} = \begin{pmatrix} 1 & 1.4 \\ 1.4 & 4 \end{pmatrix}. \tag{2.11}$$

Panel (b) shows the marginal distribution of x_1, while Panel (c) displays the conditional distribution of x_1 given $x_2 = 3$. In both cases, the empirical distributions computed from the samples closely match the theoretical values. For the conditional distribution, the empirical distribution is computed using samples (x_1, x_2), where x_2 falls within $[2.9, 3.1]$. These results confirm that the conditional distribution can have a different mean than the marginal distribution, while its variance is always less than or equal to the marginal variance.

□

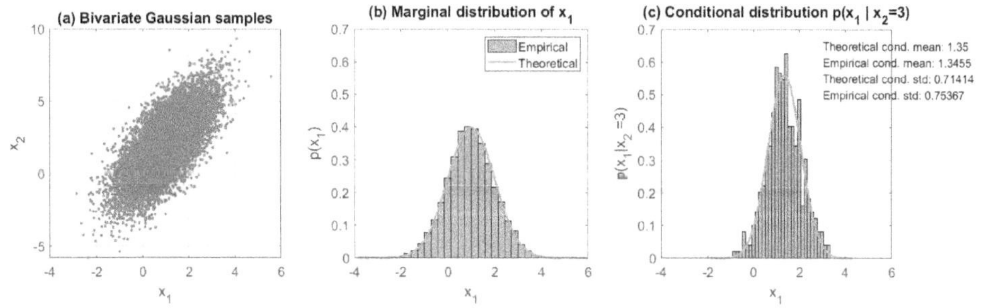

FIGURE 2.1: Numerical validation of marginal and conditional distributions of a bivariate Gaussian in (2.11). Panel (a): scatter plot of samples from the joint distribution $p(x_1, x_2)$. Panel (b): marginal distribution $p(x_1)$. Panel (c): conditional distribution $p(x_1 \mid x_2 = 3)$.

For a Gaussian random variable, the covariance matrix is a natural and intuitive measure of uncertainty. As will be rigorously justified in Chapter 8 (from an information-theoretic perspective), the determinant of the covariance (which scales with the volume of uncertainty hyperellipsoids) is the unique way to quantify uncertainty for Gaussian distributions. Therefore, the result in (2.10) indicates that the additional information provided by $\mathbf{x_2}$ helps reduce the uncertainty in $\mathbf{x_1}$. It is also worthwhile to remark that if $\mathbf{x_2}$ is regarded as the observational information while $\mathbf{x_1}$ is the forecast from a separate model, then (2.10) leads to the essence of data assimilation, in which observational data are incorporated into the model forecast to facilitate the reduction of the uncertainty and the correction of the model forecast error. The details of data assimilation and uncertainty quantification will be presented in Chapters 7 and 8.

Finally, it is worth highlighting that, according to the maximum entropy principle, the Gaussian distribution is the most unbiased choice when only the mean and variance are known. Details will be presented in Chapter 8. In addition, the central limit theorem ensures that the distribution of a normalized version of the sample mean converges to a standard Gaussian distribution, regardless of the original distributions of these samples (given finite variance). This universality explains why Gaussian noise, measurement errors, and aggregate effects frequently appear normal.

2.1.4 Moments

Moments are expected values of powers of a random variable and provide coarse-grained statistical descriptions of a probability distribution [209]. They are particularly useful in studying stochastic or chaotic systems, where the time evolution of statistics (e.g., mean, variance) is often of primary interest. Moment equations, which are ordinary differential equations (ODEs), are typically more tractable to solve numerically than the full governing equation of the distribution (e.g., the Fokker-Planck equation, a partial differential equation; see Chapter 6).

Moments can be categorized into raw, central, and standardized types, each with distinct interpretations. For clarity, we define them below for scalar random variables unless stated otherwise.

1. Raw Moments

The k-th raw moment of a random variable X ($k \geq 1$) is the expected value of X^k:

$$\mu'_k = \mathbb{E}[X^k]. \tag{2.12}$$

The mean (μ) is the first raw moment ($k = 1$) and measures central tendency:

$$\text{Mean:} \qquad \mu = \mathbb{E}[X]. \tag{2.13}$$

2. Central Moments

The k-th central moment ($k \geq 2$) quantifies deviations from the mean:

$$\mu_k = \mathbb{E}[(X - \mu)^k]. \tag{2.14}$$

The variance (σ^2) is the second central moment ($k = 2$) and measures spread:

$$\text{Variance:} \qquad \sigma^2 = \mathbb{E}[(X - \mu)^2]. \tag{2.15}$$

The square root of variance, σ, is called the standard deviation. Dimension-wise, the variance has the order of the square of the random variable. Therefore, the standard deviation, which has the same unit as the mean, is widely used in practice to quantify the relative amplitude of the sample deviations compared to the mean. For Gaussian distributions, 95% of the probability mass lies within $[\mu - 1.96\sigma, \mu + 1.96\sigma]$, which is commonly referred to as the 95% confidence interval and is used to represent estimation uncertainty. In practice, this interval is often approximated by $[\mu - 2\sigma, \mu + 2\sigma]$ for simplicity, with negligible loss of accuracy.

For a multi-dimensional random variable, the second-order central moment is given by the covariance (2.8).

3. Standardized Moments

The k-th standardized moment ($k \geq 3$) is the central moment normalized by the k-th power of the standard deviation to eliminate the dimension:

$$\tilde{\mu}_k = \frac{\mu_k}{\sigma^k}. \tag{2.16}$$

The standardized moments are used for scale-free comparisons. In particular, the third-order and fourth-order standardized moments are

$$\begin{aligned} &\text{Skewness:} \qquad \mu_3/\sigma^3, \\ &\text{Kurtosis:} \qquad \mu_4/\sigma^4, \end{aligned} \tag{2.17}$$

which measures the asymmetry and peakedness of a distribution, respectively. When the skewness is zero, the PDF is symmetric. A positive (negative) skewness commonly indicates that the tail is on the right (left) side of the distribution. A PDF with kurtosis set to the standard value of 3 exhibits the same tail behavior as the Gaussian distribution. When the kurtosis is smaller (larger) than 3, the random variable produces fewer (more) extreme outliers than the corresponding Gaussian counterpart. Therefore, the associated PDF has lighter (heavier) tails than the Gaussian distribution. Sometimes, excess kurtosis is used, which is defined as kurtosis minus 3.

2.1.5 Simple non-Gaussian distributions

A non-Gaussian distribution is any probability distribution that does not follow the Gaussian form. Many non-Gaussian distributions in practice are skewed, possess fat tails, or have multimodality. One common way to determine whether a distribution is Gaussian or not is to compute the skewness and kurtosis (2.17).

Example 2.4 (Non-Gaussian distributions). The t-distribution [251] is a continuous probability distribution that generalizes the standard Gaussian distribution. Like the latter, it is symmetric around zero and bell-shaped. However, the t-distribution has heavier tails, and the amount of probability mass in the tails is controlled by the parameter ν. The t-distribution has the following mathematical expression:

$$f(x \mid \nu) = \frac{\Gamma\left(\frac{\nu+1}{2}\right)}{\sqrt{\nu\pi}\,\Gamma\left(\frac{\nu}{2}\right)} \left(1 + \frac{x^2}{\nu}\right)^{-\frac{\nu+1}{2}}, \qquad x \in \mathbb{R}, \tag{2.18}$$

where $\nu > 0$ is the degrees of freedom (controls tail heaviness) and Γ is the Gamma function, defined for $z > 0$ by

$$\Gamma(z) = \int_0^\infty t^{z-1} e^{-t}\, \mathrm{d}t. \tag{2.19}$$

For $\nu = 1$, the Student's t-distribution becomes the standard Cauchy distribution, which has very "fat" tails; whereas for $\nu \to \infty$ it becomes the standard Gaussian distribution $\mathcal{N}(0, 1)$. The t-distribution is often used to model heavy-tailed processes, such as extreme temperature anomalies.

The Gamma distribution is a two-parameter family of continuous probability distributions

$$p(x) = \frac{1}{\Gamma(k)\theta^k} x^{k-1} e^{-x/\theta}, \qquad x \in [0, \infty), \tag{2.20}$$

where k and θ are the shape and scale parameters, respectively, and the gamma function, Γ, has been defined in (2.19). The Gamma distribution is used for positive and skewed variables. Extremely positive events often occur in these variables, which corresponds to the one-sided fat tail on the positive side. Daily rainfall and wave heights often fit a Gamma distribution.

Figure 2.2 shows a Student's t-distribution (with $\nu = 5$) and a Gamma distribution (with $k = 2.0$ and $\theta = 1.0$), both of which are non-Gaussian distributions with fat tails. The kurtosis of these distributions exceeds 3, which is the value for Gaussians. Note that fat tails are often visualized by comparison with Gaussian distributions, and these distributions are frequently plotted on a logarithmic scale, where Gaussian distributions appear as parabolas. □

These non-Gaussian distributions can be used to generate samples and noise. For example, random noises satisfying a Gamma distribution can be used to create noisy observations that preserve the positivity of rainfall data. In contrast, the standard Gaussian noise will easily break this constraint and lead to non-physical outcomes. On the other hand, time series or sequences that satisfy these non-Gaussian distributions can be constructed using many efficient methods. For example, Markov chain Monte Carlo, discussed later in this chapter, is a natural way to generate a chain in which each point is sampled from a specified target distribution. In addition, it will be shown in Chapter 10 that a systematic method exists for developing simple linear stochastic differential equations, the equilibrium distribution of which can have any given non-Gaussian characteristics (even those without analytic formulae). Temporal correlation can be easily incorporated into generating these time series by including this information in the design of the differential equations, thereby helping capture more physics. These simple models also provide efficient reduced-order models for conducting various tasks, including data assimilation, uncertainty quantification, and forecasting in nature. Intermittency, regime switching, and extreme events are all crucial topics in nature that are accompanied by strongly non-Gaussian distributions. Modeling and predicting these features will be the topics to discuss in the subsequent chapters.

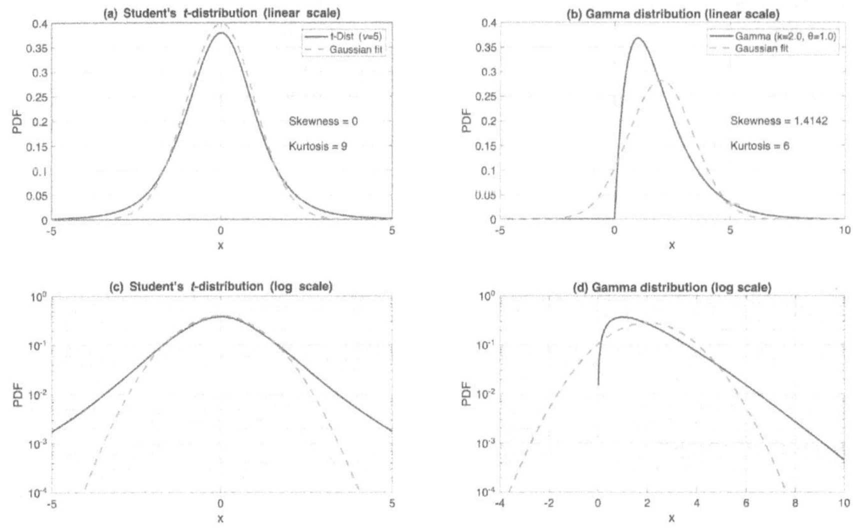

FIGURE 2.2: Student t-distribution (Panels (a) and (c)) and Gamma distribution (Panels (b) and (d)) together with their Gaussian fit. The top row shows these distributions in the linear scale while the bottom row shows them in the logarithm scale.

2.1.6 Gaussian mixture distributions

A Gaussian mixture distribution, also known as a Gaussian mixture model (GMM), is a probabilistic model that represents a probability distribution as a weighted sum of multiple Gaussian distributions [225]. It provides a powerful way to characterize complex (especially multi-modal) data that a single Gaussian cannot adequately describe. A GMM with K components is defined as:

$$p(\mathbf{x}) = \sum_{k=1}^{K} \pi_k \mathcal{N}(\mathbf{x} \mid \boldsymbol{\mu}_k, \boldsymbol{\Sigma}_k), \tag{2.21}$$

where π_k is he mixing weight (probability) of the k-th Gaussian with $\sum_{k=1}^{K} \pi_k = 1$ and $\mathcal{N}(\mathbf{x} \mid \boldsymbol{\mu}_k, \boldsymbol{\Sigma}_k)$ is the k-th Gaussian component with mean $\boldsymbol{\mu}_k$ and covariance $\boldsymbol{\Sigma}_k$. The parameters $\boldsymbol{\mu}_k$ and $\boldsymbol{\Sigma}_k$ can be determined via an expectation-maximization algorithm that alternates between updating these parameters and assigning each data point (a similar procedure will be discussed in Chapter 3 when the k-means clustering algorithm is introduced). The Bayesian information criterion (BIC), cross-validation, and other criteria can be used to select K.

GMM has many applications. GMMs are universal approximators of smooth probability densities, meaning that with enough components, they can approximate any continuous distribution arbitrarily well. In other words, any non-Gaussian distribution can be approximated by GMMs. This is similar to how a Fourier series can approximate any periodic function by summing sine waves. However, in practice, the number of required Gaussians may be very large for highly complex distributions.

Example 2.5 (GMM for smooth bimodal density estimation). There are two groups of people. One group has an average height of 165 cm (with a variance of 25 cm^2). Another group has an average height of 180 cm (with a variance of 36 cm^2). A GMM with two Gaussians can model this scenario effectively. Note that, when the sample size N is small, empirical histograms of the data may appear non-smooth and noisy due to sampling variability. In contrast, the GMM provides a smooth probability density estimate by leveraging parametric Gaussian components. See the three panels in Figure 2.3 with an increased number of samples. The smoothness in the approximate distribution is often preferred since it facilitates many subsequent mathematical operations. For example, smoothness allows us to take the derivatives. The differentiable GMM PDF advances gradient-based optimization (e.g., in maximum likelihood estimation). Smoothness also ensures the stable computation of probabilities over intervals. In addition, smoothness facilitates model generalization, which prevents overfitting sparse samples while preserving the true underlying structure.

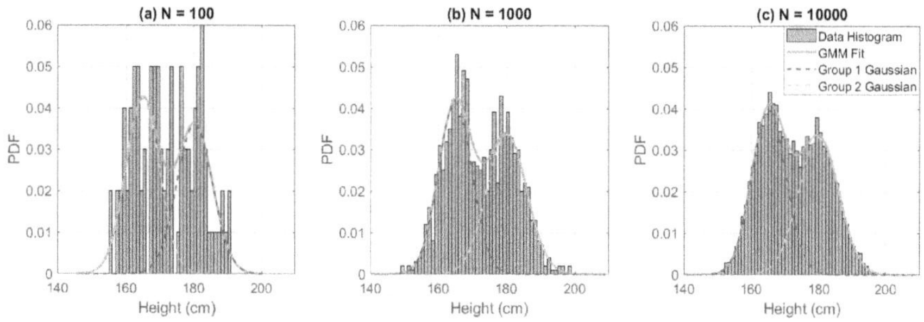

FIGURE 2.3: Using GMM to obtain a smooth estimate of the sample distribution, which exhibits bimodal features. The samples come from two groups. One group has an average height of 165 cm (with a variance of 25 cm^2). Another group has an average height of 180 cm (with a variance of 36 cm^2). The number of samples N consists of $N/2$ samples for each group.

□

Example 2.6 (GMMs and kernel density estimation). Kernel density estimation (KDE) is a method to estimate the PDF of a random variable by placing a kernel (usually a Gaussian) at each data point and summing their contributions [73]. For a dataset $\{\mathbf{x}_1, \mathbf{x}_2, \ldots, \mathbf{x}_N\}$ with N data points, where each $\mathbf{x}_i$ is d-dimensional, the KDE estimate is

$$\hat{f}(\mathbf{x}) = \frac{1}{N|\mathbf{H}|^{1/2}} \sum_{i=1}^{N} K\left(\mathbf{H}^{-1/2}(\mathbf{x} - \mathbf{x}_i)\right), \tag{2.22}$$

where K is the kernel function and the positive definite matrix $\mathbf{H}$ is the bandwidth matrix that controls smoothness. There are several existing rules to determine $\mathbf{H}$, depending on the situations. The Gaussian kernel is widely used. In such a case, the KDE simplifies to

$$\hat{f}(\mathbf{x}) = \frac{1}{N(2\pi)^{d/2}|\mathbf{H}|^{1/2}} \sum_{i=1}^{N} \exp\left(-\frac{1}{2}(\mathbf{x} - \mathbf{x}_i)^{\mathrm{T}}\mathbf{H}^{-1}(\mathbf{x} - \mathbf{x}_i)\right). \tag{2.23}$$

GMMs and KDE are both density estimation techniques, but they differ in how they model probability distributions. First, instead of placing a kernel at every data point (in total N) as in the KDE, GMM uses a small number (K) of Gaussians to approximate the PDF. If $K = N$ and $\pi_k = \frac{1}{N}$, then GMM reduces to KDE with Gaussian kernels. However, typically, $K \ll N$, which makes GMM more efficient for large datasets. Second, as KDE is non-parametric (i.e., it does not assume that the data follow a specific distribution), it is more flexible and best suited for exploratory analysis when no prior structure is assumed.

However, it is computationally heavier. In contrast, GMM is a parametric (mixture of Gaussians) model, which is efficient for large datasets. It is best suited for clustered data or when interpretability is a priority. GMM is also preferred for building generative models, which means new data points need to be sampled. In addition, KDE suffers from the so-called curse of dimensionality and becomes impractical as the dimension of each data point increases. In contrast, GMMs can mitigate this by leveraging low-rank covariance structures (e.g., diagonal $\mathbf{\Sigma}_k$).

□

2.2 Bayes' theorem

Bayes' Theorem is one of the most fundamental formulae in probability theory that describes the probability of an event based on prior knowledge of conditions related to the event [95, 32]. The theorem is stated mathematically as:

$$p(\mathbf{x} \mid \mathbf{y}) = \frac{p(\mathbf{x})p(\mathbf{y} \mid \mathbf{x})}{p(\mathbf{y})}. \tag{2.24}$$

The goal is to compute the conditional distribution $p(\mathbf{x} \mid \mathbf{y})$, which is the distribution of $\mathbf{x}$ conditioned on the known (or observed) event $\mathbf{y}$. This is also known as the posterior distribution. It can be computed based on the prior distribution $p(\mathbf{x})$ and the likelihood $p(\mathbf{y} \mid \mathbf{x})$. The prior distribution is the initial probability of $\mathbf{x}$ without knowing any information about $\mathbf{y}$. In contrast, the likelihood represents the probability of event $\mathbf{y}$ occurring given that $\mathbf{x}$ is true. Finally, $p(\mathbf{y})$ is the marginal probability, which is the total probability of event $\mathbf{y}$ occurring.

Let us use a simple example to sort out the various probability distributions mentioned above.

Example 2.7 (Tropical cyclone prediction using sea surface temperature (SST)). Historical data show that 5% of days in a given tropical region have cyclone formation ($p(\text{Cyclone}) = 0.05$). Cyclones are more likely when SST exceeds 28°C. Observations show that if a cyclone forms, there's a 90% chance SST $>$ 28°C ($p(\text{SST} > 28^o\text{C} \mid \text{Cyclone}) = 0.9$). Even without a cyclone, SST $>$ 28°C happens 30% of the time ($p(\text{SST} > 28^o\text{C} \mid \text{No Cyclone}) = 0.3$). Now, let us ask the following question. If today's SST is $>$ 28°C, what is the probability a cyclone will form ($p(\text{Cyclone} \mid \text{SST} > 28^o\text{C})$)?

We can translate the above statement into the Bayesian framework. Our target variable $\mathbf{x}$ in (2.24) is the cyclone while our conditional variable $\mathbf{y}$ is the SST exceeding 28°C. The historical data gives us the prior distribution:

$$p(\mathbf{x} = 1) = 0.05. \tag{2.25}$$

The observational evidence provides the likelihood:

$$p(\mathbf{y} \mid \mathbf{x} = 1) = 0.9 \qquad \text{and} \qquad p(\mathbf{y} \mid \mathbf{x} = 0) = 0.3. \tag{2.26}$$

Our goal is to solve $p(\mathbf{x} = 1|\mathbf{y})$, which according to Bayesian formula is

$$p(\mathbf{x} = 1 \mid \mathbf{y}) = \frac{p(\mathbf{x} = 1)p(\mathbf{y} \mid \mathbf{x} = 1)}{p(\mathbf{y})} \tag{2.27}$$

The numerator is clearly $0.05 \times 0.9 = 0.045$. The denominator is the total probability that the SST exceeds 28°C. This probability consists of two parts, with and without cyclone:

$$\begin{aligned} p(\mathbf{y}) &= p(\mathbf{y} \mid \mathbf{x} = 1)p(\mathbf{x} = 1) + p(\mathbf{y} \mid \mathbf{x} = 0)p(\mathbf{x} = 0) \\ &= 0.9 \times 0.05 + 0.3 \times 0.95 = 0.045 + 0.285 = 0.33. \end{aligned} \tag{2.28}$$

Therefore, plugging (2.28) into (2.27) yields

$$p(\mathbf{x} = 1 \mid \mathbf{y}) = \frac{0.045}{0.33} = 0.136. \tag{2.29}$$

It could be interesting to look at the contingency table (also called a confusion matrix), which helps understand marginal, joint, and conditional distributions. See Table 2.1. The

TABLE 2.1: Contingency table of joint and marginal distributions of cyclones and SSTs.

	Cyclone occurs	No cyclone	Total
SST > 28°C	4.5%	28.5%	33%
SST ≤ 28°C	0.5%	66.5%	67%
Total	5%	95%	100%

values in the first two rows and first two columns are the joint probability. For example, 4.5% is the joint distribution of cyclone occurring and SST > 28°C. The last row is the total probability (i.e., marginal distribution) of cyclone occurs (5%) and no cyclone (95%) regardless of SST. Similarly, the last column is the total probability of SST > 28°C (33%) or ≤ 28°C (67%) regardless of the cyclone occurrence.

The table does not include the likelihoods or conditional distributions. Nevertheless, they can be computed from the joint and marginal distributions. If a cyclone occurs, SST > 28°C happens 90% of the time:

$$p(\text{SST} > 28^o\text{C} \mid \text{Cyclone}) = \frac{p(\text{SST} > 28^o\text{C}, \text{Cyclone})}{p(\text{Cyclone})} = \frac{4.5\%}{5\%} = 90\%. \tag{2.30}$$

What is interesting to see is the following. If a cyclone forms, there is a 90% chance SST > 28°C. However, this does not mean cyclone occurrence and SST exceeding 28°C are highly correlated since the reverse situation has a completely different probability. In fact, even though SST > 28°C occurs in 33% of days, cyclones only happen on 4.5% of those days. This explains why "SST > 28°C" alone isn't enough to panic about cyclones. Understanding these relationships can help avoid false alarms. Furthermore, the 0.5% in the bottom-left entry shows rare cases where cyclones form despite cooler SSTs (≤ 28°C). □

Example 2.8 (Soft clustering with GMMs: A Bayesian probabilistic approach). Clustering seeks to infer the underlying distributions that generate the data and to determine which distribution each data point is most likely drawn from. GMMs can be used for probabilistic clustering, where each Gaussian component represents a cluster. Different from hard clustering (e.g., k-means, which assigns each point to exactly one cluster), GMMs allow soft clustering by computing the probability (or responsibility) that a data point $\mathbf{x}_i$ belongs to cluster k. This is derived via Bayes' theorem:

$$\gamma_{ik} = p(z_i = k \mid \mathbf{x}_i) = \frac{\pi_k \mathcal{N}(\mathbf{x}_i \mid \boldsymbol{\mu}_k, \boldsymbol{\Sigma}_k)}{\sum_{j=1}^{K} \pi_j \mathcal{N}(\mathbf{x}_i \mid \boldsymbol{\mu}_j, \boldsymbol{\Sigma}_j)} \tag{2.31}$$

where z_i indicates the cluster assignment and $\gamma_{ik} \in [0, 1]$ is a soft assignment with $\sum_{k=1}^{K} \gamma_{ik} = 1$.

Figure 2.4 compares the GMM soft clustering with the classical k-means hard clustering. The latter will be discussed in Chapter 3. In both methods, $K = 2$ clusters are used. Using the GMM soft clustering, points near the boundary of two Gaussians have $\gamma_{ik} \approx 0.5$ (ambiguous membership). Points far from boundaries show $\gamma_{ik} \approx 1$ (clear membership). □

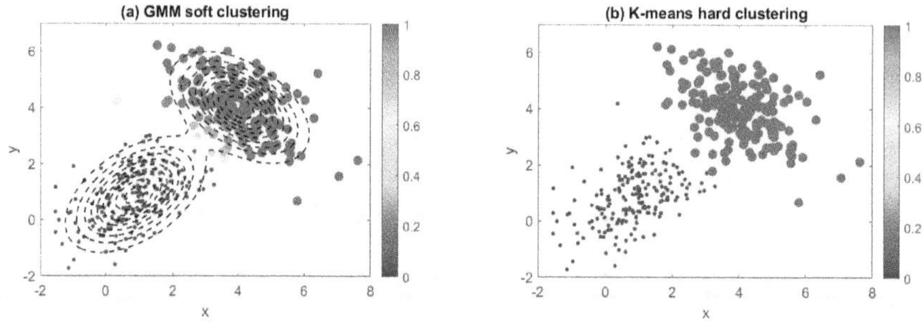

FIGURE 2.4: Comparison of GMM soft clustering (probabilistic assignments; Panel (a)) and k-means hard clustering (deterministic assignments; Panel (b)). The size and shade of the dots in Panel (a) both indicate the probability of each data point belong to the two clusters.

Bayes' theorem has wide applications. It will appear throughout this book. It serves as the foundational basis for data assimilation, uncertainty quantification, and various machine learning methods.

2.3 Monte Carlo methods

The Monte Carlo method comprises a broad class of computational algorithms that use repeated random sampling to generate numerical results [159, 194, 230]. It is recognized as one of the most important ten algorithms of the 20th century. The Monte Carlo method is particularly useful when dealing with problems where deterministic solutions are difficult or impossible to obtain. In addition to being powerful in handling problems with inherent uncertainty, it can also transform fully deterministic problems into forms that facilitate the use of random sampling techniques to solve.

2.3.1 Classical application scenarios

To demonstrate Monte Carlo methods, we start with a scenario in which we sample from a known distribution and transform the samples to approximate a target distribution. In this simple illustration, the transformation involves only algebraic squaring of Gaussian variables. However, as shown in later chapters, Monte Carlo methods become particularly powerful when approximating probability distributions arising from stochastic differential equations (SDEs). In such cases, the required transformations are far more complex, essentially reducing to a numerical solution of these SDEs.

Example 2.9 (Approximating chi-squared distributions by Monte Carlo simulations). The chi-squared distribution [2] (denoted χ^2, where k is the degrees of freedom) is a continuous probability distribution of the sum of squares of k independent standard normal random variables:

$$X = Z_1^2 + Z_2^2 + \cdots + Z_k^2 \quad \text{where} \quad Z_i \sim \mathcal{N}(0,1). \tag{2.32}$$

Its support is $X \in [0, +\infty)$. Its shape is right-skewed for small k, approaching normality as k increases. The PDF is given by

$$p(x;k) = \frac{1}{2^{k/2}\Gamma(k/2)} x^{k/2-1} e^{-x/2}, \quad x \geq 0, \tag{2.33}$$

where $\Gamma(\cdot)$ is the gamma function defined in (2.19). Its mean and variance are given by $\mathbb{E}[X] = k$ and $\text{Var}(X) = 2k$, respectively.

If $X_1 \sim \chi^2_{k_1}$ and $X_2 \sim \chi^2_{k_2}$ are independent, then

$$X_1 + X_2 \sim \chi^2_{k_1+k_2} \tag{2.34}$$

The chi-squared distribution is vital for goodness-of-fit tests (e.g., Pearson's test), independence tests in contingency tables, and variance estimation for normal data. It also supports likelihood-ratio model comparisons and regression diagnostics. Its additive property (for sums of squared normals) and dependence on degrees of freedom make it essential for statistical inference.

Figure 2.5 uses the standard Monte Carlo method to approximate the chi-squared distribution χ^2_1, which equals the square of a standard Gaussian random variable. In this Monte Carlo simulation, N samples are generated from the standard Gaussian distribution and then squared. The resulting samples approximate the χ^2_1 distribution, converging to the theoretical density $p(x;1)$ from (2.33) as N increases. □

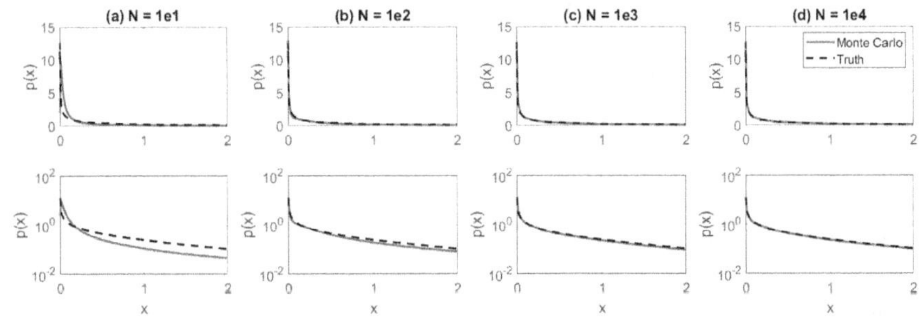

FIGURE 2.5: Monte Carlo approximation of the chi-squared distribution χ^2_1 using N samples. Top panels: PDFs in linear scale. Bottom panels: PDFs in logarithm scales, highlighting tail behavior. As N increases, the approximation converges to the theoretical χ^2_1 distribution (black dashed line).

In addition to approximating PDFs, Monte Carlo methods are widely used to estimate a target quantity r associated with complex systems where analytical solutions are intractable. Let us assume that a random variable X can be easily generated with a low computational cost. In addition, this random variable satisfies $E[X] = r$. By creating a sequence of independent and identically distributed (i.i.d.) samples $X_1, X_2, X_3, \ldots$ from the same distribution as X, the law of large numbers guarantees that the empirical average:

$$\hat{r}_N = \frac{1}{N} \sum_{i=1}^{N} X_i \tag{2.35}$$

converges to r as $N \to \infty$. Moreover, the central limit theorem provides quantitative error bounds and confidence intervals for $\hat{r}_N - r$, which forms the theoretical foundation of Monte Carlo estimation.

To build intuition before presenting the formal theory, consider the following straightforward Monte Carlo application.

Example 2.10 (Using Monte Carlo to compute the area of the unit circle). Assume we cannot compute the irrational number π directly. Our goal is to estimate the area of the unit circle (theoretical value: π) using a Monte Carlo simulation.

We begin by illustrating the method using a simple geometric probability setting. Consider the unit circle enclosed by a square of side length 2 (area = 4), as shown in Panel (a) of Figure 2.6. A natural question to ask is that, if we randomly scatter a point within this square, what is the probability it lands inside the circle? The answer is straightforward. The probability is given by the ratio of the area of the circle (π) to the area of the square (4).

This important insight motivates us to adopt the following Monte Carlo method:

1. We generate N independent random points (a_i, b_i), where each coordinate a_i and b_i is drawn independently from Unif$(-1, 1)$. This ensures uniform coverage of the square domain $[-1, 1] \times [-1, 1]$.
2. For each point, we test whether it lies within the unit circle by checking if $a_i^2 + b_i^2 \leq 1$.
3. The fraction of points inside the circle, multiplied by the square's area (4), gives our estimate of the unit circle's area (π).

It is worth highlighting that the method never explicitly computes π. All operations are elementary: uniform sampling, squaring, summation, and comparison. The value π emerges naturally from the ratio of counts.

The accuracy of the Monte Carlo procedure depends fundamentally on the sample size N. To assess its performance, we repeat the above experiment K times, each using N samples to estimate the area of the unit circle. Due to the finite value of N, these K independent estimates will exhibit variability: their average reveals whether the method is unbiased (converging to π as $K \to \infty$), while their variance quantifies the stability of the estimation. Note that, in practice, we typically perform the computation once using N samples. Therefore, it is crucial to understand whether such an estimate with N samples is likely to be close to the true value or whether it fluctuates significantly depending on the random samples. A small variance among the K estimates indicates reliable performance, whereas a large variance suggests sensitivity to randomness and less dependable results.

Panel (b) of Figure 2.6 demonstrates that the average of $K = 1000$ independent Monte Carlo estimates remains unbiased even for small N, aligning with the theoretical value. However, as expected, the 95% confidence interval (mean ± 1.96 standard deviations) is wide for small N and narrows as N increases. Panel (c) shows one critical feature: the standard deviation of the estimate decays at the rate of $1/\sqrt{N}$. In other words, the ratio of the standard deviation to $1/\sqrt{N}$ is constant. This scaling law provides a theoretical prediction of estimation variance without requiring extensive trials. It thus allows the practical pre-calculation of the sample size N needed to achieve a desired precision.

□

With the numerical evidence shown in Example 2.10, we now introduce the theoretical error bounds of the Monte Carlo simulation.

Let $p(x)$ denote the true probability density function (PDF) of a scalar random variable X. The expectation of X is defined as

$$\mu_x := E(X) = \int_{-\infty}^{\infty} x p(x) \, dx,$$

which represents the theoretical mean and serves as the target quantity of interest.

Given N independent and identically distributed (i.i.d.) samples $x_1, \ldots, x_N$ drawn from

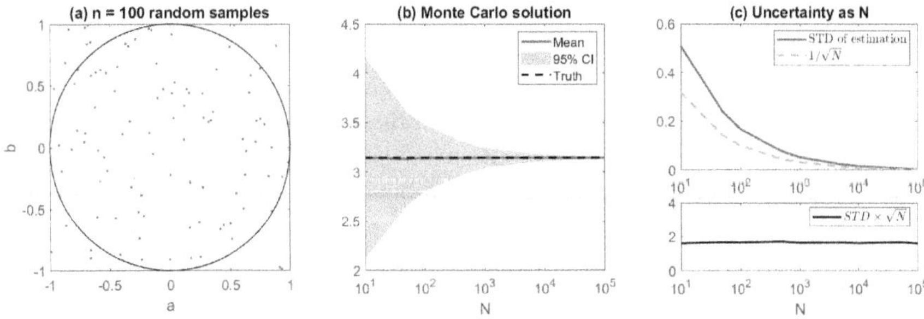

FIGURE 2.6: Using Monte Carlo to compute the area of the unit circle. Panel (a): visualization of $N = 100$ uniformly distributed random points in $[-1, 1] \times [-1, 1]$ with the unit circle shown for reference. Panel (b): convergence of the mean (solid line) and 95% confidence interval (shading area) computed from $K = 1000$ independent trials, plotted as a function of sample size N. The horizontal dashed line marks the true value (π). Panel (c): standard deviation of the estimates versus N in comparison to $1/\sqrt{N}$ (top) and the ratio of the standard deviation to $1/\sqrt{N}$.

the same distribution as X, a natural estimator of μ_x is the sample mean,

$$\bar{x} := \frac{1}{N}\sum_{i=1}^{N} x_i. \tag{2.36}$$

Because Monte Carlo simulations rely on a finite number of random samples, different realizations generally produce different values of $\bar{x}$ in (2.36). As illustrated in Figure 2.6, the variability of $\bar{x}$, that is, how much the estimate fluctuates across independent runs, depends on the sample size N. Consequently, when computed from a finite number of samples, $\bar{x}$ itself is a random variable.

A natural question is how accurate $\bar{x}$ is, on average, as an estimator of μ_x. To address this, we consider the expected difference between the estimator and the true mean,

$$E(\bar{x} - \mu_x) = E(\bar{x}) - \mu_x = E\left(\frac{1}{N}\sum_{i=1}^{N} x_i\right) - \mu_x = \frac{1}{N}\sum_{i=1}^{N} E(x_i) - \mu_x. \tag{2.37}$$

Since each x_i in (2.37) is drawn from the same distribution as X, we have $E(x_i) = \mu_x$, which implies $E(\bar{x} - \mu_x) = 0$. This result shows that the sample mean has zero expected error and is therefore an unbiased estimator of μ_x.

To further characterize the uncertainty in $\bar{x}$, we examine the variance of $\bar{x} - \mu_x$,

$$\text{Var}(\bar{x} - \mu_x) = \text{Var}(\bar{x}) - \text{Var}(\mu_x) = \text{Var}(\bar{x}) = \text{Var}\left(\frac{1}{N}\sum_{i=1}^{N} x_i\right) = \frac{1}{N^2}\text{Var}\left(\sum_{i=1}^{N} x_i\right),$$

where $\text{Var}(\mu_x) = 0$ since μ_x is a constant. Because the Monte Carlo samples are i.i.d., the variance of the sum equals the sum of the variances, yielding

$$\text{Var}(\bar{x} - \mu_x) = \frac{1}{N^2}\sum_{i=1}^{N} \text{Var}(x_i) = \frac{1}{N^2} N\sigma_x^2 = \frac{\sigma_x^2}{N}.$$

The standard error of the estimator, defined as the square root of the variance, therefore decreases at the rate $1/\sqrt{N}$, consistent with the behavior shown in Figure 2.6. Equivalently, reducing the variability of the mean estimate by a factor of 10 requires increasing the number of Monte Carlo samples by a factor of 100.

For sufficiently large N, the central limit theorem may be invoked to approximate the distribution of $\bar{x}$. Specifically, $\bar{x} - \mu_x$ converges in distribution to a normal random variable with mean 0 and variance σ_x^2/N. This result provides a practical basis for determining the sample size required to achieve a prescribed confidence level. In practice, however, the true variance σ_x^2 is typically unknown and is replaced by the sample variance when constructing error estimates or confidence intervals. This substitution introduces additional uncertainty, particularly when the sample size is small.

To complete the story, let us go back to Example (2.10). To compute σ_x, we need first to identify the random variable x_i using the notation in (2.36). The random variable in this example is whether this data point (a_i, b_i) is within a distance of 1 to the center of the domain. It takes the value one or zero, depending on whether the data point is within the unit circle. It is essentially a Bernoulli distribution but with two modified values:

$$x_i = \begin{cases} 1 \times 4, & \text{if } a_i^2 + b_i^2 \leq 1, \\ 0 \times 4, & \text{if } a_i^2 + b_i^2 > 1, \end{cases} \tag{2.38}$$

where the factor 4 is the area of the square, which needs to be multiplied to obtain the area of the circle. Numerical solution of the variance of x_i is about 1.64, which is consistent with the value in the last panel of Figure 2.6, as it can be regarded as $K = N$ independent experiments with each having only one sample.

Example 2.11 (Numerical integration). Another simple but illustrative example of Monte Carlo is numerical integration using random numbers. The method is particularly useful when the integrand is complicated such that analytic solutions are not available. Let $-\infty < a < b < \infty$ and assume our goal is to integrate a function f over the interval $[a, b]$. The integral can be rewritten as an expectation of a uniform random variable U over $[a, b]$, denoted by $\mathrm{Unif}(a, b)$,

$$\int_a^b f(x)\,\mathrm{d}x = (b-a)\int_a^b f(x)\frac{1}{b-a}\,\mathrm{d}x = (b-a)\mathbb{E}[f(U)], \tag{2.39}$$

where $p(x) := 1/(b-a)$ inside the integral is the PDF of the uniform distribution over the interval $[a, b]$. Thus, the integral can be estimated by repeatedly sampling from a $\mathrm{Unif}(a, b)$ random variable, applying f, and then taking the average (expectation).

Let us consider a simple example of estimating the following integral for illustration purpose:

$$A = \int_0^1 x^3\,\mathrm{d}x. \tag{2.40}$$

The exact solution can be found analytically, which is $A = 0.25$.

The top row of Figure 2.7 illustrates the Monte Carlo simulation based on (2.39). Panel (a) demonstrates this process: N independent samples x_i are drawn from a uniform distribution, $\mathrm{Unif}(0, 1)$, and each is evaluated by computing x_i^3 (represented by thin vertical lines). The average of these values, $\frac{1}{N}\sum_{i=1}^{N} x_i^3$, then serves as an approximation for A. Because the samples are uniformly distributed, partitioning the interval $[0, 1]$ into equi-spaced bins results in roughly equal numbers of samples per bin for large N. This is intuitively similar to the rectangle rule in numerical integration, where each bin contributes uniformly to the integral estimate. Panels (b)–(c) are similar to those in Figure 2.6, describing how the uncertainty in the estimate diminishes as N increases, following the expected $1/\sqrt{N}$ convergence rate. The standard deviation, computed from the samples x_i^3, is approximately 0.284.

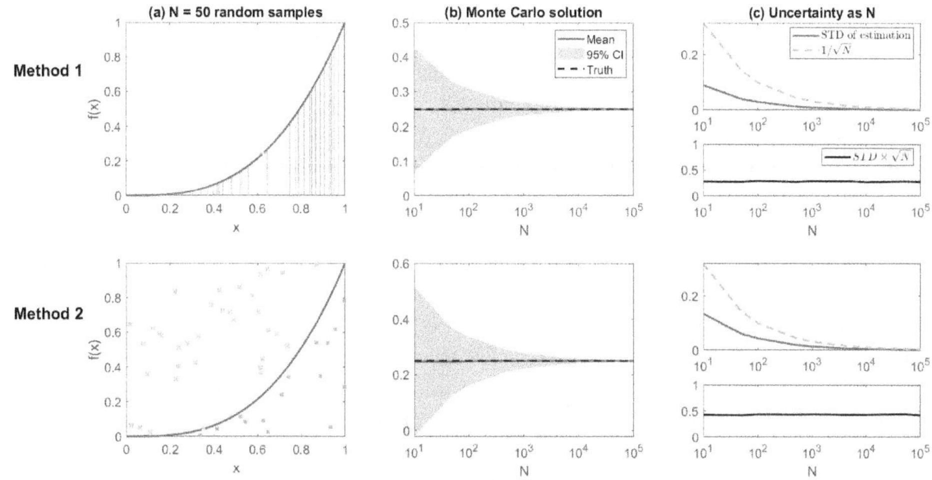

FIGURE 2.7: Using Monte Carlo methods to compute the numerical integral in (2.40). The top and bottom rows show two different methods as outlined in Example 2.11.

The bottom row of Figure 2.7 presents results using an alternative Monte Carlo method, analogous to the one in Example 2.10. Here, we scatter uniformly distributed random points across the domain $[0, 1] \times [0, 1]$ and approximate the integral by computing the ratio of points falling below the curve $y = x^3$ to the total number of points. Since the total area of the domain equals one, this ratio directly provides the integral estimate without requiring the multiplication of additional prefactors.

While this method differs conceptually from the one shown in the top row of Figure 2.7, both methods share the same $1/\sqrt{N}$ convergence rate. The main difference between these two methods lies in the standard deviation of the estimator. In this second method, the random variable satisfies a Bernoulli distribution. For each sample (x_i, y_i), we define the binary random variable:

$$z_i = \begin{cases} 1, & \text{if } y_i \leq x_i^3, \\ 0, & \text{if } y_i > x_i^3, \end{cases} \tag{2.41}$$

which gives a standard deviation of approximately 0.433.

In this example, the two methods lead to comparable standard deviations of the estimators. In practice, a Monte Carlo method with lower variance and easy sampling is always preferred to improve computational efficiency and storage. □

2.3.2 Markov chain Monte Carlo

A Markov chain is a stochastic process that transitions between states in a probabilistic manner, where the probability of moving to the next state depends only on the current state (and not on the history of states). The details of the Markov chain will be discussed in Chapter 6.

Markov chain Monte Carlo (MCMC) is a general class of methods for sampling from a prescribed target probability distribution [258, 281, 76]. The core idea of MCMC is to construct a Markov chain whose stationary (or invariant) distribution is designed to coincide with the target distribution of interest. In other words, as the number of steps tends to infinity, the probability distribution of the chain's state converges to the target distribution and remains unchanged thereafter. After simulating the chain for a sufficiently long time so that it reaches this stationary regime, the states of the chain can be treated as samples drawn from the target distribution. In many applications, MCMC is employed to generate a sequence of random samples from target distributions that are high-dimensional or otherwise difficult to sample from directly. These samples are then used to approximate the target distribution or to compute statistical quantities, such as expected values.

The Metropolis [193] (or its generalized version, Metropolis-Hastings [128]) algorithm and Gibbs sampling [113] are among the most widely used MCMC methods. The Metropolis algorithm is the first MCMC algorithm. It proposes a new state in the Markov chain and accepts/rejects it based on a probability ratio of the target distribution. It is highly flexible and works for any target distribution, even unnormalized ones. It is also easy to implement for simple problems. One potential issue with the Metropolis algorithm is that it may converge slowly in high dimensions (due to random-walk behavior). In addition, tuning the so-called proposal distribution that is used to generate samples (see below) is critical and often tricky. In contrast, Gibbs sampling samples each variable sequentially, conditional on the current values of all other variables. It has no rejection step when conditional distributions are easy to sample from. It is also efficient for hierarchical models (e.g., Bayesian networks). However, it requires knowing/being able to sample from all full conditional distributions. It may also be slow to mix if variables are highly correlated. Many advanced MCMC methods have been developed in recent years. Hamiltonian Monte Carlo and its variants [34] are often preferred for high-dimensional problems because they use gradient information to explore the space more efficiently. Variational inference or gradient-based MCMC (e.g., Langevin dynamics) [259] has also become more common due to scalability for big data or even deep learning.

In the following, we present the Metropolis algorithm and demonstrate its steps via simple examples. Let $f(x)$ be a function proportional to the target probability density $p(x)$, namely $p(x) = f(x)/Z$ with Z being an unknown normalizing constant. In practice, this normalization constant is extremely difficult to determine as it often involves solving a high-dimensional complicated integration. One of the key elements in the Metropolis algorithm (or in general MCMC) is to introduce a so-called proposal distribution $g(x' \mid x_k)$, which is used to propose the next potential sample x' based on the current one x_k. Repeatedly generating samples for $k = 1, 2, \ldots$ using this proposal distribution, we can construct a Markov chain. For the Metropolis algorithm, the proposal distribution must be symmetric, which means $g(x' \mid x_k) = g(x_k \mid x')$ for all x_k, x'. A common choice is a Gaussian centered at x_k:

$$g(x' \mid x_k) = \mathcal{N}(x' \mid x_k, \sigma^2). \tag{2.42}$$

The more general Metropolis-Hastings algorithm relaxes this symmetry requirement.

After the initialization by starting at an arbitrary point x_0, the Metropolis algorithm is as follows:

For each iteration k:

1. Generate a candidate x' for the next sample by picking from the distribution $g(x' \mid x_k)$.
2. Calculate the acceptance ratio $\alpha = f(x')/f(x_k)$, which will be used to decide whether to accept or reject the candidate. Because f is proportional to the density of p, we have that $\alpha = f(x')/f(x_k) = p(x')/p(x_k)$.
3. Accept or reject:
   (a) Generate a uniform random number $r \in [0, 1]$.
   (b) If $r \leq \alpha$, then accept the candidate by setting $x_{k+1} = x'$.
   (c) If $r > \alpha$, then reject the candidate and set $x_{k+1} = x_k$ instead.

The chain explores high-probability regions under $p(x)$ because α favors moves to higher-density areas. Notably, even if $f(x') < f(x_k)$, the candidate may still be accepted with probability α. This ensures sampling from $p(x)$ rather than optimization. In other words, MCMC characterizes the entire distribution $p(x)$, not just a local optimum. Therefore, it is fundamentally different from the gradient descent algorithm.

As can be seen from the Metropolis algorithm, the acceptance ratio α depends only on the ratio of the target densities, so the unknown normalization constant Z cancels out. The choice of g (e.g., step size σ in Gaussian proposals) affects convergence speed. Adaptive methods are often used to adjust σ as a function of k [227, 269, 19]. As a final remark, if a non-symmetric proposal g is used, then the Metropolis-Hastings correction can be implemented:

$$\alpha = \min\left(1, \frac{f(x')g(x_k \mid x')}{f(x_k)g(x' \mid x_k)}\right). \tag{2.43}$$

Example 2.12 (Metropolis algorithm for sampling from a function proportional to a bimodal distribution). Let us adopt the Metropolis algorithm to sample from the density function $p(x)$,

$$p(x) \propto f(x) = \exp\left(x + 4x^2 - 2x^3 - 4x^4\right), \tag{2.44}$$

where the normalization constant for making $f(x)$ to be a PDF is assumed to be unknown here. The profiles of $p(x)$ and $f(x)$ are shown in Panel (a) of Figure 2.8. Note that in such a simple example, the normalization constant can easily be obtained by carrying out a direct numerical integration of (2.44). However, calculating this constant becomes extremely challenging for more complicated problems. This example mainly serves to show the procedure of carrying out the MCMC in sampling from this unknown distribution.

We initialize the Metropolis algorithm with $x_0 = 0$ and use a Gaussian proposal distribution $g(x' \mid x_k) = (2\pi)^{-1/2}\exp\left(-(x' - x_k)^2/2\right)$ centered at the current sample with unit variance. The first five iterations proceed as follows (see Table 2.2): At $k = 1$, the candidate $x' = -0.6490$ is accepted with $\alpha = 2.3939$ (since $\alpha > 1$, acceptance is guaranteed). For $k = 2$, the proposed $x' = -1.4075$ yields $\alpha = 0.0114$, and the random draw $r = 0.0923$ satisfies $r > \alpha$, leading to rejection, thus $x_2 = x_1$. At $k = 3$, the candidate $x' = -1.4946$ has an extremely low acceptance ratio ($\alpha = 0.0012$), and with $r = 0.3456 > \alpha$, it is rejected again, retaining $x_3 = x_2$. Similarly, for $k = 4$, $x' = -1.2077$ ($\alpha = 0.2915$) is rejected as $r = 0.5388 > \alpha$. At $k = 5$, the candidate $x' = -0.8459$ is accepted ($\alpha = 1.3578 > 1$). Finally, for $k = 6$, the candidate $x' = -1.6978$ is rejected since $\alpha = 0.0000 < r$. This procedure demonstrates how low-probability proposals in the distribution's tails (e.g., $x' = -1.4946$) are frequently rejected, while high-density regions are preferentially sampled. Panel (b) of Figure 2.8 shows the first 100 samples in this Markov chain, with crosses and dots at the bottom of the panel showing the candidate being rejected or accepted. Panel (c) shows the reconstructed PDF based on a total of N samples with different N values. As N increases, the sample-reconstructed PDF converges to the true PDF.

□

TABLE 2.2: The first 6 sample steps using the Metropolis algorithm for (2.44).

k	x'	α	r	x_k	accept or reject?
1	−0.6490	2.3939	0.7203	−0.6490	accept
2	−1.4075	0.0114	0.0923	−0.6490	reject
3	−1.4946	0.0012	0.3456	−0.6490	reject
4	−1.2077	0.2915	0.5388	−0.6490	reject
5	−0.8459	1.3578	0.6852	−0.8459	accept
5	−1.6978	0.0000	0.8781	−0.8459	reject

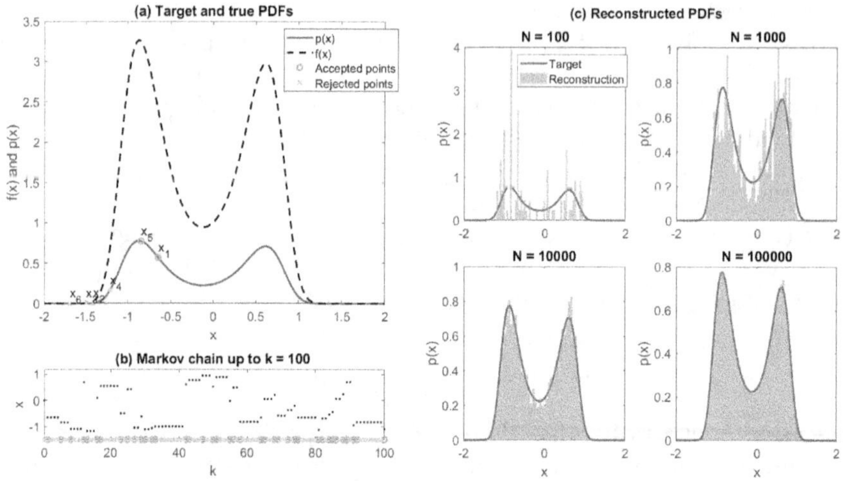

FIGURE 2.8: Metropolis algorithm for sampling from a function proportional to a bimodal distribution. Panel (a): the given function $f(x)$ (dashed curve) and the corresponding PDF $p(x)$ (solid curve) as well as the first six candidate points. If the candidate is rejected (accepted), it is marked by a cross (a circle). Panel (b): the Markov chain containing the first 100 samples. The crosses and dots at the bottom indicate the candidates being rejected and accepted, respectively. Panel (c): the reconstructed PDF for different sample sizes N, showing convergence to the true PDF as N increases.

Example 2.13 (Parameter estimation using the MCMC). One wide application of the MCMC is parameter estimation in dynamical systems. The samples from the Markov chain can be used to provide a distribution of the parameters that allows the quantification of uncertainty in parameter estimation. Since the topics of dynamical systems will be discussed in later chapters, we use a simple algebraic regression model to illustrate how the Metropolis algorithm is implemented to estimate model parameters.

Consider observed data pairs (x_i, y_i) for $i = 1, \ldots, N$, governed by:

$$y_i = ax_i + b + \varepsilon_i, \quad \varepsilon_i \sim \mathcal{N}(0, \sigma^2), \tag{2.45}$$

where ε_i represents observational noise (Figure 2.9, Panel (a)). The goal is to estimate the parameters $\boldsymbol{\theta} = (a, b, \sigma)$ using MCMC.

The target distribution $p(\boldsymbol{\theta} \mid \text{data})$ is proportional to the likelihood:

$$f(\boldsymbol{\theta}) = \prod_{i=1}^{N} \mathcal{N}(y_i \mid ax_i + b, \sigma^2). \tag{2.46}$$

The likelihood function measures how well the parameters $\boldsymbol{\theta}$ explain the observed data under the assumed Gaussian noise model, with σ controlling the spread of the observational uncertainty. Larger likelihood values correspond to parameter choices that produce model the predictions of y_i closer to the observed data.

Notably, even for this extremely simple problem, the actual distribution is already complicated, where computing the normalization constant Z is quite challenging. The Metropolis algorithm samples $\boldsymbol{\theta}$ without requiring the normalization constant Z. At each step:

1. Propose a new θ' from a symmetric Gaussian distribution $g(\theta' \mid \theta_k)$.

2. Compute $\alpha = \min\left(1, \frac{f(\theta')}{f(\theta_k)}\right)$.
3. Accept θ' if $r \sim \text{Uniform}[0,1] \leq \alpha$; otherwise, retain θ_k.

We use an adaptive MCMC [269] that adjusts the variance of the proposal distribution to maintain a 25% acceptance rate, which has been shown to be optimal for Gaussian targets. This can be done by decreasing (increasing) the variance dynamically if the acceptance rate based on the existing samples is above (below) 25%. Panels (b)–(d) in Figure 2.9 show the estimated parameters converging to their true values.

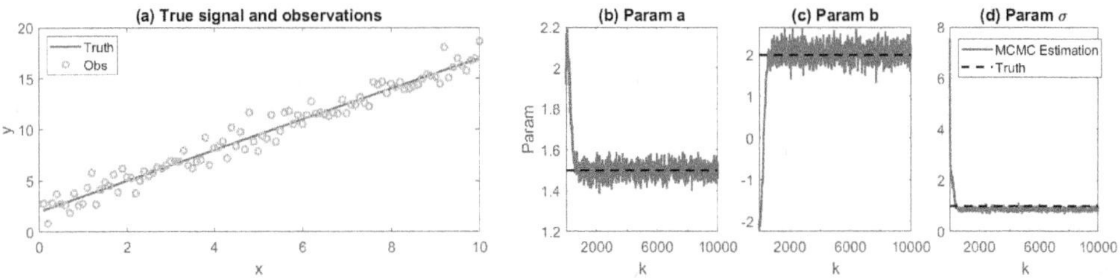

FIGURE 2.9: Using MCMC to estimate parameters in the linear model (2.45). Panel (a): true relationship between x and y (solid line) and the noisy observations of y (dots). Panels (b)–(d): the Markov chains of the estimated parameters (solid curves) from the adaptive Metropolis algorithm and the true parameter values (dashed lines).

□

2.3.3 Importance sampling

Importance sampling is a powerful technique for estimating properties of a target distribution, such as expected values, by drawing samples from a different, more tractable proposal distribution. Importance sampling proves particularly valuable in several scenarios, including when direct sampling from the target distribution $p(x)$ is computationally challenging, and when analyzing rare events (e.g., estimating probabilities of extreme weather phenomena) where conventional methods would require prohibitively many samples.

Consider estimating the expected value of $f(X)$ under $p(x)$:

$$\mathbb{E}_p[f(X)] = \int f(x)p(x)\,\mathrm{d}x. \tag{2.47}$$

While standard Monte Carlo methods approximate this by averaging $f(x_i)$ over samples $x_i \sim p(x)$, this becomes inefficient when sampling from $p(x)$ is difficult. Importance sampling circumvents this issue by introducing a proposal distribution $q(x)$ that is easier to sample from, while preserving the expectation through careful reweighting:

$$\mathbb{E}_p[f(X)] = \int f(x)\frac{p(x)}{q(x)}q(x)\,\mathrm{d}x = \mathbb{E}_q\left[f(X)\frac{p(X)}{q(X)}\right], \tag{2.48}$$

where $w(x) := \frac{p(x)}{q(x)}$ is called the importance weight. The corresponding Monte Carlo estimate becomes:

$$\hat{\mathbb{E}} = \frac{1}{N}\sum_{i=1}^{N} f(x_i)w(x_i), \quad x_i \sim q(x). \tag{2.49}$$

Notably, while direct sampling from $p(x)$ may be challenging, the method requires that $p(x)$ remains computationally tractable to evaluate, which is a necessary condition for calculating the importance weights that is fortunately satisfied in most practical applications.

Example 2.14 (Rare event probability). Estimate $P(X > 4)$ where $X \sim \mathcal{N}(0, 1)$. It is anticipated that applying the standard Monte Carlo simulation by generating samples directly from X is inefficient as it rarely happens that a sample is outside 4 standard deviations. To implement importance sampling, we adopt the following proposal function

$$q(x) = \mathcal{N}(5, 1). \tag{2.50}$$

The estimate is:

$$\hat{P}(X > 4) = \frac{1}{N} \sum_{i=1}^{N} \mathbf{1}_{x_i > 4} \cdot w(x_i), \tag{2.51}$$

where $w(x_i) = p(x_i)/q(x_i)$. Here, the indicator function $\mathbf{1}_{x_i>4}$ equals 1 if the sampled value x_i exceeds the threshold 4 and 0 otherwise, ensuring that only samples corresponding to the rare event contribute to the estimate. The importance weight $w(x_i)$ then corrects for the fact that samples are drawn from the proposal distribution $q(x)$ rather than the target distribution $p(x)$. According to Panel (a) of Figure 2.10 there is a much higher chance that the value exceeds 4 when sampled from the proposal distribution $q(x)$ (shading area). In addition, although sampling an extreme event directly from $p(x)$ is difficult, evaluating $p(x_i)$ (by plugging x_i into the explicit expression of the standard Gaussian density) is straightforward regardless of whether x_i is a normal or extreme event.

Let us set $N = 100$, which mimics the practical situation that we may not have a large number of samples. With these N samples, we can evaluate the probability $\hat{P}(X > 4)$. Since the number of samples is limited, we aim to understand the discrepancy that arises when using different random numbers, just as we did in studying the variance of the standard Monte Carlo simulation in Section 2.3. To this end, we carry out such an experiment K times using independent random numbers. The resulting mean and variance of the estimate using both importance sampling and standard Monte Carlo simulation are shown in Table 2.3. The importance sampling outperforms the standard Monte Carlo method dramatically. When $K = 100$, none of the samples in these K experiments (in total $K \times N = 10,000$ samples) hit the tail for a single time using the standard Monte Carlo simulation. This is not surprising since the true probability of hitting the tail is 3.167×10^{-5}, which on average requires 31,576 samples to have one extreme event. When K increases, the performance of the Monte Carlo simulation indeed improves. However, even with $K = 100,000$, the mean estimate is still different from the truth, and the variance is four orders larger than that of the importance sampling. The latter shows very steady behavior with different numbers of K, where the mean is close to the truth while the variance remains small, which is of the order $O(10^{-11})$. This is as expected, since most of the samples from the importance sampling are kept. Panel (b) in Figure 2.10 shows the distribution using the K independent estimators, which validates its performance in Table 2.3.

TABLE 2.3: Comparison of the importance sampling (IS) and standard Monte Carlo (MC) simulation for estimating $P(X > 4)$ where $X \sim \mathcal{N}(0, 1)$. In each experiment, $N = 100$ points are used to estimate the probability of the extreme events. In total, K independent experiments are conducted to calculate the mean and variance of the estimation.

K	IS (Mean)	IS (Var)	MC(mean)	MC(Var)
100	3.272e-5	7.140e-11	0	0
1,000	3.199e-5	7.149e-11	6.000e-5	5.970e-7
10,000	3.169e-5	7.133e-11	3.600e-5	3.587e-7
100,000	3.163e-5	7.156e-11	2.960e-5	2.971e-7
Truth	3.167e-5			

Unlike importance sampling, where the importance weight varies according to the sample values and affects the results, the standard Monte Carlo method essentially provides the solution by counting the number of samples hitting the tail divided by N. As most of the $N = 100$ samples cannot hit the tail, the majority of the individual experiments give a solution of zero, while very occasionally some experiments have one or two samples hitting the tail, giving solutions 0.01 and 0.02. These findings can be seen in Panel (c) of Figure 2.10. Therefore, even if the mean estimate with $K = 100,000$ has only 7% of the error (which is nevertheless much larger than the importance sampling), not a single individual experiment will nearly hit this value. In contrast, the K estimates from the importance sampling approach almost form a Gaussian distribution, with the mean being almost identical to the truth.

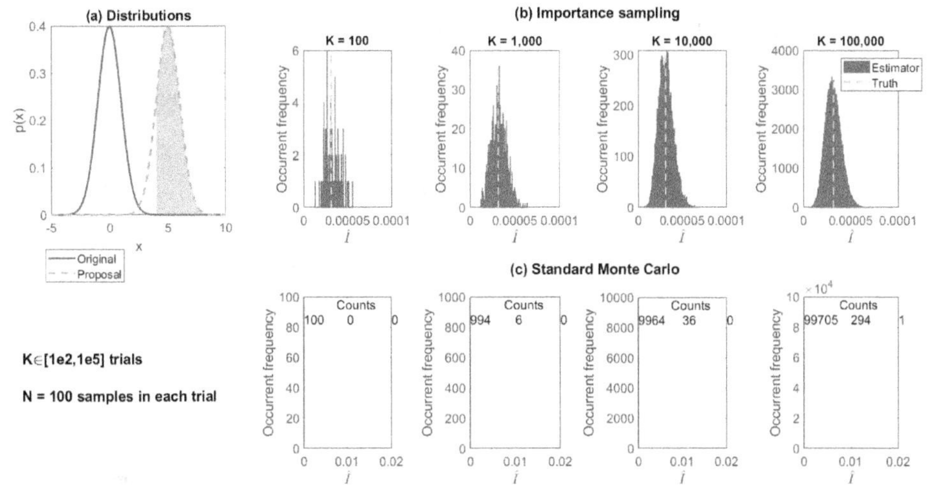

FIGURE 2.10: Comparison of using importance sampling and standard Monte Carlo methods to estimate $P(X > 4)$ where $X \sim \mathcal{N}(0, 1)$. Panel (a): the original and the proposal distributions. Panel (b): estimation using importance sampling, where K independent experiments are carried out and each experiment contains $N = 100$ samples. Panel (c): estimation using the standard Monte Carlo method.

□

Recall from (2.39) that a Monte Carlo simulation can be used to compute the numerical integration within a finite interval. However, what if the domain of integration is infinite and the integrand is complicated?

Example 2.15 (Numerical integration, a complicated function in an infinite domain). Compute the following integral:

$$\int_0^\infty e^{-x^{1.5}}\,\mathrm{d}x. \tag{2.52}$$

The standard Monte Carlo method (with uniformly distributed samples) can still be used by confining the integral in a finite interval. Usually, a large interval is preferred to reduce the mean bias in the estimation. However, as shown in Table 2.4, if this interval is too large, many samples will fall into the tail of the distribution, which contributes very little to the integral value. Each experiment yields quite different results, as shown in Figure 2.11. As a result, the variance becomes large when the interval size increases. In such cases, most individual experiments significantly underestimate the integral value. See Panel (c) of Figure 2.11.

In contrast, we can use a proposal that is an exponential function

$$q(x) = e^{-x}, \tag{2.53}$$

for which random numbers can be easily generated using most existing software. The exponential function is closer to the target function and is more efficient when sampling from it compared to the uniform distribution. As is seen in Table 2.4 and Panel (b) of Figure 2.11, the importance sampling using such a proposal has a more accurate mean estimation compared to the Monte Carlo simulation confined to $x \in [0, 5]$ and the variance of the former is only 1/20 of the latter. The distribution of the estimator is also nearly Gaussian regardless of the number of experiments K.

TABLE 2.4: Mean and variance of using the standard Monte Carlo method and importance sampling to estimate the integral in (2.52). The number of independent experiments is $K = 100,000$. In each experiment, $N = 100$ samples are used. For the Monte Carlo method, different finite intervals are adopted.

Truncated interval	Mean	Var
$x \in [0, 5]$	0.902795	2.03e-2
$x \in [0, 50]$	0.902484	2.77e-1
$x \in [0, 500]$	0.909030	2.85e0
$x \in [0, 5000]$	0.903659	2.85e1
IS	0.902734	1.10e-3
Truth	0.902745	

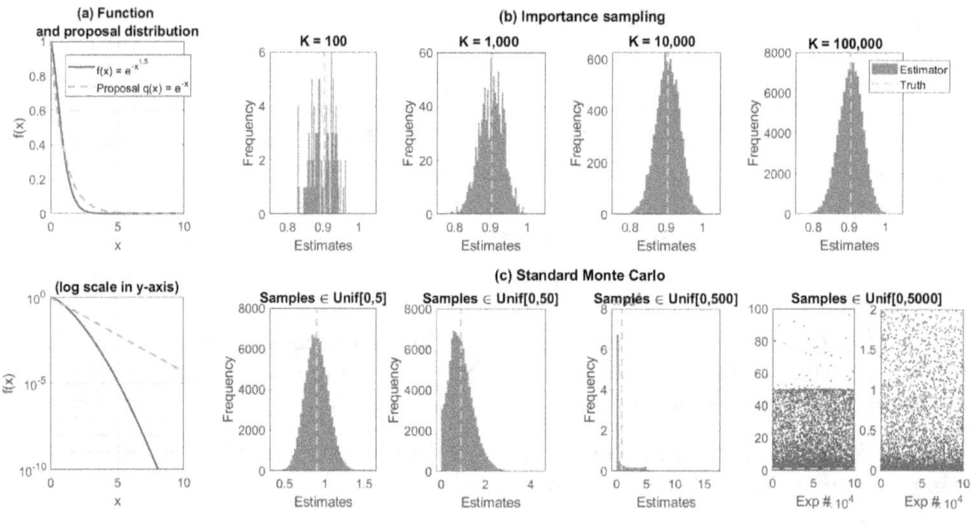

FIGURE 2.11: Comparison of using importance sampling and standard Monte Carlo methods to estimate the integral in (2.52). Panel (a): the original and the proposal distributions, in both linear and logarithmic scales. Panel (b): estimation using importance sampling, where K independent experiments are carried out and each experiment contains $N = 100$ samples. Panel (c): estimating using the standard Monte Carlo method. The last part of Panel (c) is shown using a scatter plot, as the PDF plot in this case, with a wide range of mass concentrated around the original, is nearly invisible.

□

2.3.4 Random optimization: Simulated annealing

The initial concept of the simulated annealing algorithm was first proposed by N. Metropolis et al. in 1953 [193]. In 1983, S. Kirkpatrick et al. [150] successfully introduced the annealing concept into the field of combinatorial optimization. Simulated annealing is a stochastic optimization algorithm based on a Monte Carlo iterative solving strategy, inspired by the annealing process in metallurgy, where a material is heated and slowly cooled to reduce defects. It is used to approximate global optima in complex, non-convex problems by allowing controlled uphill moves to escape local minima.

Before introducing simulated annealing, let us first explain the hill climbing algorithm [245]. Hill climbing is a simple greedy search algorithm, which is widely used in many optimization problems. The greedy algorithm makes a locally optimal choice at each step, based solely on the current state, without considering future consequences or the global structure of the problem. Such algorithms are typically efficient and easy to implement, but they do not guarantee a globally optimal solution. While simple to implement, the main drawback of hill climbing is that it can get stuck in local optima and may not necessarily find the global optimal solution. Note that both hill climbing and gradient descent [87] are iterative optimization algorithms that aim to find a local maximum (hill climbing) or minimum (gradient descent) by making incremental steps toward improving the objective function. The difference is that hill climbing evaluates neighboring solutions discretely (often in combinatorial problems), while gradient descent uses continuous gradient information to guide updates, making it more suitable for differentiable functions in continuous spaces.

In contrast to hill climbing and gradient descent, simulated annealing is a global optimization algorithm. The core idea of simulated annealing is as follows. It starts from a relatively high initial temperature. As the temperature parameter gradually decreases, it combines probabilistic jumping characteristics to randomly search for the global optimum of the objective function in the solution space. This means it can probabilistically escape local optima and eventually converge to the global optimum. The probability calculation references the annealing process in metallurgy. The temperature T serves as the control parameter, the objective function value f represents the internal energy E, and a state of the solid at temperature T corresponds to a solution x_i. The algorithm aims to gradually reduce the objective function f (internal energy E) as the control parameter T decreases, eventually approaching the global minimum (the lowest energy state at low temperatures in annealing), much like the metal annealing process.

An interesting analogy between hill climbing and simulated annealing is as follows. Hill climbing is like a sober rabbit jumping toward higher ground. It finds the nearest peak, but this peak might not be Mount Everest. This represents hill climbing, which doesn't guarantee that the local optimum is the global optimum. Simulated annealing mimics a drunk rabbit. It jumps randomly for a long time. During this period, it might move upward or stay on flat ground. However, it gradually sobers up and begins jumping toward higher ground. This represents simulated annealing.

The detailed steps of the simulated annealing algorithm are as follows:

1. Initialization. Set T_0 (initial temperature), randomly generate an initial solution x_0, and compute its objective function value $E(x_0)$.
2. Reduce the temperature: $T_k = \alpha T_0$, where $0 < \alpha < 1$ (cooling rate) decreases as the iteration step k increases.
3. Apply a random perturbation to the current solution x_k to generate a new solution x_{k+1} in its neighborhood, and compute the corresponding objective function value

$E(x_{k+1})$, then calculate the energy difference:

$$\Delta E = E(x_{k+1}) - E(x_k). \tag{2.54}$$

4. Acceptance criterion. If $\Delta E < 0$, accept the new solution as the current solution. Otherwise, accept it with probability $e^{-\Delta E/T_k}$.
5. If the termination temperature is reached, stop the algorithm; otherwise, return to step 2.

Finally, it is worth remarking that both MCMC methods and simulated annealing rely on random sampling to explore the state space and avoid purely deterministic search. However, there is an important distinction between the two. MCMC is designed to sample from a prescribed target probability distribution, whereas the primary objective of simulated annealing is optimization rather than distributional inference.

Example 2.16 (Global vs. Local Optimization). We compare hill climbing (modified for minimization) and simulated annealing in finding the global minimum of the multimodal function:

$$f(x) = x^2 + 3\sin(5x), \quad x_0 = 4. \tag{2.55}$$

Figure 2.12 shows the results for $K = 2000$ iterations. As is expected, hill climbing quickly converges to the nearest local minimum. In contrast, simulated annealing (with cooling rate $\alpha = 1 - k/K$) escapes local minima through thermal fluctuations. Large exploratory moves occur during early iterations (T high) while the solution meanders near the global minimum during late iterations ($T \to 0$). The temperature schedule balances exploration and exploitation, demonstrating the ability of simulated annealing for non-convex optimization.

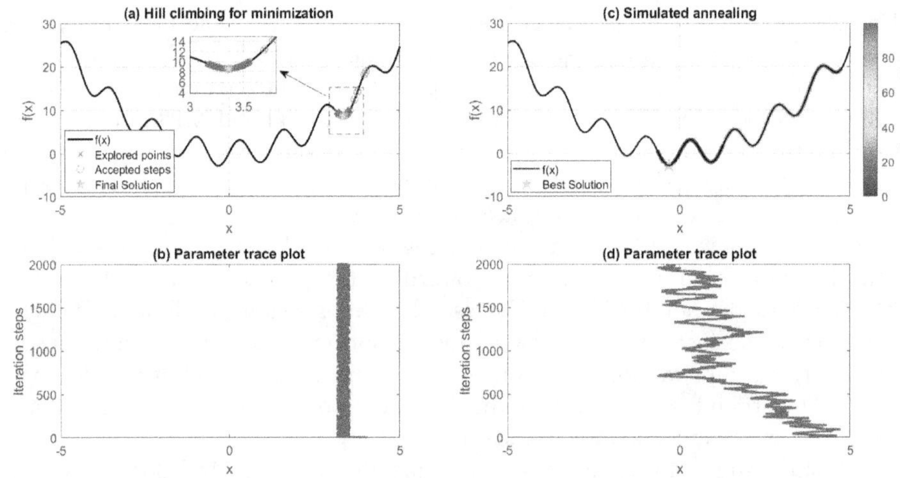

FIGURE 2.12: Optimization trajectories for hill climbing and simulated annealing on $f(x) = x^2 + 3\sin(5x)$. Hill climbing stagnates at a local minimum, while simulated annealing (with linearly decreasing temperature) reaches the global minimum. Shading indicates the annealing temperature $T(k)$.

□

2.4 Code listings for figures and examples

- `Chapter_2_Marginal_Conditional_Gaussian.m`: generates Figure 2.1 in Example 2.3, which illustrates the marginal and conditional distributions of a joint Gaussian.

- `Chapter_2_t_Gamma_Distributions.m`: generates Figure 2.2 in Example 2.4, which illustrates two non-Gaussian distributions with fat tails: the t-distribution and the Gamma distribution.

- `Chapter_2_GMM_Bimodal.m`: generates Figure 2.3 in Example 2.5, which illustrates the use of Gaussian mixture models to fit bimodal data.

- `Chapter_2_GMM_Clustering.m`: generates Figure 2.4 in Example 2.8, which compares clustering results obtained using the Gaussian mixture model and the k-means algorithm.

- `Chapter_2_Chi_Squared.m`: generates Figure 2.5 in Example 2.9, which uses Monte Carlo approximation to generate a chi-squared distribution.

- `Chapter_2_Area_Circle.m`: generates Figure 2.6 in Example 2.10, which demonstrates how Monte Carlo simulation is used to compute the area of the unit circle.

- `Chapter_2_Numerical_Integration.m`: generates Figure 2.7 in Example 2.11, which demonstrates the use of two Monte Carlo methods for numerical integration.

- `Chapter_2_MCMC_Distribution.m`: generates Figure 2.8 and Table 2.2 in Example 2.12, which demonstrates how Markov chain Monte Carlo is used to sample from a function proportional to a probability distribution.

- `Chapter_2_MCMC_Linear_Regression.m`: generates Figure 2.9 in Example 2.13, which demonstrates how Markov chain Monte Carlo can be used for parameter estimation.

- `Chapter_2_MCMC_Importance_Sampling_Rare_Event_Probability.m`: generates Figure 2.10 and Table 2.3 in Example 2.14, which illustrates the advantage of using importance sampling for computing rare-event probabilities.

- `Chapter_2_Importance_Sampling_Numerical_Integration.m`: generates Figure 2.11 and Table 2.4 in Example 2.15, which demonstrates how importance sampling facilitates numerical integration over an infinite domain.

- `Chapter_2_Simulated_Annealing.m`: generates Figure 2.12 in Example 2.16, which compares simulated annealing with hill climbing (modified for minimization) in finding the global minimum of a multimodal function.

3

Machine Learning Toolkit

3.1 Supervised and unsupervised learning

Machine learning has become a popular and powerful tool in contemporary applied math. Beyond the methods themselves, the fundamental ideas behind these techniques are advancing other disciplines.

Machine learning methods can often be categorized into supervised and unsupervised learning, depending on whether labeled outputs are available. In supervised learning, the goal is to learn a mapping from inputs to known targets using labeled data. Examples include regression, where continuous outputs are predicted, and classification, where data are assigned to discrete classes. In addition, many neural network models also belong to this category. In contrast, unsupervised learning aims to reveal the underlying data structures or patterns with no labeled training data. Examples include principal component analysis (PCA) for linear dimensionality reduction and clustering methods for grouping similar data points. While supervised learning focuses on prediction and inference given known targets, unsupervised learning emphasizes representation, structure discovery, and data exploration.

3.2 Principal component analysis

We start with a motivating example before formally introducing principal component analysis (PCA). Intuitively, PCA is a linear dimension reduction method that seeks a small number of orthogonal directions (principal components) along which the data exhibit the largest variance, thereby providing a low-dimensional representation that captures the dominant structure of the data.

Example 3.1 (A motivating example of the PCA). Figure 3.1 shows the scatter plots of (x_i, y_i) with $i = 1, \ldots, 100$ from three data sets. In Panel (a), even though each point is given by a pair of data (x_i, y_i), a clear linear relationship between the two components is observed, which is $x_i = y_i$. Therefore, we simply need one degree of freedom to fully characterize these data points. This simple case serves as a motivation for dimension reduction, though it does not yet capture the full essence of PCA, since the data lie exactly on a line with no variability orthogonal to it. Panel (b) illustrates a situation that is closely related to, but not identical to, PCA. Here, noise is added to the data, so the points no longer lie exactly on the line $y = x$, although their dominant variation remains approximately aligned with this direction. The figure visually suggests identifying a "main direction" of the data and discarding fluctuations orthogonal to it. This idea captures the essence of PCA, which aims to separate directions of large variance from those of small variance. However, the specific line $y = x$ shown in the figure should be viewed only as a

DOI: 10.1201/9781003738459-3

schematic illustration of this dominant direction. In practice, PCA determines this direction by maximizing variance rather than by fitting a regression line. Although both PCA and linear regression can lead to noise reduction, their underlying principles are fundamentally different. This subtle difference will be discussed in the following subsection when linear regression is introduced. Finally, in Panel (c), although we know that a curve can perfectly fit the data and, intuitively, we can reduce the degree of freedom from 2 to 1, there does not exist a linear relationship between x and y that fully characterizes the data. In such a case, PCA is not the best choice, but other nonlinear dimension reduction methods can be utilized to improve the results.

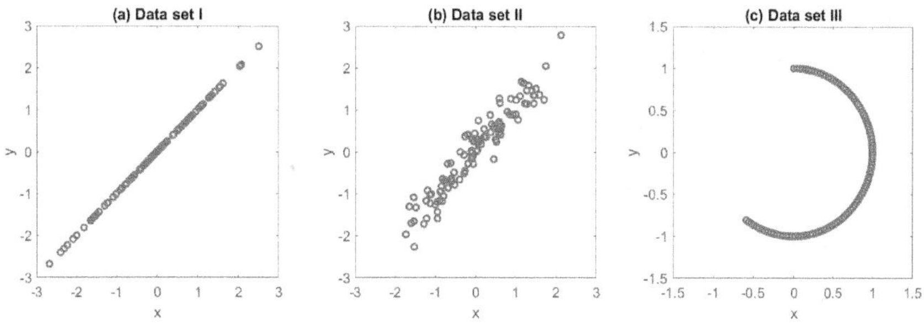

FIGURE 3.1: A motivating example of the PCA: three different data sets.

□

Now let us introduce the mathematical details of PCA. PCA is a fundamental data analysis and machine learning method with many applications [1]. It linearly transforms the data onto a new coordinate system such that the directions (principal components) capturing the largest variation in the data can be easily identified. PCA is primarily used as a dimensionality reduction technique. It also has wide applications in feature extraction, data visualization, and denoising.

Consider n data points $\mathbf{x}_1, \mathbf{x}_2, \ldots, \mathbf{x}_n$, where each $\mathbf{x}_i = (x_{i1}, \ldots, x_{ip})^{\mathrm{T}}$ is a p-dimensional vector. These p components are referred to as features. For example, if $p = 2$, we have n points in a two-dimensional plane. For the convenience of statement, let us represent the data as an $n \times p$ matrix $\mathbf{X}$:

$$\mathbf{X} = \begin{bmatrix} x_{11} & \cdots & x_{1p} \\ \vdots & \ddots & \vdots \\ x_{n1} & \cdots & x_{np} \end{bmatrix}, \tag{3.1}$$

where each row stands for one data point, while each column contains one feature. To find the principal components, the first step is to center the data, which is done by subtracting the mean of each feature:

$$\mathbf{X}_c = \mathbf{X} - \mathbf{1}\bar{\mathbf{x}}^{\mathrm{T}}, \tag{3.2}$$

where $\mathbf{1} = (1, \ldots, 1)^{\mathrm{T}}$ is an $n \times 1$ vector of ones and $\bar{\mathbf{x}} = \frac{1}{n}\mathbf{X}^{\mathrm{T}}\mathbf{1}$ contains the feature means. With the centered data, the next step is to construct a $p \times p$ covariance matrix:

$$\mathbf{C} = \frac{1}{n-1}\mathbf{X}_c^{\mathrm{T}}\mathbf{X}_c, \tag{3.3}$$

which characterizes how different features are correlated with each other. The principal components are then obtained through the eigendecomposition:

$$\mathbf{C}\mathbf{v}_i = \lambda_i \mathbf{v}_i, \qquad i = 1, \ldots, p, \tag{3.4}$$

where $\mathbf{v}_i$ are the eigenvectors (principal directions) and λ_i the corresponding eigenvalues. Usually, λ_i is used to represent the variance of each component, as the unit of λ_i is by design the square of the variable. With the eigendecomposition, the data can be projected onto the new bases, which are orthogonal to each other,

$$\mathbf{Z}_c = \mathbf{X}_c\mathbf{V}, \quad \mathbf{V} = [\mathbf{v}_1 \cdots \mathbf{v}_p]. \tag{3.5}$$

It is worth remarking that, if we do not center the data, then the first principal component may be dominated by the mean of the data rather than the true variance structure. The explained variance captured by PCA will be artificially inflated by the mean offset. Since PCA aims to extract statistical features from data, it is crucial to remove this deterministic effect due to the mean contribution.

In many practical situations, only the leading p' components, corresponding to the first p' rows of the matrix $\mathbf{Z}_c$, are kept (with $p' \ll p$):

$$\tilde{\mathbf{Z}}_c = \mathbf{X}_c\tilde{\mathbf{V}}, \quad \tilde{\mathbf{V}} = [\mathbf{v}_1 \cdots \mathbf{v}_{p'}]. \tag{3.6}$$

The reconstructed data in the original space becomes:

$$\tilde{\mathbf{X}} = \tilde{\mathbf{Z}}_c\tilde{\mathbf{V}}^{\mathrm{T}} + \mathbf{1}\bar{\mathbf{x}}^{\mathrm{T}}. \tag{3.7}$$

This low-rank approximation allows not only dimension reduction but also has wide applications, such as denoising (discarding low-variance components) and data compression (efficient storage).

Example 3.2 (PCA and multivariate Gaussian distribution). PCA and multivariate Gaussian distributions can be fundamentally linked through covariance structure and orthogonal transformation. Recall that the PDF of a multivariate Gaussian random variable $\mathbf{x}$ with mean $\boldsymbol{\mu}$ and covariance $\boldsymbol{\Sigma}$ is given by

$$p(\mathbf{x}) = \frac{1}{(2\pi)^{n/2}|\boldsymbol{\Sigma}|^{1/2}} \exp\left(-\frac{1}{2}(\mathbf{x}-\boldsymbol{\mu})^\top\boldsymbol{\Sigma}^{-1}(\mathbf{x}-\boldsymbol{\mu})\right), \tag{3.8}$$

where in this example we assume the covariance matrix $\boldsymbol{\Sigma}$ in (3.8) is non-diagonal. PCA can be used to transform the state variables into a new coordinate system such that the covariance matrix becomes diagonalized.

The matrix $\mathbf{C}$ in (3.3) using the standard PCA procedure is computed from samples. However, since the covariance of the Gaussian is already given, such information can be directly used to carry out the eigendecomposition

$$\boldsymbol{\Sigma} = \mathbf{V}\boldsymbol{\Lambda}\mathbf{V}^\top, \tag{3.9}$$

where $\mathbf{V}$ contains eigenvectors and $\boldsymbol{\Lambda} = \mathrm{diag}(\lambda_1, \ldots, \lambda_n)$ collects the eigenvalues. With these principal components given by the eigenvectors $\mathbf{V}$, we can carry out the change of variables

$$\mathbf{z} = \mathbf{V}^{\mathrm{T}}(\mathbf{x}-\boldsymbol{\mu}). \tag{3.10}$$

Note that $\mathbf{x}-\boldsymbol{\mu}$ is the original centered variable. The above manipulation leads to a transformed distribution:

$$p(\mathbf{z}) = \prod_{i=1}^{n} \frac{1}{\sqrt{2\pi\lambda_i}} \exp\left(-\frac{z_i^2}{2\lambda_i}\right), \tag{3.11}$$

which is a product of n independent one-dimensional Gaussian distributions. Essentially, PCA rotates the coordinate system to align with the Gaussian's principal axes, eliminating correlations between variables. It is also worthwhile highlighting that the total variance $\mathrm{tr}(\boldsymbol{\Sigma}) = \sum \lambda_i$ remains unchanged, which means PCA is a variance-preserving method.

Figure 3.2 shows applying PCA to find the principal axes of a two-dimensional Gaussian distribution (Panel (a)), which has the mean and covariance:

$$\boldsymbol{\mu} = \begin{bmatrix} 2 \\ 3 \end{bmatrix} \quad \text{and} \quad \boldsymbol{\Sigma} = \begin{bmatrix} 2 & 1.5 \\ 1.5 & 2 \end{bmatrix}. \tag{3.12}$$

The eigenvalues are $\lambda_1 = 3.5$ and $\lambda_2 = 0.5$. They account for 87.5% and 12.5% explained variance, respectively. Panel (b) shows the sampled data points reconstructed from the one-dimensional Gaussian distribution corresponding to the first principal component, which intuitively captures the majority of the information since PC1 contains 87.5% of the explained variance.

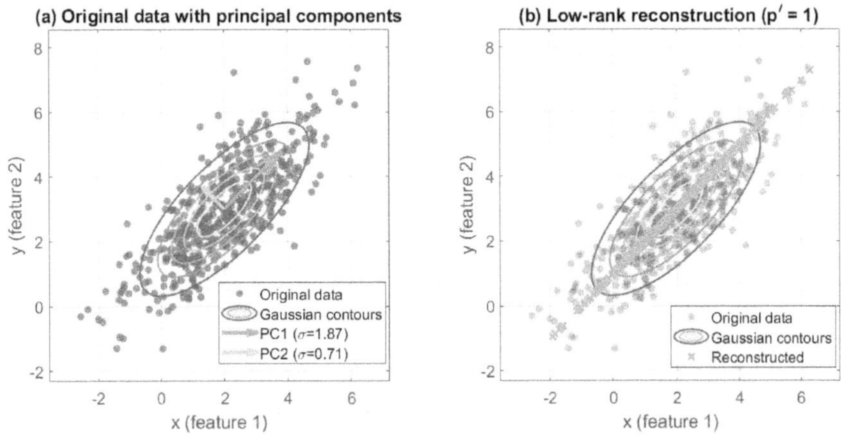

FIGURE 3.2: Principal directions of a two-dimensional Gaussian distribution and its one-dimensional reduction. Panel (a): the two principal components (PC1 and PC2) of the two-dimensional Gaussian, which are marked by arrows. The σ values are the standard deviations, which are the square root of the eigenvalues. Panel (b): the original data and the reconstructed one-dimensional reduced data.

□

Although PCA is illustrated here for static data, it can also be applied to multi-dimensional time series data to reveal the time evolution of the principal components. In addition, although standard PCA is a linear method, it has nonlinear generalizations. These topics will be discussed in the subsequent chapters.

3.3 Clustering

Clustering is a widely used unsupervised learning technique. It seeks to group data points with similar features into clusters without knowing the class labels in advance. Clustering has wide applications, including feature recognition, pattern discovery, and dimensionality reduction. Clustering methods vary in their assumptions and flexibility. Hard clustering algorithms (e.g., k-means) assign each object to exactly one cluster. Soft clustering methods, such as fuzzy c-means (which allows partial membership) and Gaussian mixture models

(GMM) (which model probabilistic membership; see also Example 2.8 in Chapter 2), capture ambiguity in data assignments. Different clustering methods have their own advantages, depending on the features of the data and application scenarios.

3.3.1 K-means algorithm

One of the most fundamental and widely used clustering algorithms is called the k-means algorithm [6, 106]. It aims to partition n observations into k clusters, where each observation belongs to the cluster with the nearest mean, referred to as the cluster centers.

Assume we have n data points: $\mathbf{x}_1, \mathbf{x}_2, \ldots, \mathbf{x}_n$ all in $\mathbb{R}^d$. The number of clusters k is predetermined (the details will be shown later). Let us denote by $C_1, \ldots, C_k$ the k clusters, which are disjoint in the k-means algorithm. Each data point must belong to one cluster. The goals of the k-means clustering algorithm are to find (1) the cluster centers (representatives) of the k clusters, denoted by $\boldsymbol{\mu}_1, \ldots, \boldsymbol{\mu}_k$, and (2) the cluster assignment for each data point. An optimization problem can be formulated to achieve these goals. One natural choice of the loss function in the optimization problem is the total distance between each data point and its corresponding cluster center, where the Euclidean norm is a commonly used distance metric. The loss function reads:

$$L = \sum_{j=1}^{k} \sum_{i \in C_j} \|\mathbf{x}_i - \boldsymbol{\mu}_j\|^2, \tag{3.13}$$

where $\|\cdot\|$ is the standard L^2-norm in the Euclidean space. The index i represents the data point while j stands for the cluster. Unfortunately, finding the exact solution to this optimization problem is computationally an NP-hard problem. Nevertheless, efficient heuristic algorithms converge quickly to a sufficiently skillful local optimum. One commonly used heuristic method is the expectation-maximization (EM) algorithm, which is an iterative refinement procedure. After randomly selecting k points representing the initial choice of the cluster centers, the k-means algorithm becomes an iterative process that alternates between the following two steps:

- In the E-step, assign each data point to the nearest cluster center. Mathematically, for each data point $\mathbf{x}_i$, it is assigned to cluster C_j if:

$$C_j = \arg\min_j \|\mathbf{x}_i - \boldsymbol{\mu}_j\|^2. \tag{3.14}$$

- In the M-step, recompute the cluster center based on the current cluster assignments. As will be shown below, the cluster center is the centroid of the data belonging to this cluster:

$$\boldsymbol{\mu}_j = \frac{1}{|C_j|} \sum_{\mathbf{x}_i \in C_j} \mathbf{x}_i. \tag{3.15}$$

  where $|C_j|$ is the number of points in the cluster C_j.

Essentially, (3.14) and (3.15) seek the optimal partition for a fixed set of representatives and the optimal representatives for a fixed partition, respectively.

Example 3.3 (Simple illustration of the iterative process in the k-means algorithm). A simple example is shown in Figure 3.3, where 20 scatter points in the two-dimensional space need to be clustered into two groups. Panels (a)–(d) in Figure 3.3 illustrate this iterative procedure of finding the cluster centers and the cluster assignments to each data set. The results in each panel are shown after assigning each data point to a cluster (namely, the E-step). As shown in Panel (a), with the new assignments, the cluster centers from the previous step are no longer suitable. Therefore, the two centers are recomputed (in the M-step) based on the assignments shown in Panel (a). Then, with the newly computed cluster centers, an update of the assignments is shown in Panel (b). We notice a significant drop in the loss from Panel (a) to Panel (b). A further improvement is found in Panel (c). After that, additional iterations no longer change the results (Panel (d)). For such a simple problem, the convergence is fast. Panel (e) includes the loss as a function of the number of clusters. As anticipated, the total loss decreases as the number of clusters increases. Yet, determining the optimal number of clusters should be based on a balance between model fit and complexity to avoid overfitting. Exploiting the Akaike Information Criterion (AIC), the optimal number of clusters is identified as $k = 2$ (details will be discussed below). □

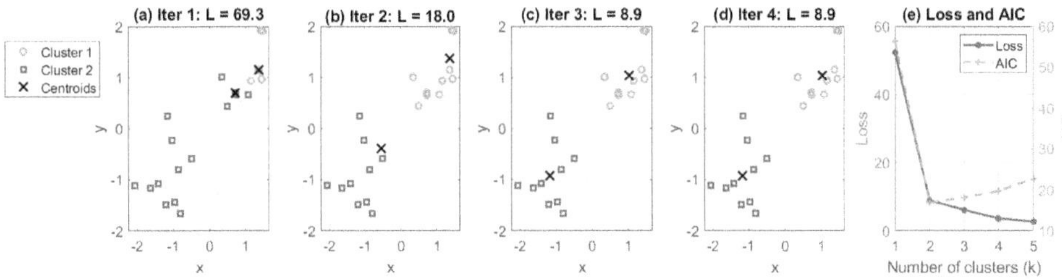

FIGURE 3.3: Clustering 20 scatter points in the two-dimensional space. Panels (a)–(d): the iterative process of the k-means algorithm using two clusters. The results in each panel are shown after assigning the cluster assignment to each data point (namely, the E-step). Panel (e): the total loss function (left y-axis) and the Akaike Information Criterion (AIC; right y-axis) as a function of the number of clusters.

It is seen in Figure 3.3 that the loss function in the k-means algorithm is non-increasing during the iterative process. This fact can be shown by simple mathematical analysis. Let $C'_1, \ldots, C'_k$ be the current clusters, with representatives $\boldsymbol{\mu}'_1, \ldots, \boldsymbol{\mu}'_k$. After the M-Step , the new representatives are $\boldsymbol{\mu}''_1, \ldots, \boldsymbol{\mu}''_k$. As the M-Step finds the optimal representatives given the current cluster assignment, they satisfy

$$\sum_{j=1}^{k} \sum_{i \in C'_j} \|\mathbf{x}_i - \boldsymbol{\mu}''_j\|^2 \leq \sum_{j=1}^{k} \sum_{i \in C'_j} \|\mathbf{x}_i - \boldsymbol{\mu}'_j\|^2.$$

After the E-Step, the updated clusters are $C''_1, \ldots, C''_k$. Similarly, as the E-Step finds the optimal cluster assignments with the given representatives, they satisfy

$$\sum_{j=1}^{k} \sum_{i \in C''_j} \|\mathbf{x}_i - \boldsymbol{\mu}''_j\|^2 \leq \sum_{j=1}^{k} \sum_{i \in C'_j} \|\mathbf{x}_i - \boldsymbol{\mu}''_j\|^2.$$

Combining these two inequalities gives

$$\sum_{j=1}^{k} \sum_{i \in C''_j} \|\mathbf{x}_i - \boldsymbol{\mu}''_j\|^2 \leq \sum_{j=1}^{k} \sum_{i \in C'_j} \|\mathbf{x}_i - \boldsymbol{\mu}'_j\|^2,$$

which means the sequence of loss function values is monotone non-increasing after each iteration. Since the loss function is bounded from below by 0. Hence, it converges.

We have claimed in the algorithm (the M-Step (3.15)) that the cluster center is precisely the centroid of the data within this cluster. Now we show this mathematically. To this end, we rewrite the k-means loss function in the scalar form:

$$\begin{aligned}\sum_{j=1}^{k}\sum_{i\in C_j}\|\mathbf{x}_i-\boldsymbol{\mu}_j\|^2&=\sum_{j=1}^{k}\sum_{i\in C_j}\sum_{m=1}^{d}(x_{i,m}-\mu_{j,m})^2\\&=\sum_{m=1}^{d}\sum_{j=1}^{k}\left(\sum_{i\in C_j}(x_{i,m}-\mu_{j,m})^2\right),\end{aligned}\tag{3.16}$$

where the index m represents the component of the vector $\mathbf{x}_i$. Note that the expression in the bracket is a quadratic function in $\mu_{i,m}$:

$$q_{i,m}(\mu_{i,m})=\underbrace{(|C_i|)}_{\tilde{a}}\mu_{i,m}^2+\underbrace{\left(-2\sum_{j\in C_i}x_{j,m}\right)}_{\tilde{b}}\mu_{i,m}+\underbrace{\left(\sum_{j\in C_i}x_{j,m}^2\right)}_{\tilde{c}}.\tag{3.17}$$

Recall the fact from elementary algebra that the unique minimum of $\tilde{q}(\tilde{x})=\tilde{a}\tilde{x}^2+\tilde{b}\tilde{x}+\tilde{c}$ for $\tilde{a}>0$ is at $\tilde{x}^*=-\tilde{b}/(2\tilde{a})$. Therefore, the minimum of $q_{i,m}$ in (3.17) is obtained at

$$\mu_{i,m}^*=-\frac{-2\sum_{j\in C_i}x_{j,m}}{2|C_i|}=\frac{1}{|C_i|}\sum_{j\in C_i}x_{j,m},$$

which is the centroid of the m-th dimension of the data. Since each term in the sum of (3.16) is strictly minimized at $\boldsymbol{\mu}_1^*,\ldots,\boldsymbol{\mu}_k^*$, so is the total loss function.

Next, recall in the above discussions that we have specified the number of clusters k before carrying out the k-means algorithm. There exist different ways to determine k [153].

One simple but widely used method is the so-called elbow method [86]. This can be done by plotting the loss function against k. Then choose the k at the elbow as the optimal value of k. The elbow method is subjective and empirical. However, the elbow point indicates that the rate of decrease in loss sharply changes, implying that adding more clusters (beyond this point) yields diminishing returns. This elbow point has been used to determine the optimal number of clusters. See the solid curve in Panel (e) of Figure 3.3, which indicates the optimal $k=2$.

Information criteria are also popular methods for determining the optimal k [8]. They are more theoretically sound. Among various information criteria, the Bayesian Information Criterion (BIC) [286] and the Akaike Information Criterion (AIC) [9] are commonly employed. These information criteria balance the model fit and the complexity. For example,

$$\text{AIC}=-\log(L)+2\cdot k.\tag{3.18}$$

Here we use information criteria in a heuristic sense, treating the k-means loss as a measure of data misfit and penalizing model complexity through the number of clusters. This means even though increasing the number of clusters will reduce the total loss, the model becomes more complicated. Therefore, a penalty term is imposed to prevent overfitting. The dashed curve in Panel (e) of Figure 3.3 shows the AIC. Its turning point is at $k=2$ as well, indicating the optimal choice of the number of clusters.

Finally, we want to highlight that k-means and many other clustering algorithms may suffer from the so-called curse of dimensionality [90, 247]. In other words, they may not perform well for high-dimensional problems. See the following example.

Example 3.4 (Applying the k-means algorithm in high-dimensional cases). Let us consider two clustering problems with dimensions 2 and 100, respectively. The data in the first dimension is generated from a Gaussian mixture distribution, with two components $\mathcal{N}(2, 1)$ and $\mathcal{N}(-2, 1)$. The data in the other dimensions are all independently generated from the one-dimensional Gaussian distribution $\mathcal{N}(0, 4)$. For consistency, we use the same random number seeds to generate the samples of the two-dimensional data and the first two dimensions of the 100-dimensional data. We set 2 clusters due to the nature of the Gaussian mixture distribution used in the first dimension. The clustering results are shown in Figure 3.4. As is consistent with observations, in the two-dimensional case (Panel (a)), the k-means algorithm works well. However, even though visualization reveals that in the second case, the clustering results should be based on the first dimension, as the remaining 99 dimensions are all Gaussian white noise, the k-means algorithm cannot provide a reasonable separation of the data points. Intuitively, this is because, in the 100-dimensional setting, the contribution of the informative first dimension to the total Euclidean loss becomes negligible compared to that of the noisy dimensions.

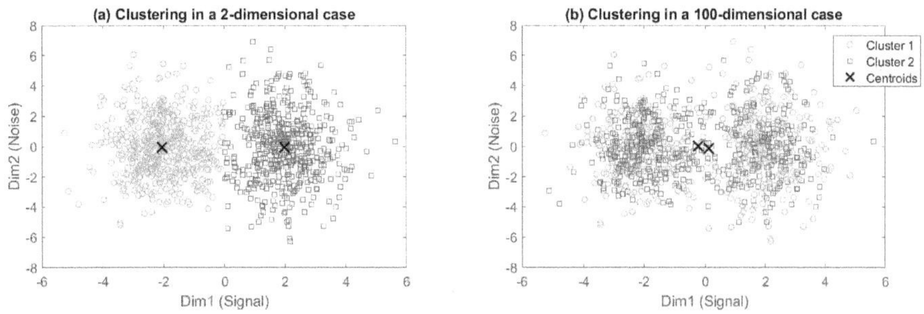

FIGURE 3.4: Two clustering problems with dimensions 2 (Panel (a)) and 100 (Panel (b)), respectively. In the second case, only the data components in the first two dimensions are shown. For consistency, we use the same random number seeds to generate the samples of the two-dimensional data and the first two dimensions of the 100-dimensional data.

□

To provide a more systematic explanation of why the k-means algorithm performs poorly in high-dimensional settings, let us consider points uniformly sampled from the d-dimensional unit hypercube $[0, 1]^d$. Consider two arbitrary data points $\mathbf{x} = (x_1, \ldots, x_d)^\mathrm{T}$ and $\mathbf{y} = (y_1, \ldots, y_d)^\mathrm{T}$, where each component x_i and y_i satisfies a uniform distribution Unif$(0, 1)$. The Euclidean distance squared is

$$\|\mathbf{x} - \mathbf{y}\|^2 = \sum_{i=1}^{d} (x_i - y_i)^2. \tag{3.19}$$

Since all x_i and y_i are i.i.d, the expected distance squared is:

$$\mu := \mathbb{E}\big[\|\mathbf{x} - \mathbf{y}\|^2\big] = d \cdot \mathbb{E}\big[(x_1 - y_1)^2\big] = \frac{1}{6} \cdot d. \tag{3.20}$$

Similarly, the variance of the distance squared is:

$$\sigma^2 := \mathrm{Var}\big[\|\mathbf{x} - \mathbf{y}\|^2\big] = d \cdot \mathrm{Var}\big[(x_1 - y_1)^2\big] = \frac{7}{180} \cdot d. \tag{3.21}$$

The relative deviation of distances shrinks as the dimension d grows

$$\frac{\sigma}{\mu} = C \cdot \frac{1}{\sqrt{d}} \to 0 \qquad \text{as } d \to \infty, \tag{3.22}$$

where the constant $C = 6\sqrt{\frac{7}{180}}$. The result in (3.22) indicates that, in high dimensions, distances between random points converge to almost the same value. Since the k-means algorithm relies on comparing distances to determine cluster assignments, if the distances between one data point and all others (including all the centroids) are nearly identical, the cluster assignments become random.

3.3.2 Fuzzy clustering: Fuzzy c-means and Gaussian mixture models

While k-means assigns each data point to exactly one cluster (hard clustering), fuzzy clustering methods allow for soft assignments, where data points can belong to multiple clusters with varying degrees of membership. This is particularly useful when clusters overlap or when uncertainty exists in the classification. Below, we present two commonly used fuzzy clustering methods. We will then compare these methods with k-means to understand the scenarios each technique is most appropriate for.

Fuzzy c-means (FCM) algorithm.

FCM generalizes the standard k-means algorithm by introducing a membership matrix $U = [u_{ij}]$, where $u_{ij} \in [0, 1]$ represents the degree to which data point $\mathbf{x}_i$ belongs to cluster j [35, 93]. The algorithm minimizes the following objective function:

$$L_{\text{FCM}} = \sum_{j=1}^{k} \sum_{i=1}^{n} u_{ij}^m \|\mathbf{x}_i - \boldsymbol{\mu}_j\|^2, \tag{3.23}$$

where $m > 1$ is a fuzziness parameter (typically $m = 2$). Similar to the k-means method, the FCM algorithm also finds the solution iteratively. It alternatively updates:

- Membership values:

$$u_{ij} = \left(\sum_{l=1}^{k} \left(\frac{\|\mathbf{x}_i - \boldsymbol{\mu}_j\|}{\|\mathbf{x}_i - \boldsymbol{\mu}_l\|} \right)^{2/(m-1)} \right)^{-1}, \tag{3.24}$$

  where the weight u_{ij} is small (large) if the distance from the data point $\mathbf{x}_i$ to the j-th centroid is large (small).

- Cluster centers (weighted centroids):

$$\boldsymbol{\mu}_j = \frac{\sum_{i=1}^{n} u_{ij}^m \mathbf{x}_i}{\sum_{i=1}^{n} u_{ij}^m}. \tag{3.25}$$

Different from k-means, FCM provides probabilistic interpretations of cluster assignments but remains sensitive to initialization and high-dimensional noise.

Gaussian mixture models (GMMs)

Using GMMs for clustering has been briefly mentioned in Chapter 2. Recall that GMMs treat a clustering task as a density estimation problem, assuming data points are drawn from a mixture of k Gaussian distributions. Each cluster is characterized by a mean $\boldsymbol{\mu}_j$ and covariance matrix $\boldsymbol{\Sigma}_j$, and a mixing weight π_j, which is the prior probability of cluster j. The likelihood of a data point $\mathbf{x}_i$ is:

$$p(\mathbf{x}_i) = \sum_{j=1}^{k} \pi_j \mathcal{N}(\mathbf{x}_i \mid \boldsymbol{\mu}_j, \boldsymbol{\Sigma}_j). \tag{3.26}$$

The parameters of a GMM are typically estimated using the EM algorithm [291], which proceeds iteratively as follows:

- (E-step) Compute posterior memberships of each data point $\mathbf{x}_i$ to the mixture component j: $\gamma_{ij} = \frac{\pi_j \mathcal{N}(\mathbf{x}_i|\boldsymbol{\mu}_j,\boldsymbol{\Sigma}_j)}{\sum_{l=1}^{k} \pi_l \mathcal{N}(\mathbf{x}_i|\boldsymbol{\mu}_l,\boldsymbol{\Sigma}_l)}$, which is analogous to the soft assignment in (3.24).
- (M-step) Update the parameters $\boldsymbol{\mu}_j$, $\boldsymbol{\Sigma}_j$, and π_j by maximizing the expected log-likelihood weighted by the posterior memberships.

Note that GMMs offer flexibility in cluster shape (via $\boldsymbol{\Sigma}_j$) and principled probabilistic foundations but require more computational effort than k-means or FCM.

It is worth highlighting the following differences between these two fuzzy clustering algorithms. First, FCM memberships are heuristic, while GMM memberships derive from probability theory. Second, GMMs provide elliptical clusters; FCM assumes spherical clusters like k-means. Third, both GMMs and FCM suffer from the curse of dimensionality. However, GMMs can regularize $\boldsymbol{\Sigma}_j$ to mitigate this issue, for example, by imposing diagonal constraints for the covariance matrices.

Example 3.5 (Comparison between k-means, FCM, and GMM). Figure 3.5 compares the clustering results of k-means, FCM, and GMM in three different scenarios. In the first scenario, data points are well separated into two groups. Therefore, the three methods essentially provide the same clustering results, which are intuitive. However, when data becomes more mixed, as is shown in the second row of the figure, the clustering results using the three methods become distinct. Since k-means is a hard clustering method, it provides a clear boundary between the two clusters. However, since the data points near the boundary are closely clustered, and even a small perturbation can change the cluster to which they belong, the FCM yields more intuitive results by assigning membership degrees to these data points. The GMM provides very distinct clustering results, which is due to the outlier that makes the GMM with two mixture members cover it well. Finally, if the data distribution exhibits elliptical structures, the k-means and FCM can provide similar clustering results as the GMM. However, FCM assigns membership degrees especially to those closer to the other cluster. □

3.4 Classification

Classification aims to predict the correct label of a given input data from the labeled training data. Different from PCA and clustering, classification is a supervised learning method that requires the use of a set of training data to learn a model and then use the model for prediction. While there exist many classification methods, we will only focus on two simple ones. We will use the logistic regression to present the general idea of classification and illustrate the training process, which will also be used in the neural network, as will be discussed in the following subsection. Then we will state the k-nearest neighbor method, using which to highlight the critical concept of overfitting and underfitting.

3.4.1 Logistic regression

Logistic regression is one of the simplest and most fundamental classification algorithms for binary classification [260]. The two classes are usually labeled 0 and 1. Logistic regression predicts the probability that an input $\mathbf{x}$ belongs to one of two classes. Different from linear

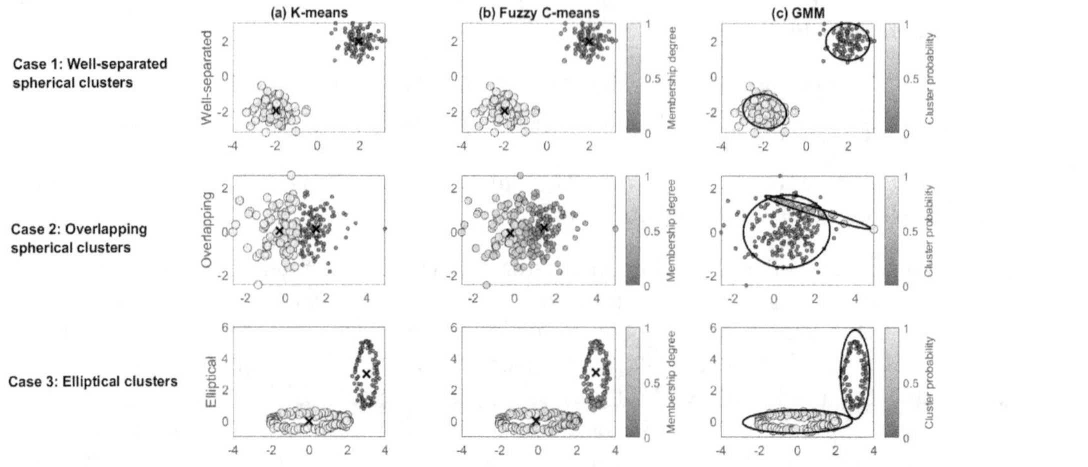

FIGURE 3.5: Comparison the clustering results of k-means, FCM, and GMM in three different scenarios. The cluster centers in k-means and FCM are marked by cross while the clusters of the GMM is marked by circles. Dots with different sizes and shades indicate the clusters they belong to.

regression, whose outputs take continuous real values, logistic regression first forms a linear combination of the input features and then passes this quantity through the so-called sigmoid function, which maps any real-valued input to the interval $(0, 1)$. As a result, the model output can be naturally interpreted as a probability.

The logistic regression model is defined as:

$$P(y = 1|\mathbf{x}) = \sigma(\mathbf{w}^{\mathrm{T}}\mathbf{x} + b), \tag{3.27}$$

where $\mathbf{w}$ is the weight vector, b is the bias term, and $\sigma(z)$ is the sigmoid function [123]:

$$\sigma(z) = \frac{1}{1 + e^{-z}}. \tag{3.28}$$

Equation (3.27) specifies the conditional probability that the class label y equals 1, given the input features $\mathbf{x}$. Note that the sigmoid function does not output the label y itself. Instead, it produces a probability value in the interval $(0, 1)$. The training procedure aims to determine the parameters $\mathbf{w}$ and b. Once the model is trained, a new data point $\mathbf{x}$ is mapped through (3.27) to obtain a predicted probability $p = P(y = 1|\mathbf{x})$. A class label is then assigned by thresholding this probability. If $p > 0.5$, the data point is classified as class 1. Otherwise, it is classified as class 0. The threshold 0.5 is commonly used when false positives and false negatives carry equal costs, though other thresholds may be more appropriate for unbalanced data sets.

Note that the sigmoid is essential because it maps any input $z \in (-\infty, \infty)$ to $(0, 1)$. Therefore, the output can be interpreted as a valid probability. In addition, the derivative of the sigmoid function can be expressed as a simple function of the sigmoid function itself,

$$\sigma'(z) = \sigma(z)(1 - \sigma(z)). \tag{3.29}$$

This property is crucial for gradient descent optimization during the training process, which is the fundamental training algorithm not only for logistic regression but also for neural networks and many other machine learning methods. Figure 3.6 shows the profile of the sigmoid function and its derivative. The sigmoid function will appear again when we discuss the neural network below.

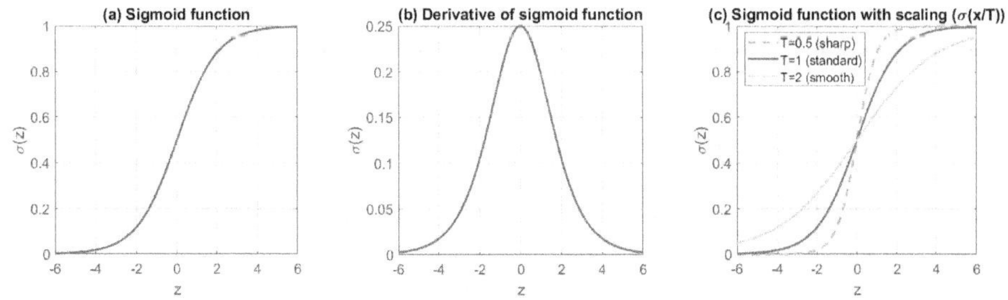

FIGURE 3.6: Sigmoid function and its derivative.

To train a logistic regression model, we introduce a loss function that quantifies the discrepancy between the model predictions and the observed class labels. In binary classification, the binary cross-entropy (BCE) loss is commonly used. Let $y_i \in \{0,1\}$ denote the true class label of the i-th data point, and let

$$\hat{y}_i := P(y_i = 1 \mid \mathbf{x}_i) \tag{3.30}$$

denote the predicted probability produced by the logistic regression model. The BCE loss is defined as

$$\mathcal{L}(\mathbf{w}, b) = -\frac{1}{N}\sum_{i=1}^{N}\Big[y_i \log(\hat{y}_i) + (1-y_i)\log(1-\hat{y}_i)\Big], \tag{3.31}$$

where N is the number of training samples.

This loss function can be derived from a maximum likelihood perspective. Under the logistic regression model, each label y_i is assumed to follow a Bernoulli distribution conditioned on $\mathbf{x}_i$, with success probability $\hat{y}_i$. Therefore, the conditional likelihood of observing y_i given $\mathbf{x}_i$ is

$$P(y_i \mid \mathbf{x}_i) = \hat{y}_i^{\,y_i}(1-\hat{y}_i)^{\,1-y_i}. \tag{3.32}$$

Assuming that the data points are independent, the likelihood of the entire data set is

$$\mathcal{L}_{\text{lik}}(\mathbf{w}, b) = \prod_{i=1}^{N} P(y_i \mid \mathbf{x}_i) = \prod_{i=1}^{N} \hat{y}_i^{\,y_i}(1-\hat{y}_i)^{\,1-y_i}. \tag{3.33}$$

Taking the logarithm of this likelihood and multiplying by $-1/N$ yields the binary cross-entropy loss defined above. Minimizing the BCE loss is therefore equivalent to maximizing the log-likelihood of the observed labels under the logistic regression model.

Gradient descent is commonly used for training [45]. During the training process, the weight update rule is:

$$w_j \leftarrow w_j - \alpha\frac{\partial \mathcal{L}}{\partial w_j}, \tag{3.34}$$

where α is the learning rate, which may be chosen as a fixed constant or adjusted adaptively during training. The gradient is given by

$$\frac{\partial \mathcal{L}}{\partial w_j} = \frac{1}{N}\sum_{i=1}^{N}(\hat{y}_i - y_i)x_{i,j}. \tag{3.35}$$

Example 3.6 (Spam detection). Let us use the logistic regression to predict whether an email is spam ($y = 1$) based on the number of exclamation marks (x). The training data set is $(x, y) = \{(1,0), (3,0), (5,1), (7,1)\}$.

We start by arbitrarily setting the initial values of the parameters to be $w = 0.5$, $b = -2.0$, and the learning rate $\alpha = 0.1$ (for weight update). We compute the forward pass for $x = 5$:

$$\begin{aligned} z &= w \cdot x + b = 0.5 \times 5 - 2.0 = 0.5, \\ \hat{y} &= \sigma(0.5) \approx 0.622. \end{aligned} \tag{3.36}$$

Since the true label of this data point is $y = 1$, the loss function has the value:

$$\mathcal{L} = -\log(0.622) \approx 0.474. \tag{3.37}$$

Then we apply the backward pass:

$$\begin{aligned} \frac{\partial \mathcal{L}}{\partial w} &= (\hat{y} - y) \cdot x = (0.622 - 1) \times 5 \approx -1.89, \\ w_{new} &= 0.5 - 0.1 \times (-1.89) \approx 0.689. \end{aligned} \tag{3.38}$$

After 2000 iterations, we obtain $w^* \approx 2.7642$ and $b^* \approx -11.0567$, which give the decision boundary:

$$x_{decision} = \frac{-b^*}{w^*} \approx 4.00. \tag{3.39}$$

See Figure 3.7. Note that the input data x (the number of exclamation marks) is symmetric ($x = 1, 3, 5$ and 7) regarding the detected decision boundary, which provides an intuitive validation of the classification result. □

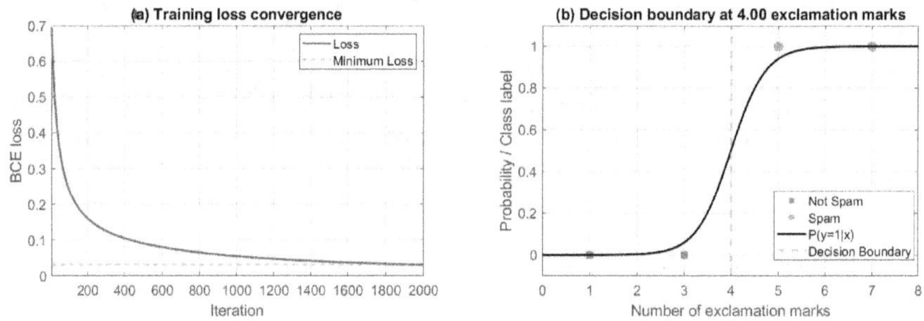

FIGURE 3.7: Logistic regression for spam detection. Panel (a): convergence of the training loss. Panel (b): training sample points and the decision boundary after training.

3.4.2 K-nearest neighbors

K-nearest neighbors (KNN) is another popular and one of the simplest classification methods [104]. Its classification criterion is straightforward. A class label is assigned based on a majority vote among its k nearest neighbors. The KNN algorithm is part of a family of the so-called lazy learning models [293], because it does not have a formal training process, but instead stores a training dataset. Therefore, all the computation occurs when a classification is being made.

The implementation of the KNN is relatively straightforward. Given a new data point, the first step is to compute the distance between the new data point and all training data points. Then, select the k training points closest to this new point. The last step is to calculate the majority voting, based on which we assign the class that appears most frequently among the k neighbors.

KNN has several advantages. First, it is easy to implement, as was seen from the above procedure. Second, it adapts easily. As new training samples are added, the algorithm adjusts to account for any new data since all training data is stored in memory. Third, unlike many other machine learning algorithms that demand many hyperparameters and are more subjective, KNN only requires one hyperparameter k, which makes it more objective. However, it is also worth commenting that KNN often does not scale well. This is because KNN is a lazy algorithm, which means it takes up more memory and data storage compared to other classifiers. The value of k can also largely impact the algorithm. Lower values of k can overfit the data, whereas higher values of k tend to "smooth out" the prediction values.

Example 3.7 (KNN and its overfitting and underfitting). The dataset consists of two classes of interlocking half-moon (semicircular) with random perturbations in $\mathbb{R}^2$. Each class has $N = 250$ samples:

$$\begin{aligned} \text{Class 1:}\quad \mathbf{X}_1 &= \begin{bmatrix} \cos(t) \\ \sin(t) \end{bmatrix} + \boldsymbol{\epsilon}, \\ \text{Class 0:}\quad \mathbf{X}_2 &= \begin{bmatrix} 1-\cos(t) \\ 1-\sin(t) \end{bmatrix} + \boldsymbol{\epsilon}, \end{aligned} \tag{3.40}$$

where the noise satisfies a Gaussian distribution $\boldsymbol{\epsilon} \sim \mathcal{N}(\mathbf{0}, r^2\mathbf{I}_2)$ with $r = 0.15$ while the parameter t is uniformly distributed along the semicircle $t = \left(0, \frac{\pi}{249}, \frac{2\pi}{249}, \ldots, \pi\right)^{\mathrm{T}}$.

Figure 3.8 shows the KNN results with three k values: $k = 1$, $k = 10$ and $k = 200$. We divide the domain $[-1.5, 2.5] \times [-1, 2]$ into a uniformed mesh containing 200×200 grid points. Then we compute the KNN at each grid to determine the class it belongs to. In this manner, we can identify the decision boundary, which is represented by the solid black curves. This visualization facilitates the judgment of whether a data point is misclassified. The top row of this figure shows the labeled reference data, which are stored by the KNN algorithm and used to induce the decision boundary, while the bottom row illustrates the classification of unseen test data. Note that in KNN, there is no explicit training process in the sense of fitting model parameters. Instead, the labeled reference data are directly used at prediction time to determine the class of new data points.

When $k = 1$, the decision boundary is complicated. This is not surprising, as each labeled data point needs to be put exactly into the corresponding class. Due to the noise, many data points overlap, which leads to a complicated decision boundary. Because the decision boundary (and the classification) is primarily determined by the noise, this is often called overfitting [129]. On the other extreme, when $k = 200$, determining the class of a new data point requires the averaged information of its nearby 200 points. Noise is certainly smoothed out. However, the valuable information is dissipated as well. The decision boundary has an oversimplified profile, leading to an unskillful model, known as underfitting. With an appropriate choice of k, which is $k = 10$ here, the decision boundary captures enough information while mitigating the effect from noise.

Note that when overfitting occurs, the classification accuracy evaluated on the reference (in-sample) data is very high. However, because the induced decision boundary is overly sensitive to noise, the method generalizes poorly, and the accuracy on unseen test data is significantly lower. Since the results lack good generalization skills, the accuracy using the testing data is much lower. For underfitting, the decision boundary is overly smooth and fails to capture essential structure in the data, resulting in low accuracy on both the reference data and the test data.

□

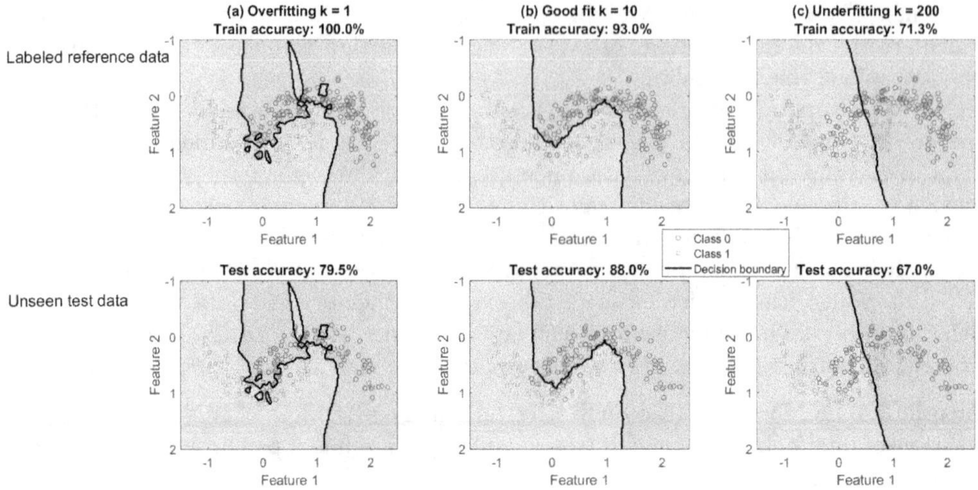

FIGURE 3.8: KNN with $k = 1$ (Panel (a)), $k = 10$ (Panel (b)) and $k = 200$ (Panel (c)). The top row shows the results of determining the decision boundary using the training data while the bottom row illustrates the classification for the unseen test data using the decision boundary from the top row. In this context, we use the term "training data" solely to denote the labeled reference dataset, rather than a dataset used to train a parametric model.

3.5 Regression

Regression is a fundamental machine learning technique for modeling relationships between input variables and continuous output targets. Parametric regression methods assume that the underlying relationship between inputs and outputs can be described by a model with a fixed number of parameters, which is independent of the size of the dataset [127]. The simplest form, linear regression, assumes a linear relationship and provides interpretable parameters. On the other hand, nonparametric methods do not take a specific functional form. The effective model complexity can grow with the amount of data. Therefore, these methods can adapt more flexibly to complex and nonlinear relationships. While there are many nonparametric methods, we will focus on the Gaussian Process Regression (GPR), which provides a flexible Bayesian method by modeling functions as distributions over possible solutions, explicitly capturing uncertainty and smoothness.

3.5.1 Linear regression

Given a data set $\{y_i, x_{i1}, x_{i2}, \dots, x_{ip}\}_{i=1}^{n}$ of n data points, a linear regression model assumes that the relationship between the dependent variable y and the vector of regressors x is linear [107]. This relationship is modeled through a residual term or error variable. Thus, the model takes the form,

$$y_i = \theta_0 + \theta_1 x_{i1} + \cdots + \theta_p x_{ip} + \varepsilon_i = \mathbf{x}_i^{\mathrm{T}} \boldsymbol{\theta} + \varepsilon_i, \qquad i = 1, \dots, n, \tag{3.41}$$

where ε_i is the residual term, $\boldsymbol{\theta} = (\theta_0, \theta_1, \dots, \theta_p)^{\mathrm{T}}$ collects all the parameters and $\mathbf{x}_i = (1, x_{i1}, x_{i2}, \dots, x_{ip})^{\mathrm{T}}$ includes the intercept term. These n equations are often stacked together and written in matrix notation as

$$\mathbf{y} = \mathbf{X}\boldsymbol{\theta} + \boldsymbol{\varepsilon}, \tag{3.42}$$

where

$$\mathbf{y} = \begin{bmatrix} y_1 \\ y_2 \\ \vdots \\ y_n \end{bmatrix}, \quad \mathbf{X} = \begin{bmatrix} \mathbf{x}_1 \\ \mathbf{x}_2 \\ \vdots \\ \mathbf{x}_n \end{bmatrix} = \begin{bmatrix} 1 & x_{11} & \cdots & x_{1p} \\ 1 & x_{21} & \cdots & x_{2p} \\ \vdots & \vdots & \ddots & \vdots \\ 1 & x_{n1} & \cdots & x_{np} \end{bmatrix}, \quad \boldsymbol{\theta} = \begin{bmatrix} \theta_0 \\ \theta_1 \\ \theta_2 \\ \vdots \\ \theta_p \end{bmatrix}, \quad \boldsymbol{\varepsilon} = \begin{bmatrix} \varepsilon_1 \\ \varepsilon_2 \\ \vdots \\ \varepsilon_n \end{bmatrix}. \tag{3.43}$$

The goal is to estimate the coefficients $\boldsymbol{\theta}$. Note that $\mathbf{X}$ is of size $n \times p$, which is usually not a square matrix and often $n \gg p$. The solution $\boldsymbol{\theta}$ is given by the least squares [112]. Minimizing the error term ϵ is equivalent to minimizing

$$\begin{aligned} L &= \|\mathbf{X}\boldsymbol{\theta} - \mathbf{y}\|^2 \\ &= (\mathbf{X}\boldsymbol{\theta} - \mathbf{y})^{\mathrm{T}}(\mathbf{X}\boldsymbol{\theta} - \mathbf{y}) \\ &= \mathbf{y}^{\mathrm{T}}\mathbf{y} - \mathbf{y}^{\mathrm{T}}\mathbf{X}\boldsymbol{\theta} - \boldsymbol{\theta}^{\mathrm{T}}\mathbf{X}^{\mathrm{T}}\mathbf{y} + \boldsymbol{\theta}^{\mathrm{T}}\mathbf{X}^{\mathrm{T}}\mathbf{X}\boldsymbol{\theta}. \end{aligned} \tag{3.44}$$

As the loss is convex, the optimal solution lies at the point where the gradient is zero. The gradient of the loss function is

$$\frac{\partial L}{\partial \boldsymbol{\theta}} = -2\mathbf{X}^{\mathrm{T}}\mathbf{y} + 2\mathbf{X}^{\mathrm{T}}\mathbf{X}\boldsymbol{\theta}. \tag{3.45}$$

Setting the gradient to zero produces the optimum parameter:

$$\begin{aligned} -2\mathbf{X}^{\mathrm{T}}\mathbf{y} + 2\mathbf{X}^{\mathrm{T}}\mathbf{X}\boldsymbol{\theta} &= 0 \\ \mathbf{X}^{\mathrm{T}}\mathbf{X}\boldsymbol{\theta} &= \mathbf{X}^{\mathrm{T}}\mathbf{y} \\ \boldsymbol{\theta} &= (\mathbf{X}^{\mathrm{T}}\mathbf{X})^{-1}\mathbf{X}^{\mathrm{T}}\mathbf{y}. \end{aligned} \tag{3.46}$$

It is worth highlighting that, although the above derivation was presented assuming that the independent variables x_i enter the model linearly, the same procedure applies when the predictors are nonlinear functions of the inputs. In this case, the nonlinearity is absorbed into the design matrix $\mathbf{X}$ through appropriately defined basis functions, while the model remains linear in the parameters $\boldsymbol{\theta}$. As a result, the least-squares solution is still obtained by solving (3.46). In other words, linear regression refers to linearity with respect to the parameters rather than the input variables, and as long as $\boldsymbol{\theta}$ appears linearly in the model, the regression framework and solution remain unchanged.

In many cases, changes of variables can be used to transform problems involving nonlinear functions of parameters into a linear regression framework.

Example 3.8 (Change of variables to fit into linear regression framework). Consider the following scalar equation:

$$y = ax + e^b x^2 + c\sin(2\pi t + \phi)x^3 + \epsilon, \tag{3.47}$$

where the parameters are $\boldsymbol{\theta} = (a, b, c, \phi)$. Although e^b and $c\sin(2\pi t + \phi)$ contain nonlinear functions of b and ϕ, the following manipulations can be carried out to fit this into a standard linear regression problem. First, introduce a new variable $\tilde{b} = e^b$, which builds a one-to-one map between b and $\tilde{b}$. Therefore, $e^b x^2 = \tilde{b}x^2$. Next, introduce two additional variables $\tilde{c}_1$ and $\tilde{c}_2$

$$c\sin(2\pi t + \phi) = c\sin(2\pi t)\cos(\phi) + c\sin(\phi)\cos(2\pi t) := \tilde{c}_1 \sin(2\pi t) + \tilde{c}_2 \cos(2\pi t),$$

where $\tilde{c}_1 = c\cos(\phi)$ and $\tilde{c}_2 = c\sin(\phi)$. Therefore, $\phi = \arctan(\tilde{c}_2/\tilde{c}_1)$ and $c = \tilde{c}_1/\cos(\phi)$ can be uniquely determined once $\tilde{c}_1$ and $\tilde{c}_2$ are estimated. With the new set of parameters $\tilde{\boldsymbol{\theta}} = (a, \tilde{b}, \tilde{c}_1, \tilde{c}_2)^\mathtt{T}$, the equation used for the regression problem becomes

$$y = ax + \tilde{b}x^2 + \tilde{c}_1 \sin(2\pi t)x^3 + \tilde{c}_2 \cos(2\pi t)x^3 + \epsilon, \tag{3.48}$$

where the deterministic part of the right-hand side becomes a linear function of the new parameters $\tilde{\boldsymbol{\theta}}$. This example demonstrates that time-periodic functions, including seasonality in many ocean and atmospheric applications, can be naturally handled in such a parameter estimation framework. □

Both PCA and linear regression can be used to find the optimal linear fit of scatter data and serve as denoising methods. However, the optimization criteria differ, resulting in distinct outcomes.

Example 3.9 (Comparing PCA and linear regression for signal denoising). Let us assume we have data pairs (y_i, x_i) for $i = 1, \ldots, n$. The true relationship between the two variables is $y = 0.8x + 1$. Gaussian noise is added to the observed values of y_i. The goal is to compare PCA and linear regression in denoising the signal. Panels (a)–(b) of Figure 3.9 show the noisy data, the true linear relationship, and the identified linear model using linear regression and PCA, respectively. The fundamental difference between these methods is the following. Linear regression minimizes the vertical distance between the data points and the regression line. This is due to the loss function (3.44) used in linear regression, where the truth and the estimate are evaluated at the same x value. In contrast, PCA aims to find the principal axis (direction of maximum variance), while the perpendicular direction is regarded as noise. Therefore, it minimizes the orthogonal distance from data points to the principal axis, which is given by the length of the perpendicular line drawn from the data points to the principal axis. As shown in Panel (c), as the number of data points increases, the linear regression converges to the truth, as expected, due to the design of this example. In contrast, the principal direction from PCA may deviate from the slope of the linear regression since it maximizes variance rather than minimizes vertical error. In reality, since we often lack prior knowledge of the noise structure, both methods have applications in various scenarios. □

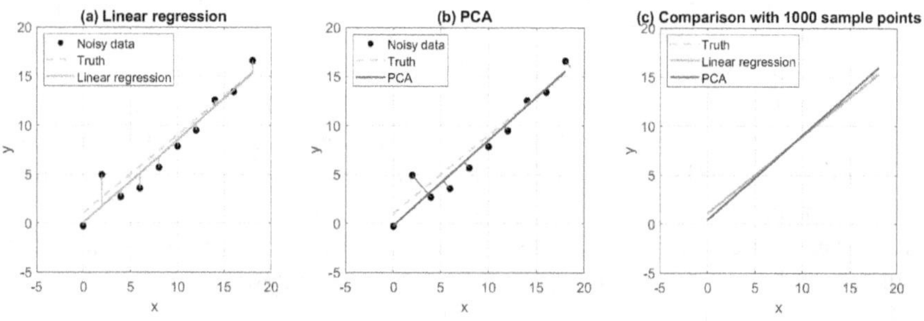

FIGURE 3.9: Comparing PCA and linear regression for signal denoising. Panel (a): illustration of using linear regression for signal denoising. Panel (b): illustration of using PCA for signal denoising. Panel (c): the resulting linear fits using the two methods with a sufficient number of samples (1000 samples).

3.5.2 Gaussian process regression

A Gaussian process (GP) is a supervised learning method used for regression and probabilistic classification [178], and it can be regarded as an infinite-dimensional generalization of multivariate Gaussian distributions. Gaussian process regression (GPR) is a nonparametric Bayesian inference method [240, 285]. Different from parametric approaches, which infer a probability distribution over a finite set of model parameters, GPs infer a distribution over functions. Specifically, a GP assigns a prior distribution over functions before any data are observed, which is fully specified by a mean function and a covariance function, known as the kernel. The kernel includes prior knowledge about the target function, such as smoothness and correlations between inputs. It quantifies how function values at different input locations are statistically related. From a predictive perspective, GPR can also be viewed as a kernel-based interpolation method. The predictions at a new input location are obtained by weighting the contributions of all observed data points, where the weights are determined by their kernel-evaluated correlations with the target point. Observations that are farther from the target point typically have weaker correlations and therefore contribute less to the prediction. This leads to smooth interpolation or extrapolation of the data. After observing input–output data, the prior distribution is updated via Bayes' rule to obtain a posterior distribution over functions, which naturally provides both predictive estimates and associated uncertainty.

To simplify notation, let us consider one-dimensional cases. The framework can be easily generalized to higher dimensions. Before observing any output data, the function values at the observed input locations are assumed to follow a multivariate Gaussian distribution,

$$p(\mathbf{f} \mid \mathbf{x}) = \mathcal{N}(\mathbf{f} \mid \mathbf{m}, \mathbf{K}), \tag{3.49}$$

where $\mathbf{x} = (x_1, \ldots, x_n)^{\mathrm{T}}$ represents the n input locations, $\mathbf{f} = (f(x_1), \ldots, f(x_n))^{\mathrm{T}}$ contains the corresponding (unknown) function values, and $\mathbf{m} = (m(x_1), \ldots, m(x_n))^{\mathrm{T}}$ is the mean function evaluated at these inputs. The positive definite matrix $\mathbf{K}$ is the kernel (covariance) matrix, whose (i, j)-th entry quantifies the prior dependence between the function values at x_i and x_j. For centered data, as in many practical cases, the mean function is set to $m(\mathbf{x}) = 0$, which will also be adopted hereafter.

The distribution $p(\mathbf{f} \mid \mathbf{x})$ thus defines the prior distribution over function values at the observed inputs. The kernel $\mathbf{K}$ encodes prior assumptions about the function, such as smoothness and correlation length scales. Therefore, if two inputs x_i and x_j are considered similar by the kernel, their corresponding function values $f(x_i)$ and $f(x_j)$ are expected to be similar as well. Importantly, at this stage no output observations are used, and no learning from data has yet occurred.

Given observed data, the goal of GPR is to predict the function values at new input locations $\mathbf{x}_*$, denoted by $f(\mathbf{x}_*)$. See Panel (a) of Figure 3.10 for a schematic illustration, where the dots indicate the observed input locations $\mathbf{x}$ and the shaded region represents plausible functions drawn from the Gaussian process. The function values at the new inputs $\mathbf{x}_*$ to be predicted are marked by rectangles.

Next, we consider the joint prior distribution of the function values at the observed and new input locations,

$$\begin{bmatrix} \mathbf{f} \\ \mathbf{f}_* \end{bmatrix} \sim \mathcal{N}\left(\begin{bmatrix} \mathbf{0} \\ \mathbf{0} \end{bmatrix}, \begin{bmatrix} \mathbf{K} & \mathbf{K}_* \\ \mathbf{K}_*^{\mathrm{T}} & \mathbf{K}_{**} \end{bmatrix} \right), \tag{3.50}$$

where $\mathbf{K} = K(\mathbf{x}, \mathbf{x})$, $\mathbf{K}_* = K(\mathbf{x}, \mathbf{x}_*)$, and $\mathbf{K}_{**} = K(\mathbf{x}_*, \mathbf{x}_*)$. This expression specifies the joint prior over the function values before any output data are taken into account.

While the joint distribution describes $p(\mathbf{f}, \mathbf{f}_* \mid \mathbf{x}, \mathbf{x}_*)$, regression requires the conditional distribution of $\mathbf{f}_*$ given $\mathbf{f}$. Using standard results for multivariate Gaussian distributions (see Chapter 2), we obtain

$$\mathbf{f}_* \mid \mathbf{f}, \mathbf{x}, \mathbf{x}_* \sim \mathcal{N}\left(\mathbf{K}_*^{\mathrm{T}} \mathbf{K}^{-1} \mathbf{f}, \mathbf{K}_{**} - \mathbf{K}_*^{\mathrm{T}} \mathbf{K}^{-1} \mathbf{K}_*\right). \tag{3.51}$$

In realistic scenarios, the true function values are not observed directly. Instead, we observe noisy outputs $y = f(x) + \epsilon$, where ϵ is additive i.i.d. Gaussian noise with variance

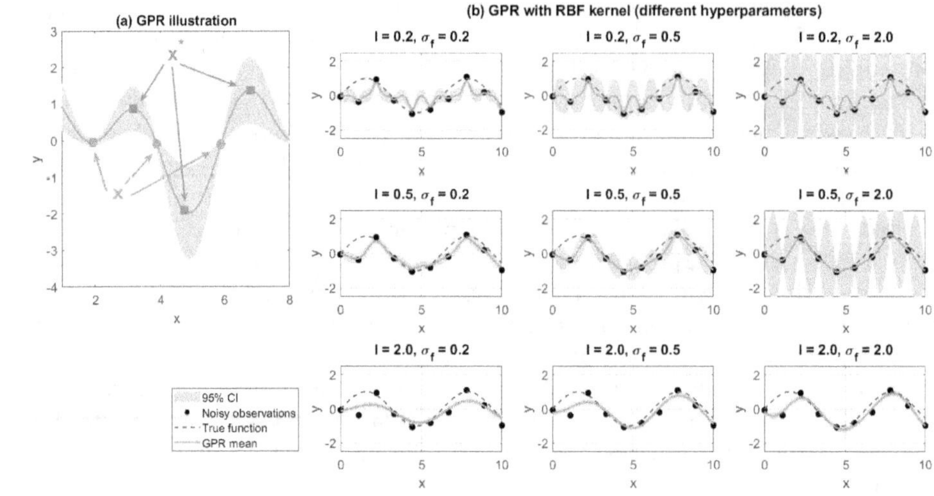

FIGURE 3.10: Gaussian process regression (GPR). Panel (a): a schematic illustration of GPR, where the dots and the shading areas are the input data $\mathbf{x}$ and the estimated function $\mathbf{f}$, respectively. The function values $f(\mathbf{x}_*)$ to be estimated at a few new points $\mathbf{x}_*$ are marked by rectangles. Panel (b): the results of GPR with different hyperparameters using an RBF kernel.

σ_n^2. Incorporating this observation model, the prior covariance of the observed outputs becomes $\mathrm{Cov}(\mathbf{y}) = \mathbf{K} + \sigma_n^2\mathbf{I}$, where $\mathbf{y} = (y_1, \ldots, y_n)^{\mathrm{T}}$. The joint distribution of the observed outputs and the function values at new inputs is then

$$\begin{pmatrix} \mathbf{y} \\ \mathbf{f}_* \end{pmatrix} \sim \mathcal{N}\left(\mathbf{0}, \begin{bmatrix} \mathbf{K} + \sigma_n^2\mathbf{I} & \mathbf{K}_* \\ \mathbf{K}_*^{\mathrm{T}} & \mathbf{K}_{**} \end{bmatrix}\right). \tag{3.52}$$

Conditioning this joint distribution on the observed data $\mathbf{y}$ yields the posterior predictive distribution, which represents the outcome of training the GPR model:

$$\mathbf{f}_* \mid \mathbf{x}, \mathbf{y}, \mathbf{x}_* \sim \mathcal{N}\left(\bar{\mathbf{f}}_*, \mathrm{Cov}(\mathbf{f}_*)\right), \tag{3.53}$$

where

$$\begin{aligned} \bar{\mathbf{f}}_* &= \mathbb{E}[\mathbf{f}_* \mid \mathbf{x}, \mathbf{y}, \mathbf{x}_*] = \mathbf{K}_*^{\mathrm{T}}[\mathbf{K} + \sigma_n^2\mathbf{I}]^{-1}\mathbf{y}, \\ \mathrm{Cov}(\mathbf{f}) &= \mathbf{K}_{**} - \mathbf{K}_*^{\mathrm{T}}[\mathbf{K} + \sigma_n^2\mathbf{I}]^{-1}\mathbf{K}_*. \end{aligned} \tag{3.54}$$

Here, the mean $\bar{\mathbf{f}}_*$ provides the predicted function values, while the covariance $\mathrm{Cov}(\mathbf{f}_*)$ quantifies the remaining uncertainty after conditioning on the data. Notably, the predictive uncertainty depends only on the input locations and the kernel structure, whereas the observed outputs $\mathbf{y}$ influence only the posterior mean.

From this perspective, GPR can be interpreted as a kernel-based interpolation method. Predictions at new points are obtained by combining information from all observed data points, with weights determined by kernel-based correlations. Points that are farther away from the target input are less correlated and therefore contribute less to the prediction.

Example 3.10 (RBF kernel). The radial basis function (RBF) kernel (also called the squared exponential or Gaussian kernel) is a widely used kernel function $\mathbf{K}$ in GPR [268]. It is defined as:

$$K(\mathbf{x}, \mathbf{x}') = \sigma_f^2 \exp\left(-\frac{\|\mathbf{x} - \mathbf{x}'\|^2}{2l^2}\right), \tag{3.55}$$

where $\mathbf{x}, \mathbf{x}'$ are input vectors. The length-scale hyperparameter l determines how quickly the correlation decays with distance. A larger l corresponds to a long-range correlation, which makes the function smoother. In contrast, the RBF kernel with a smaller l often leads to more wiggly functions. The output-scale hyperparameter σ_f determines the amplitude of function variations. A larger σ_f allows bigger peaks/valleys while a smaller σ_f constrains variations closer to the mean. This hyperparameter directly scales the covariance magnitude $\mathrm{Cov}(f(\mathbf{x})) = \sigma_f^2$.

Panel (b) of Figure 3.10 shows the results of GPR with different hyperparameters using an RBF kernel. It is observed that, from top to bottom, the function becomes smoother as l increases. Likewise, from the left to the right column, the amplitude of the function variance increases as a response to the amplification of the hyperparameter σ_f.

□

3.6 Neural networks

Neural networks are a powerful class of machine learning models inspired by the structure and function of the human brain [221, 205, 192]. Composed of interconnected layers of simple processing units called neurons, these networks learn to recognize patterns and make decisions by adjusting their connections based on data. Their ability to automatically learn hierarchical representations, from raw inputs to high-level features, makes them remarkably flexible and powerful. Neural networks have wide applications from image recognition to language translation to forecasting complex systems.

3.6.1 Basic neural network architectures

Example 3.11 (A single-layer perceptron). Let us begin with a single-layer perceptron, a fundamental building block of a neural network. It is one of the simplest neural networks, mainly used to classify input data into two categories.

The architecture presented in Panel (a) of Figure 3.11 is a single-layer perceptron with two inputs x_1, x_2 and one output y. In such a network, each input x_1, x_2 is multiplied by a weight w_1, w_2. Then all the weighted inputs are added together with a bias b, and the sum is passed through an activation function f that gives the output y:

$$y = f(w_1 x_1 + w_2 x_2 + b). \tag{3.56}$$

The activation function f is used to turn an unbounded input into an output that has a predictable form. A commonly used activation function is the sigmoid function (3.28). Recall that one desirable property of the sigmoid function is that its derivative can be expressed as a function of f. See (3.29). This property is crucial to train the neural networks, as will be discussed in the next example.

As a simple illustration, let us assume our 2-input neuron uses the sigmoid activation function and has the following parameters:

$$w = [w_1, w_2] = [0, 1] \qquad \text{and} \qquad b = 4. \tag{3.57}$$

Now, let us give the neuron an input of $x = [2, 3]$. Then

$$y = f(w_1 x_1 + w_2 x_2 + b) = f(0 \cdot 2 + 1 \cdot 3 + 4) = f(7) = 0.999. \tag{3.58}$$

This process of passing inputs forward to get an output is known as feedforward. □

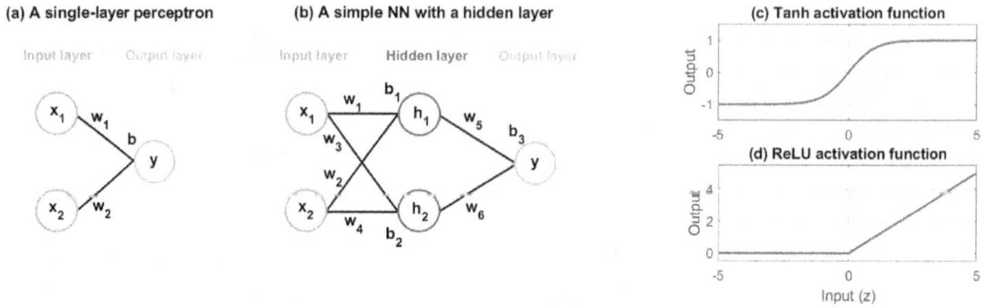

FIGURE 3.11: Panel (a): the architecture of a single-layer perceptron. Panel (b): the architecture of a simple neural network with one hidden layer. Panel (c): tanh and ReLU activation functions.

Example 3.12 (Combining neurons into a neural network). A typical neural network is nothing more than a bunch of neurons connected together. Consider the neural network presented in Panel (b) of Figure 3.11. This network has 2 inputs (x_1 and x_2), an additional hidden layer with 2 neurons (h_1 and h_2), which differs from the simple perceptron, and an output layer with 1 neuron (y). Here, a hidden layer is any layer between the input (first) layer and output (last) layer. Neural networks can have multiple hidden layers. In general, a neural network can have any number of layers with any number of neurons in those layers. The basic idea stays the same: feed the input(s) forward through the neurons in the network to get the output(s) at the end.

Going back to the example presented in Panel (b) of Figure 3.11, assume the bias from input to hidden layer $b_1 = b_2 = 0$ and the bias from hidden to output layer $b_3 = 0$. Let's give the neuron the same input as in the previous example, i.e., $x = [2, 3]$, and the weights $[w_1, w_2] = [w_3, w_4] = [w_5, w_6] = [0, 1]$. Then,

$$\begin{aligned} h_1 &= f(w_1x_1 + w_2x_2 + b_1) = f(0 \cdot 2 + 1 \cdot 3 + 0) = f(3) = 0.9526, \\ h_2 &= f(w_3x_1 + w_4x_2 + b_2) = f(0 \cdot 2 + 1 \cdot 3 + 0) = f(3) = 0.9526, \\ y &= f(w_5h_1 + w_6h_2 + b_3) = f(0 \cdot h_1 + 1 \cdot h_2 + 0) = f(0.9526) = 0.7216. \end{aligned} \tag{3.59}$$

□

3.6.2 Training a neural network

According to Examples 3.11–3.12, once the parameters of weights and biases are specified, the neural network operates by passing information forward from the input layer to the hidden layers and then to the output layer. The operations are simple addition, multiplication, and function evaluations. Therefore, neural networks, with the given parameters, can typically run quite quickly, which enables them to be applied to many practical problems and outperform traditional solvers in terms of computational efficiency. In practice, a neural network may have millions of neurons and parameters. A systematic way of training a neural network to determine its parameters is crucial.

Example 3.13 (Step-by-step training of a neural network). Let us still use the neural network architecture in Panel (b) of Figure 3.11. We need to determine the parameters in the neural network:

$$\boldsymbol{\theta} = (w_1, w_2, w_3, w_4, w_5, w_6, b_1, b_2, b_3). \tag{3.60}$$

A loss function can be used to quantify how skillfully the neural network is doing. One commonly used loss function is the mean squared error (MSE) loss:

$$L := \text{MSE} = \frac{1}{n}\sum_{i=1}^{n}(y_{truth} - y_{pred})^2, \tag{3.61}$$

where n is the number of samples in the training data set and y_{truth} is the true output, while y_{pred} is the output using the current estimation of the parameters.

We train a network, aiming to minimize its loss, which corresponds to a better prediction. Consider a training data with $n = 4$ samples, shown in Table 3.1.

TABLE 3.1: The training data set for a neural network.

Name	x_1: Weight (minus 135lb)	x_2: Height (minus 66in)	y: Gender
Sookie	-2	-1	1
Ray	15	10	0
George	37	6	0
Amy	-15	-6	1

If our network always outputs 0 (male), then the predictions of Sookie and Amy are wrong, which leads to

$$\text{MSE} = \frac{1}{n}\sum_{i=1}^{n}(y_{true} - y_{pred})^2 = \frac{1}{4}(1 + 0 + 0 + 1) = 0.5 \tag{3.62}$$

Below, for the sake of simplicity in presenting the training procedure, let us assume we only have Charlotte in our training dataset. Then the mean squared error loss is just Sookie's squared error:

$$L := \text{MSE} = \frac{1}{1}\sum_{i=1}^{1}(y_{true} - y_{pred})^2 = (1 - y_{pred})^2 \tag{3.63}$$

The loss function $L(\boldsymbol{\theta}) = L(w_1, w_2, w_3, w_4, w_5, w_6, b_1, b_2, b_3)$. To determine the parameters, we make use of the fact that the loss reaches the minimum if $\partial L/\partial\boldsymbol{\theta} = 0$.

Let's focus on the partial derivative in terms of $\partial L/\partial w_1$ using the chain rule:

$$\frac{\partial L}{\partial w_1} = \frac{\partial L}{\partial y_{pred}} \cdot \frac{\partial y_{pred}}{\partial w_1}. \tag{3.64}$$

Since $L = (1 - y_{pred})^2$, we have

$$\frac{\partial L}{\partial y_{pred}} = \frac{\partial(1 - y_{pred})^2}{\partial y_{pred}} = -2(1 - y_{pred}). \tag{3.65}$$

To compute $\partial y_{pred}/\partial w_1$, recall $y_{pred} = f(w_5h_1 + w_6h_2 + b_3)$. Since w_1 only affects h_1 (not h_2), we can write

$$\frac{\partial y_{pred}}{\partial w_1} = \frac{\partial y_{pred}}{\partial h_1} \cdot \frac{\partial h_1}{\partial w_1},$$

where

$$\frac{\partial y_{pred}}{\partial h_1} = w_5 \cdot f'(w_5h_1 + w_6h_2 + b_3), \tag{3.66}$$

and

$$\frac{\partial h_1}{\partial w_1} = x_1 \cdot f'(w_1x_1 + w_2x_2 + b_1). \tag{3.67}$$

This system of calculating partial derivatives (3.64)–(3.67) by working backward is known as backpropagation, or "backprop".

Let us initialize all the weights to 1 and all the biases to 0. If we do a feedforward pass through the network, we get:

$$\begin{aligned} h_1 &= f(w_1x_1 + w_2x_2 + b_1) = f(-2-1+0) = 0.0474, \\ h_2 &= f(w_3x_1 + w_4x_2 + b_2) = f(-2-1+0) = 0.0474, \\ y &= f(w_5h_1 + w_6h_2 + b_3) = f(0.0474 + 0.0474 + 0) = 0.524. \end{aligned} \tag{3.68}$$

The network outputs $y_{pred} = 0.524$, which doesn't strongly favor Male (0) or Female (1). Let's calculate $\partial L/\partial w_1$:

$$\frac{\partial L}{\partial w_1} = \frac{\partial L}{\partial y_{pred}} \cdot \frac{\partial y_{pred}}{\partial h_1} \cdot \frac{\partial h_1}{\partial w_1} = -0.952 \cdot 0.249 \cdot (-0.0904) = 0.0214. \tag{3.69}$$

With $\partial L/\partial w_1$ in hand, we can use an optimization algorithm called gradient descent or its stochastic version that tells us how to change our weights and biases to minimize loss,

$$w_1 \leftarrow w_1 - \eta \frac{\partial L}{\partial w_1}, \tag{3.70}$$

where η is a constant called the learning rate that controls how fast we train. If we apply this process to every weight and bias in the network, the loss will gradually decrease, and our network will improve.

To summarize, training a neural network typically consists of two phases:

1. A forward phase, where the input is passed completely through the network.
2. A backward phase, where gradients are backpropagated and weights are updated.

□

3.6.3 The role of weights and biases

From the above discussions, we have seen that neural networks usually contain two distinct sets of parameters: weights and biases. Weights determine the strength of connections between neurons. However, biases are equally important. They adjust the activation function to allow the neural network to characterize a broader range of functions and patterns.

Example 3.14 (The necessity of including both weight and bias parameters). Let us revisit the perceptron in Panel (a) of Figure 3.11. Assume our goal is to classify whether a student passes (1) or fails (0) an exam based on two factors: x_1 (hours studied) and x_2 (previous exam score). Both x_1 and x_2 are scaled to $[0, 1]$ for the convenience of discussion. The output of the perceptron is defined as

$$y = \begin{cases} 1 & \text{if } (w_1x_1 + w_2x_2 + b) \geq 0, \\ 0 & \text{otherwise.} \end{cases} \tag{3.71}$$

Note that this step function is a nonlinear activation function.

Let us start by considering a case with no bias term, namely $b = 0$. Obviously, w_1 and w_2 should both be positive to reflect the fact that students who are more hardworking or have a more solid background are more likely to pass the exam. Let us take $w_1 = 1$ and $w_2 = 0.5$. For a student whose $x_1 = 0.8$ and $x_2 = 0.6$, the output will be

$$y = 1 \cdot 0.8 + 0.5 \cdot 0.6 + 0 = 1.1 \geq 0. \tag{3.72}$$

The student will pass the exam, which seems to make sense. However, for another student whose $x_1 = 0$ and $x_2 = 0$, which means that the student did not spend any time preparing for the exam and received poor scores previously, the output will still be

$$y = 0 + 0 + 0 = 0 \geq 0, \tag{3.73}$$

and thus, according to the activation function, this student will also pass the exam. The above simple argument shows the problem in the neural network without the bias term.

Now let us consider the case with a bias $b = -0.7$ while the weights remain the same as above. Now the output for the first student will be reduced to 0.4. Nevertheless, the student will still pass the exam. For the second student, the score will be -0.7, indicating a failure on the exam. The bias essentially shifts the threshold so that the neuron only activates when the weighted input is strong enough. □

We can also explain the necessity of including the bias term based on a simple geometric interpretation. The decision boundary is $w_1 x_1 + w_2 x_2 + b = 0$ (even though the activation function is nonlinear). Without the bias term, the boundary must pass through the origin $(0, 0)$. With the bias term, the boundary can shift away from the origin, which provides a better separation. Overall, weights determine the importance of each input. Bias acts like a threshold, which determines whether the neuron should activate even in the absence of input signals. Without bias, the model is too rigid, which cannot represent cases where all inputs are zero, but the output should be negative. In other words, it cannot find the optimal linear decision boundary.

3.6.4 The role of nonlinear activation functions

A crucial feature in neural network design is that the activation functions are typically nonlinear. This is because these nonlinear activation functions allow the network to learn complex patterns, such as curves, discontinuities, and interactions between features. If we used only linear activations, the entire network would collapse into a single linear transformation, regardless of the number of layers it had. This can be seen by the following simple example x(input) $\to h$(hidden) $\to y$(output) with

$$\begin{aligned} &\text{from the input layer to the hidden layer} \qquad h = W_1 x + b_1, \\ &\text{from the hidden layer to the output layer} \qquad y = W_2 h + b_2. \end{aligned} \tag{3.74}$$

Plugging the first equation into the second yields

$$y = W_2(W_1 x + b_1) + b_2 = (W_1 W_2)x + (W_2 b_1 + b_2), \tag{3.75}$$

which is equivalent to one linear layer

$$y = Wx + b \qquad \text{with } W = W_1 W_2, \quad \text{and} \quad b = W_2 b_1 + b_2. \tag{3.76}$$

Therefore, without nonlinearity, depth adds no expressive power.

Example 3.15 (nonlinear activation functions). In the previous example, we have seen that the sigmoid function served as the activation function for both the hidden and output layers. Other commonly used nonlinear activation functions include tanh and ReLU (Rectified linear unit) [51]. The profiles of tanh and ReLU are shown in Panel (c) of Figure 3.11. The properties of these nonlinear activation functions are summarized as follows.

Sigmoid function maps inputs into a 0 to 1 range. It is particularly useful for probability outputs. It has smooth gradients and interpretable outputs (e.g., for binary classification). However, sigmoid activations suffer from the so-called vanishing gradient problem. In fact, when the input is very large or very small, the output saturates near 1 or 0, which causes the gradient to be close to zero. In deep networks, this leads to extremely small gradients being backpropagated to earlier layers, which significantly slows or even prevents effective learning. In addition, sigmoid outputs are not zero-centered, which may further slow convergence. Note that zero-centered activation functions make learning easier because they allow weight updates to move naturally in both positive and negative directions, which leads to faster and more stable training. Furthermore, the function is computationally more expensive than simpler alternatives. Thus, sigmoid activations are typically used only in the output layer for binary classification, rather than in hidden layers of deep networks.

Tanh is another widely used nonlinear activation function. It is similar to sigmoid but maps inputs into a -1 to 1 range. Therefore, the output is zero-centered. It also has stronger gradients than sigmoid, which helps with convergence due to symmetry around zero. However, the tanh activation function still suffers from vanishing gradients at extreme values. Thus, it is sometimes used in hidden layers since it is better than sigmoid for the vanishing gradient issue, but is often replaced by ReLU nowadays.

To further clarify the zero-centered and non-zero-centered activation function, consider a single neuron with input x, weight w, and activation function $\phi(\cdot)$,

$$z_0 = wx, \qquad z_i = w_i a_{i-1}, \qquad a_i = \phi(z_i). \tag{3.77}$$

Let L be a loss function depending on the activation a. By the chain rule, the gradient of the loss with respect to the weight is

$$\frac{\partial L}{\partial w_i} = \frac{\partial L}{\partial a_i} \cdot \phi'(z_i) \cdot a_{i-1}. \tag{3.78}$$

If the activation a is always positive, as is the case for non-zero-centered activation functions such as the sigmoid, the gradient $\partial L/\partial w$ tends to have a consistent sign, as both $\phi'(z_i)$ and a_{i-1} will be positive, leading to biased and inefficient weight updates. In contrast, when the activation function is zero-centered, such as the hyperbolic tangent, the activation a can take both positive and negative values. As a result, the gradient updates are more balanced, allowing gradient descent to move more directly toward the optimum and improving convergence.

Finally, ReLU outputs zero for negative inputs and linear for positive inputs ($f(x) = \max(0, x)$). It avoids vanishing gradients for positive inputs and is highly efficient in computations. Therefore, it works well in deep networks. However, ReLU can suffer from the so-called dying ReLU problem, in which neurons become inactive and output zero for all inputs due to consistently negative pre-activations. ReLU is the default choice for hidden layers in most deep learning models. It has several variants, such as leaky ReLU or parametric ReLU, which can mitigate the issues of the dying neurons. □

Now let us see how nonlinear activations allow neural networks to model nonlinear relationships.

Example 3.16 (The necessity of nonlinear activation functions). Let us assume our data come from the following simple quadratic relationship:

$$y = x^2. \tag{3.79}$$

If we build a neural network with all activation functions being linear, then the best we can do is a straight line

$$y = mx + b, \tag{3.80}$$

which obviously fails to capture the curve and is unskillful in predicting y given x values.

In contrast, using nonlinear activations (e.g., ReLU), a 2-layer network can approximate the parabola:

$$y = w_2\text{ReLU}(w_1 x + b_1) + b_2 \tag{3.81}$$

With enough neurons, the network with ReLU activations can piecewise approximate the curve. □

It is worth highlighting that not all nonlinear functions are equally effective as activation functions in neural networks. Appropriate choices, including ReLU, sigmoid, and tanh functions, have some common properties. First, their nonlinearity breaks linear transformations, which allows neural networks to capture complex patterns. Second, they are (mostly) differentiable, which is essential for applying gradient-based optimization to train the neural networks, though some (like ReLU) have limited non-differentiable points (e.g., at zero). Third, their monotonicity (or piecewise monotonicity, in ReLU's case) helps maintain stable gradient directions during backpropagation, avoiding erratic weight updates. These properties maintain a balance between expressiveness and trainability, making them practical for deep learning.

3.7 Machine learning with limited data

In machine learning, larger training data sets are often preferred to extract key underlying features. However, in many practical problems, the amount of training data is quite limited. While dealing with limited data on a case-by-case basis, the general principle is to provide additional information beyond the training data to help build machine learning models. Putting this into a Bayesian framework, we can always propose useful prior models and then use data to further adjust them. The prior models may have a different format, may not necessarily be explicit models. They can be additional knowledge or constraints that reduce the degree of freedom in training the machine learning models. In this subsection, we provide some fundamental ideas for building machine learning models with limited data. Many relevant and broader techniques will be developed in the subsequent chapters when dynamical and stochastic tools are introduced.

3.7.1 Transfer learning

Transfer learning is a commonly used technique in machine learning and statistical inference [279]. It leverages knowledge gained from one task to enhance performance and improve learning efficiency on another task, assuming they are somewhat related. It is typically used in situations where the training data is insufficient for the second task.

Instead of showing an example using a neural network, the details of which can be tedious even for a simple one, we use a regression example to deliver the idea of transfer learning.

Example 3.17 (Transfer learning using linear regression). We have two related tasks. The source task is to predict house prices based on their size, while the target task is to predict office rental prices based on their size. After training the source model (house prices) with sufficient data, we aim to transfer some knowledge from the source model to help train the target model (office prices) with very little data.

We have data for house prices in the form of (x, y), where x is size and y is price: $\{(1000, 200), (1500, 300), (2000, 400)\}$. We fit a linear regression model:

$$y = ax + b. \tag{3.82}$$

The optimal parameters by applying least squares are $a = 0.2$ and $b = 0$.

Now, suppose we have only one data point for office rentals: $\{(1200, 180)\}$. Instead of training from scratch, we transfer the learned weight $a = 0.2$ and only fine-tune the bias b. Therefore, the target model is

$$y = 0.2x + b. \tag{3.83}$$

Using the single data point $\{(1200, 180)\}$ to determine the parameter b, which is $b = -60$ and leads to the modified model

$$y = 0.2x - 60. \tag{3.84}$$

Note that if we trained only on the target data (with only one data point), we could set:

$$y = 0.15x \qquad \left(\text{with } 0.15 = \frac{180}{1200}\right). \tag{3.85}$$

However, this model is overfit to the single data point.

Transfer learning reuses parameters (a) from a source model and fine-tunes remaining parameters (b) on the target task. This is particularly helpful when the source and target tasks are related (both predict prices based on size) and the target data is limited (here, just one data point). □

In neural networks, transfer learning aims to repurpose a trained model for a new task. For example, a model pre-trained on general image recognition (e.g., identifying cats/dogs) can be adapted for a specific task (e.g., diagnosing X-rays). Usually, we keep early layers frozen (they detect universal features like edges) while we replace the final layer (e.g., swapping a cat/dog classifier for a disease/no disease one), and fine-tune later layers lightly if needed. This way, the model leverages prior knowledge while adapting to new data efficiently, which is especially ideal when labeled data is scarce.

Example 3.18 (Transfer learning for simple 1D classification tasks). To demonstrate transfer learning in neural networks, let us consider the following simple 1D classification tasks:

- Source task: Classify $x \in [0, 1]$ as "low" ($y = 0$ if $x < 0.5$) or "high" ($y = 1$ if $x \geq 0.5$).
- Target task: Classify $x \in [0, 1]$ as "left" ($y = 0$ if $x < 0.6$) or "right" ($y = 1$ if $x \geq 0.6$).

We use the following neural network architecture, which is a minimal network with only one hidden neuron:

$$\begin{aligned} h &= f(W_1 x + b_1) \quad \text{(Hidden layer)} \\ y &= f(W_2 h + b_2) \quad \text{(Output)} \end{aligned}$$

where $f(\cdot)$ is the sigmoid function shown in Panel (c) of Figure 3.11.

We first pretrain the neural network based on the source task (with sufficient training data). The detailed training process follows exactly those described in Example 3.13, so it is omitted here. Let us assume that after training on the source task, we obtain the following parameters:

$$W_1 = 10.2, \quad b_1 = -5.1 \quad \Rightarrow \quad h = f(W_1 x + b_1) \approx \begin{cases} 0, & x < 0.5, \\ 1, & x \geq 0.5, \end{cases}$$

$$W_2 = 4, \quad b_2 = -2.0 \quad \Rightarrow \quad y = f(W_2 h + b_2) \approx \begin{cases} 0, & h = 0, \\ 1, & h = 1, \end{cases}$$

which recovers the exact true decision boundary for the source task.

Now, let us apply transfer learning for the target task. To do so, we freeze W_1 and b_1. We then initialize new output layer weights $\theta = (W_2^{\text{new}}, b_2^{\text{new}})$ and use the training data for the target data to update these two parameters. Let us use the following data points for the target task: $(x, y) = \{(0.4, 0), (0.5, 0), (0.7, 1), (0.8, 1)\}$. We initialize a new output layer with:

$$W_2^{\text{new}} = 1.5, \quad b_2^{\text{new}} = -0.5. \tag{3.86}$$

The loss function is

$$\mathcal{L} = \frac{1}{2}(y_{\text{pred}} - y_{\text{true}})^2, \tag{3.87}$$

and the gradients are:

$$\begin{aligned} \frac{\partial \mathcal{L}}{\partial W_2^{\text{new}}} &= (y_{\text{pred}} - y_{\text{true}}) \cdot f'(z) \cdot h, \\ \frac{\partial \mathcal{L}}{\partial b_2^{\text{new}}} &= (y_{\text{pred}} - y_{\text{true}}) \cdot f'(z), \end{aligned} \tag{3.88}$$

where $z = W_2^{\text{new}} h + b_2^{\text{new}}$. Note that we train the network with momentum $\beta = 0.9$ and learning rate $\eta = 0.05$:

$$\begin{aligned} v_t &= \beta v_{t-1} + \eta \nabla_\theta \mathcal{L}(\theta_t), \\ \theta_{t+1} &= \theta_t - v_t. \end{aligned} \tag{3.89}$$

The learning rate η controls the step size in gradient descent. Small steps ensure stability, avoiding overshooting in parameter updates. However, small learning rates can exacerbate slow convergence when gradients are already small due to vanishing gradients (e.g., when $f'(z) \approx 0$). To address this, we use momentum ($\beta = 0.9$), which accumulates past gradients to maintain update velocity. This helps the model escape flat regions where $f'(z) \approx 0$, smoothes noisy gradient estimates and accelerates convergence in stable regions (without increasing η).

The results are shown in Figure 3.12. Panel (a) shows the convergence of the MSE loss, the weight parameter W_2^{new}, and the bias parameter b_2^{new}. Panel (b) shows the prediction of the resulting network after transfer learning, where we consider all possible values of the input x. Model decision boundary $x = 0.57$ is quite close to the actual boundary $x = 0.6$, especially since we only have four training data.

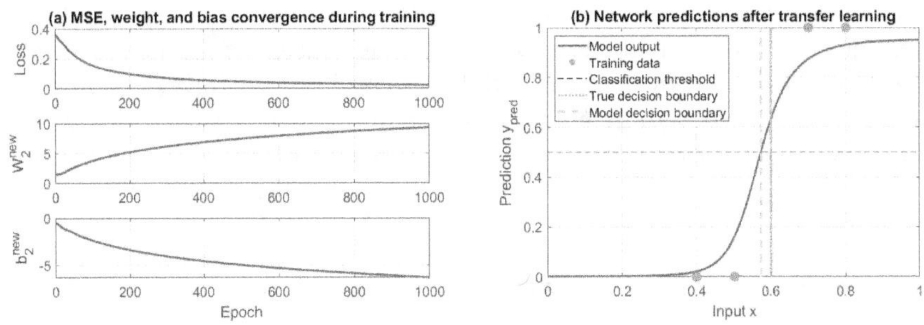

FIGURE 3.12: Transfer learning for the simple classification problems. Panel (a): the convergence of the MSE loss, the weight parameter W_2^{new}, and the bias parameter b_2^{new}. Panel (b): the prediction of the resulting network after transfer learning, where we consider all possible values of the input x.

□

3.7.2 Entropy-constrained learning

Entropy is also widely used as a constraint for learning with limited data [31]. Again, we use a simple example to illustrate the idea and will connect it with neural network applications at the end of this subsection.

Example 3.19 (Estimating the bias (probability of heads) of a fair coin using just 2 flips). The standard method for this problem is the maximum likelihood estimate. Let p denote the probability of observing heads. If we observed [Heads, Tails], which is denoted by $[1, 0]$, then the likelihood is

$$L(p) = p \cdot (1 - p). \tag{3.90}$$

Taking the derivative of L with respect to p and setting it to zero gives the maximum likelihood estimate

$$L'(p) = 1 - p - p = 0 \qquad \Rightarrow \qquad p = 0.5. \tag{3.91}$$

This is intuitive. However, the results may not be reliable as we have only two data points! In fact, if by chance the two flips give us [Heads, Heads], which corresponds to $[1, 1]$, then the likelihood becomes

$$L(p) = p^2. \tag{3.92}$$

The maximum of L is obtained at $p = 1$. This leads to a biased coin that always gives heads due to the limited training data.

To overcome the issue with limited data, we can incorporate additional knowledge into the optimization problem. The idea here is to restrict p to values where the entropy $H(p)$ is sufficiently high, preventing extreme certainty. The details of entropy and its usage in measuring uncertainty will be discussed in Chapter 8. Here, we simply need to use the basic formula of entropy in such a case,

$$H(p) = -p \log p - (1 - p) \log(1 - p). \tag{3.93}$$

Entropy is a measurement of chaos. It is maximized at $p = 0.5$ (corresponding to the maximum of disorder; $H(0.5) \approx 0.693$) and minimized at $p = 0$ or $p = 1$ (corresponding to rare events). We can impose a constraint, which requires $H(p) > 0.5$ to allow moderate uncertainty in this problem. This narrows the range of solutions to $p \in [0.11, 0.89]$, which excludes the overconfident estimates (e.g., $p = 0$ or $p = 1$). Since a coin with two flips can only give $p = 0, 0.5$ or 1, this constraint already leads to the solution $p = 0.5$. □

In neural networks, entropy constraints are used to calibrate uncertainty by penalizing low-entropy predictions on small datasets. They are also used for semi-supervised learning, which encourages high entropy on unlabeled data to avoid overfitting. In addition, Bayesian neural networks use entropy to quantify epistemic uncertainty. The entropy can be a hard constraint, or it can be part of the loss function, serving as a penalty.

3.8 Code listings for figures and examples

- `Chapter_3_PCA_Gaussian_Dimension_Reduction.m`: generates Figure 3.2 in Example 3.2, which applies principal component analysis to find the principal component of a two-dimensional Gaussian distribution.

- `Chapter_3_PCA_Motivation.m`: generates Figure 3.1 in Example 3.1, which provides motivations for the principal component analysis.

- `Chapter_3_K_Means_Standard.m`: generates Figure 3.3 in Example 3.3, which shows the iterative process of the k-means algorithm as well as determining the optimal number of clusters using the elbow method and the AIC.

- `Chapter_3_K_Means_High_Dimension.m`: generates Figure 3.4 in Example 3.4, which shows the failure of the k-means algorithm in handling high-dimensional data.

- `Chapter_3_Clustering_Comparison.m`: generates Figure 3.5 in Example 3.5, which compares clustering results of k-means, fuzzy c-means, and Gaussian mixture model in three different scenarios.

- `Chapter_3_Logistic_Regression.m`: generates Figure 3.7 in Example 3.6, which uses the logistic regression to predict whether an email is spam based on the number of exclamation marks.

- `Chapter_3_KNN.m`: generates Figure 3.8 in Example 3.7, which illustrates the use of the k-nearest neighbors method for classification and the issues with underfitting and overfitting.

- `Chapter_3_Linear_Regression_PCA_Comparison.m`: generates Figure 3.9 in Example 3.9, which illustrates the difference in denoising in using the linear regression and the principal component analysis.

- `Chapter_3_GPR.m`: generates Panel (b) of Figure 3.10 in Example 3.10, which shows the Gaussian process regression with different hyperparameters using a radial basis function kernel.

- `Chapter_3_Transfer_Learning.m`: generates Figure 3.12 in Example 3.18, which shows the use of transfer learning for a simple classification problem.

4

Dynamical Toolkit

4.1 Ordinary differential equations

Ordinary differential equations (ODEs) are the foundations of modeling continuous change in mathematics, science, and engineering [126, 272]. From predicting the growth of populations to designing control systems in robotics, ODEs provide the tools to describe how quantities evolve based on their current state. In this section, we will provide a brief overview of ODEs. For more details, we refer the reader to, for example, [249, 121, 135].

Let u denote a scalar state variable and t represent time. In dynamical systems, time is always a fundamental independent variable.

Example 4.1 (Solving a linear ODE: constant velocity to position). One of the simplest ODEs is

$$\frac{\mathrm{d}u}{\mathrm{d}t} = c, \tag{4.1}$$

where c is a constant. If u represents velocity, $\mathrm{d}u/\mathrm{d}t$ is the acceleration. According to Newton's second law, constant acceleration corresponds to motion under a constant force. Thus, this equation models motion under constant acceleration. Given an initial condition $u(0)$, the solution $u(t)$ is derived as follows:

$$\begin{aligned} \mathrm{d}u &= c\,\mathrm{d}t \quad \text{(separation of variables)} \\ \int_{u(0)}^{u(t)} \mathrm{d}u' &= \int_0^t c\,\mathrm{d}s \quad \text{(integration on both sides)} \\ u(t) - u(0) &= ct \quad \text{(evaluating the integral)} \\ u(t) &= ct + u(0). \end{aligned} \tag{4.2}$$

□

Example 4.2 (Difference between algebraic equations and ODEs). Algebraic equations describe static relationships between variables. Their solutions are numbers (or vectors) rather than functions. For example, the equation $u^2 - 5u + 6 = 0$ has the solutions $u = 2$ or $u = 3$. In contrast, ODEs describe dynamic processes involving rates of change, and their solutions are functions. A simple example is the one in Example 4.1, whose solution is $u(t) = ct + \text{constant}$, representing a family of curves.

While algebraic equations determine equilibrium or static states, ODEs track how quantities evolve over time, capturing the dynamical behavior of the underlying system. Solving ODEs typically involves integration or numerical approximation.

Many real-world problems involve quantities that evolve over time or space due to rates of change (e.g., velocity, growth, decay). Algebraic equations alone cannot describe how a system transitions from one state to another. Therefore, ODEs are essential when the change itself is part of the unknown. □

DOI: 10.1201/9781003738459-4

4.1.1 General first-order linear ODEs

The general form of a first-order linear ODE is

$$\frac{du}{dt} + a(t)u = b(t). \tag{4.3}$$

Here, "first-order" indicates that the highest derivative is du/dt, and "linear" means u appears only linearly, excluding nonlinear terms such as u^2 or u^3.

Attempting to solve this equation using the same method as (4.2) leads to a complication:

$$\begin{aligned} du + a(t)u\,dt &= b(t)\,dt \\ \int_{u(0)}^{u(t)} du' + \int_0^t a(s)u\,ds &= \int_0^t b(s)\,ds. \end{aligned} \tag{4.4}$$

Without knowing $u(t)$ explicitly, the integral $\int_0^t a(s)u\,ds$ cannot be evaluated directly.

If $b(t) \equiv 0$, the equation becomes homogeneous (meaning no external forcing terms independent of the unknown function u) and can be solved via separation of variables:

$$\begin{aligned} \frac{du}{u} &= -a(t)\,dt \\ \ln u(t) - \ln u(0) &= -\int_0^t a(s)\,ds \\ u(t) &= u(0)\exp\left(-\int_0^t a(s)\,ds\right), \end{aligned} \tag{4.5}$$

where the last step uses the property $e^{A+B} = e^A e^B$.

Returning to the inhomogeneous case (4.3), we employ an integrating factor. Define

$$\Psi(t) = \exp\left(\int_0^t a(s)\,ds\right). \tag{4.6}$$

Multiplying the equation by $\Psi(t)$ gives

$$\Psi\frac{du}{dt} + \Psi a(t)u = \Psi b(t). \tag{4.7}$$

Since $\Psi'(t) = \Psi(t)a(t)$, the left-hand side becomes

$$\frac{d}{dt}\left(\Psi(t)u(t)\right) = \Psi(t)b(t). \tag{4.8}$$

Integrating from 0 to t yields

$$\Psi(t)u(t) - \Psi(0)u(0) = \int_0^t \Psi(s)b(s)\,ds. \tag{4.9}$$

Thus,

$$u(t) = \frac{\Psi(0)}{\Psi(t)}u(0) + \frac{1}{\Psi(t)}\int_0^t \Psi(s)b(s)\,ds. \tag{4.10}$$

This provides the general solution to the first-order linear ODE.

In addition to solving the equation, it is essential to intuitively understand the role of different components in such a first-order ODE.

Example 4.3 (The effect of damping, forcing, and their balance). Consider the homogeneous equation with constant $a(t) \equiv a > 0$:

$$\frac{\mathrm{d}u}{\mathrm{d}t} = -au. \tag{4.11}$$

The solution is $u(t) = u(0)e^{-at}$, exhibiting exponential decay. The term $-au$ acts as a damping mechanism, driving the solution toward the equilibrium $u = 0$. If $a < 0$, the solution grows exponentially, and the equilibrium becomes unstable.

On the other hand, if $a(t) \equiv 0$, equation (4.3) reduces to

$$\frac{\mathrm{d}u}{\mathrm{d}t} = b(t), \tag{4.12}$$

where $b(t)$ represents external forcing. The solution is

$$u(t) = u(0) + \int_0^t b(s)\,\mathrm{d}s. \tag{4.13}$$

For constant b, this becomes $u(t) = u(0) + bt$, which grows linearly in time.

When both damping and forcing are present,

$$\frac{\mathrm{d}u}{\mathrm{d}t} = -au + b, \tag{4.14}$$

their competition can lead to a balance. If $a > 0$ and b are constants, the solution converges to a steady-state (equilibrium) solution as $t \to \infty$. Since $\mathrm{d}u/\mathrm{d}t = 0$ at steady state,

$$0 = -au + b, \tag{4.15}$$

which gives the equilibrium value

$$u(\infty) = \frac{b}{a}. \tag{4.16}$$

Thus, in the long-term limit, the effects of damping and forcing exactly balance each other. □

4.1.2 High-order linear ODEs

An n-th order ODE can be rewritten as a system of n first-order ODEs by introducing new variables representing successive derivatives. This transformation is purely structural and applies to both linear and nonlinear ODEs. It explains the fundamental importance of studying first-order systems.

Example 4.4 (Transforming a second-order ODE into a first-order system). Consider the second-order linear ODE:

$$\frac{\mathrm{d}^2u}{\mathrm{d}t^2} + c_1(t)\frac{\mathrm{d}u}{\mathrm{d}t} + c_0(t)u = c_2(t). \tag{4.17}$$

Introduce a new variable $v = \mathrm{d}u/\mathrm{d}t$. Then $\mathrm{d}v/\mathrm{d}t = \mathrm{d}^2u/\mathrm{d}t^2$, and substituting into (4.17) gives

$$\frac{\mathrm{d}v}{\mathrm{d}t} = -c_0(t)u - c_1(t)v + c_2(t). \tag{4.18}$$

Therefore, (4.17) is equivalent to the first-order system

$$\begin{aligned}\frac{\mathrm{d}u}{\mathrm{d}t} &= v,\\ \frac{\mathrm{d}v}{\mathrm{d}t} &= -c_0(t)u - c_1(t)v + c_2(t).\end{aligned} \tag{4.19}$$

This transformation shows that higher-order ODEs can always be reformulated as first-order systems, which form the foundation for modern dynamical systems theory and numerical methods. □

4.1.3 Solving first-order linear systems

Consider the linear system

$$\frac{\mathrm{d}\mathbf{u}}{\mathrm{d}t} + \mathbf{A}\mathbf{u} = \mathbf{f}, \tag{4.20}$$

where $\mathbf{u} \in \mathbb{R}^n$ and $\mathbf{A}$ is a constant $n \times n$ matrix. Assume that $\mathbf{A}$ is diagonalizable: $\mathbf{A} = \mathbf{C}\mathbf{\Lambda}\mathbf{C}^{-1}$, where $\mathbf{\Lambda}$ is diagonal and $\mathbf{C}$ contains the eigenvectors. Multiplying the equation by $\mathbf{C}^{-1}$ gives

$$\mathbf{C}^{-1}\frac{\mathrm{d}\mathbf{u}}{\mathrm{d}t} + \mathbf{\Lambda}\mathbf{C}^{-1}\mathbf{u} = \mathbf{C}^{-1}\mathbf{f}. \tag{4.21}$$

Define the new variable $\mathbf{v} = \mathbf{C}^{-1}\mathbf{u}$. Then

$$\frac{\mathrm{d}\mathbf{v}}{\mathrm{d}t} = \mathbf{C}^{-1}\frac{\mathrm{d}\mathbf{u}}{\mathrm{d}t}, \tag{4.22}$$

and the system becomes

$$\frac{\mathrm{d}\mathbf{v}}{\mathrm{d}t} + \mathbf{\Lambda}\mathbf{v} = \mathbf{C}^{-1}\mathbf{f}. \tag{4.23}$$

Since $\mathbf{\Lambda}$ is diagonal, the system decouples into n independent scalar equations, which can be solved component-wise. The original solution is recovered via

$$\mathbf{u} = \mathbf{C}\mathbf{v}. \tag{4.24}$$

Example 4.5 (Linear oscillators). The coupled system

$$\begin{aligned}\frac{\mathrm{d}u_1}{\mathrm{d}t} &= -au_1 + \omega u_2 + f_1,\\ \frac{\mathrm{d}u_2}{\mathrm{d}t} &= -\omega u_1 - au_2 + f_2,\end{aligned} \tag{4.25}$$

with $a > 0$ is called a linear oscillator. Here, ω determines the oscillation frequency, a is the damping, and f is the external forcing.

Denote $\mathbf{u} = (u_1, u_2)^{\mathrm{T}}$ and $\mathbf{f} = (f_1, f_2)^{\mathrm{T}}$. Then (4.25) can be written as

$$\frac{\mathrm{d}\mathbf{u}}{\mathrm{d}t} + \mathbf{A}\mathbf{u} = \mathbf{f}, \qquad \mathbf{A} = \begin{pmatrix} a & -\omega \\ \omega & a \end{pmatrix}. \tag{4.26}$$

The eigen-decomposition is $\mathbf{A} = \mathbf{C}\mathbf{\Lambda}\mathbf{C}^{-1}$, where

$$\mathbf{\Lambda} = \begin{pmatrix} a + i\omega & \\ & a - i\omega \end{pmatrix}, \qquad \mathbf{C} = \begin{pmatrix} 1 & 1 \\ -i & i \end{pmatrix}. \tag{4.27}$$

Let $\mathbf{v} = \mathbf{C}^{-1}\mathbf{u}$. Then

$$\frac{\mathrm{d}\mathbf{v}}{\mathrm{d}t} + \mathbf{\Lambda}\mathbf{v} = \mathbf{C}^{-1}\mathbf{f} = \begin{pmatrix} \frac{f_1 + if_2}{2} \\ \frac{f_1 - if_2}{2} \end{pmatrix}. \tag{4.28}$$

Hence (for constant f_1, f_2),

$$v_1(t) = v_1(0)e^{-(a+i\omega)t} + \frac{f_1 + if_2}{2(a+i\omega)}\left(1 - e^{-(a+i\omega)t}\right),$$
$$v_2(t) = v_2(0)e^{-(a-i\omega)t} + \frac{f_1 - if_2}{2(a-i\omega)}\left(1 - e^{-(a-i\omega)t}\right).$$

Transforming back via $\mathbf{u} = \mathbf{C}\mathbf{v}$ yields

$$\mathbf{u}(t) = e^{-at}\begin{pmatrix} A\cos\omega t + B\sin\omega t \\ -A\sin\omega t + B\cos\omega t \end{pmatrix} + \frac{1}{a^2+\omega^2}\begin{pmatrix} af_1 + \omega f_2 \\ -\omega f_1 + af_2 \end{pmatrix}, \tag{4.29}$$

where A, B are determined by $\mathbf{u}(0)$. The first term represents a damped oscillation originating from the initial state, while the second term is the steady-state response induced by the forcing.

The linear oscillation (4.25) also admits a complex representation. Define $u = u_1 + iu_2$ and $f = f_1 + if_2$. Then

$$\frac{du}{dt} = (-a - i\omega)u + f, \tag{4.30}$$

so

$$u(t) = u(0)e^{(-a-i\omega)t} + \frac{f}{a+i\omega}\left(1 - e^{(-a-i\omega)t}\right). \tag{4.31}$$

where the constant C is determined by the initial condition $u(0)$. The homogeneous term $Ce^{(-a-i\omega)t}$ describes damped oscillations arising from the initial state, while the particular solution $\frac{f}{a+i\omega}$ represents the steady-state response.

Linear oscillators are essential tools for modeling complex systems. As will be seen in later chapters, adding suitable stochastic noise allows linear oscillators to realistically capture large-scale features of many natural phenomena identified through data-driven mode decomposition methods. In this setting, they characterize not only the dynamics but also the associated statistics. Moreover, although many natural phenomena involve complex fields with many degrees of freedom, linear oscillators can still be used effectively. In particular, they can model each component of the system separately, such as individual Fourier modes. □

4.1.4 Simple nonlinear ODEs

We now focus on nonlinear ODEs with constant coefficients. Recall the first-order linear ODE in (4.3), whose equilibrium $u_* = b/a$ (when $a \neq 0$) is asymptotically stable if $a > 0$, unstable if $a < 0$, and marginal when $a = 0$.

For nonlinear systems

$$\frac{du}{dt} = F(u), \tag{4.32}$$

the behavior is richer: multiple equilibria may exist, and stability becomes a local property. An equilibrium u^* is stable if nearby solutions converge to it, but global behavior may involve other attractors or divergence. Small parameter changes can qualitatively alter system dynamics.

Example 4.6 (Multiple equilibria in a cubic nonlinear model). Consider the cubic system:

$$\frac{du}{dt} = f + au + bu^2 - cu^3, \tag{4.33}$$

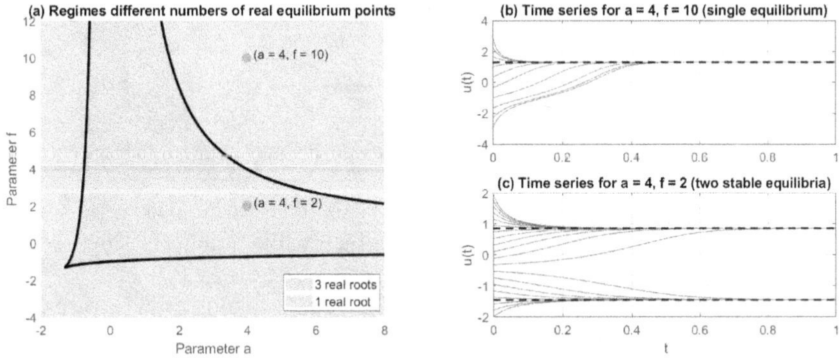

FIGURE 4.1: Regimes of the cubic model (4.33) with $b = -4$, $c = 4$. Panel (a): partition of the (a, f)-plane into regions with different numbers of real equilibria. Panel (b): trajectories starting from different initial conditions with the parameters $a = 4$ and $f = 10$ in the regime with only one equilibrium. Panel (c): trajectories starting from different initial conditions with the parameters $a = 4$ and $f = 2$ in the regime with three equilibria (two being stable). The stable equilibria in Panels (b)–(c) are shown by the horizontal dashed lines.

where a controls the linear growth/decay rate, f represents constant external forcing, while the terms bu^2 and cu^3 introduce nonlinear effects. Let us fix the parameters $b = -4$ and $c = 4$ and vary a and f. The equilibria can be found by solving

$$f + au - 4u^2 - 4u^3 = 0. \tag{4.34}$$

This cubic equation can yield either one real root (plus two complex conjugate roots), or three real roots, depending on parameters (a, f). The discriminant is

$$\Delta = -18abcf - 4b^3 f + b^2 a^2 + 4ca^3 - 27c^2 f^2. \tag{4.35}$$

When $\Delta > 0$, there exist three real roots. In contrast, when $\Delta < 0$, only one real root exists.

As a case study, we take $a = 4$ and $f = 10$. In such a case, strong forcing f dominates the dynamics, suppressing multistability, and therefore these parameters lead to a single equilibrium. The system has one stable equilibrium at $r_1 \approx 1.2868$ (and two complex conjugate roots $r_{2,3} \approx -1.1434 \pm 0.7971i$). All trajectories converge to r_1, regardless of initial conditions (see Panel (b) of Figure 4.1).

On the other hand, with $a = 4$ and $f = 2$, there exist three equilibria. Correspondingly, the system admits three real roots:

$$r_1 \approx -1.4516 \text{ (stable)}, \quad r_2 \approx -0.4030 \text{ (unstable)}, \quad r_3 \approx 0.8546 \text{ (stable)}. \tag{4.36}$$

If the initial condition is exactly equal to r_2, the solution remains at this value for all time. However, even an arbitrarily small perturbation will cause the trajectory to move away from r_2, which is why it is referred to as an unstable equilibrium. For any initial condition other than r_2, trajectories converge to either of the two stable equilibria r_1 or r_3, depending on the initial condition (see Panel (c) of Figure 4.1).

□

Example 4.7 (Predator-prey model). The classic predator-prey model (also known as the Lotka-Volterra model) describes interacting populations [50, 138]. The model consists of a pair of first-order nonlinear differential equations:

$$\begin{aligned}\frac{dx}{dt} &= \alpha x - \beta xy,\\ \frac{dy}{dt} &= \delta xy - \gamma y.\end{aligned} \tag{4.37}$$

All variables and parameters are positive and real. The variable x represents the prey population density (e.g., rabbits per square kilometer), while y denotes the predator density (e.g., foxes per square kilometer).

The derivatives dx/dt and dy/dt quantify the instantaneous growth rates. In the prey equation, the term αx describes exponential growth assuming unlimited resources, while $-\beta xy$ models predation losses proportional to encounter rates. The predator equation contains a gain term δxy linking predator reproduction to prey consumption, offset by natural mortality $-\gamma y$. The asymmetry in coefficients ($\beta \neq \delta$) reflects energy transfer inefficiencies between trophic levels. The model makes three key biological assumptions: (1) prey populations grow exponentially in the absence of predators, (2) predation rate is proportional to both population densities, and (3) predator mortality follows exponential decay in the absence of prey.

This system behaves as a nonlinear oscillator: although the dynamics are nonlinear, solutions evolve along closed orbits in the phase plane due to the nonlinear coupling between species. The phase portrait shown in Figure 4.2 depicts trajectories in the state space spanned by prey and predator populations. These closed orbits correspond to sustained oscillations whose amplitudes depend on initial conditions. In particular, predator peaks follow prey peaks, reflecting the delayed response of predators to changes in prey abundance.

FIGURE 4.2: Phase portrait (left) and time evolution (right) of the predator-prey system.

□

4.2 Bifurcation in dynamical systems

A bifurcation occurs in a dynamical system when a small change in a parameter leads to a qualitative change in its behavior, such as the creation or destruction of equilibrium points or a shift in their stability [217, 82, 77]. In many real systems, parameters naturally vary due to external forcing or gradual environmental changes. For example, increasing the driving force in a mechanical system may cause a previously stable equilibrium to become unstable. In ecology, gradually increasing harvesting pressure can suddenly drive a population from a stable positive level to extinction. In Earth science, slow changes in radiative forcing may push the system past a tipping point, triggering a transition to a qualitatively different regime. In each case, the parameter changes smoothly, but the system response changes abruptly. Bifurcations are fundamental to understanding how systems transition between different dynamical regimes. They provide a framework for predicting critical transitions in systems ranging from mechanical oscillators to biological networks. Understanding them helps in analyzing stability and long-term behavior under parameter variations. There are many types of bifurcations. We provide three commonly seen bifurcations to present the ideas.

4.2.1 Saddle-node bifurcation

The saddle-node bifurcation is one of the simplest bifurcations, where two fixed points collide and annihilate [164].

Example 4.8 (Saddle-node bifurcation). Consider the one-dimensional system:

$$\frac{\mathrm{d}x}{\mathrm{d}t} = r + x^2, \tag{4.38}$$

where r is a parameter. For $r < 0$, there are two equilibrium points: $x^* = \pm\sqrt{-r}$. The positive fixed point is unstable, meaning that trajectories starting nearby move away from it under small perturbations, while the negative one is stable, meaning that nearby trajectories are attracted toward it. At $r = 0$, the equilibria merge into a single semistable equilibrium (stable from one side and unstable from the other), and for $r > 0$, no real equilibria exist. Thus, $r = 0$ is the bifurcation point where the system undergoes a sudden change. See the first row of Figure 4.3 for a numerical illustration. □

4.2.2 Transcritical bifurcation

In a transcritical bifurcation, two equilibrium points exchange stability [250].

Example 4.9 (Transcritical bifurcation). Consider the following system:

$$\frac{\mathrm{d}x}{\mathrm{d}t} = rx - x^2. \tag{4.39}$$

Here, the fixed points are $x^* = 0$ and $x^* = r$. For $r < 0$, $x^* = 0$ is stable, and $x^* = r$ is unstable. At $r = 0$, they coincide, and for $r > 0$, the stabilities swap: $x^* = 0$ becomes unstable, while $x^* = r$ becomes stable. See the second row of Figure 4.3 for a numerical illustration. □

4.2.3 Pitchfork bifurcation

A pitchfork bifurcation occurs when a single equilibrium point splits into three as a parameter crosses a critical value. This bifurcation is common in symmetric systems and comes in two forms: supercritical (stable) and subcritical (unstable). Pitchfork bifurcations arise in systems with inherent symmetry, such as buckling beams or magnetic phase transitions [283].

Example 4.10 (Supercritical pitchfork bifurcation). Consider the system:

$$\frac{\mathrm{d}x}{\mathrm{d}t} = rx - x^3, \tag{4.40}$$

where r is a parameter. For $r < 0$, the only equilibrium is $x^* = 0$, which is stable. At $r = 0$, the system loses linear stability, and for $r > 0$, two new stable equilibria appear at

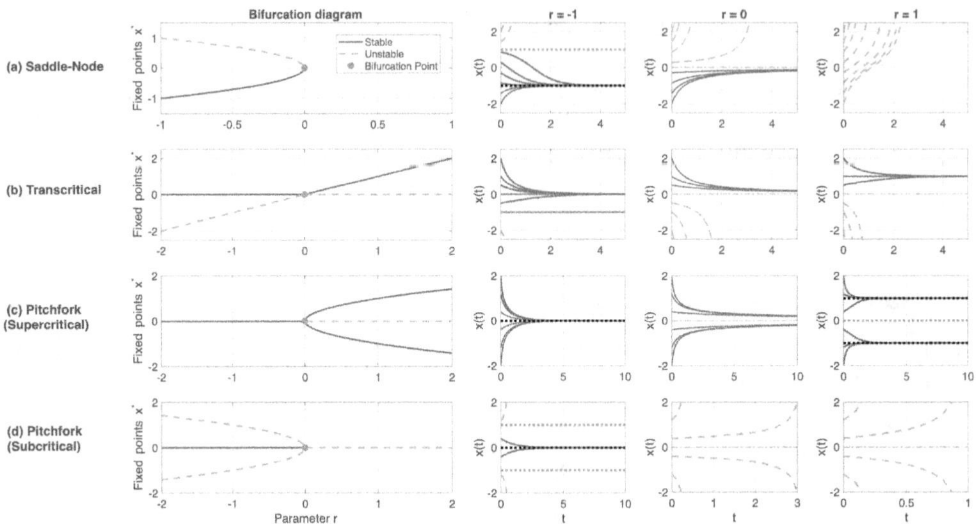

FIGURE 4.3: Bifurcation examples of different types. Row (a): saddle-node bifurcation. Row (b): Transcritical bifurcation. Row (c): Supercritical pitchfork bifurcation. Row (d): Subcritical pitchfork bifurcation. The first column shows the bifurcation diagram. The solid curves represent the stable fixed points, with their value (y-axis) and the corresponding parameter choice r. The dashed curves show the unstable equilibria (fixed points). The black dots show the bifurcation point. The other three columns show the trajectories starting from different initial conditions with three r choices. The dotted horizontal lines show the fixed point with black and gray being the stable and unstable ones, respectively. The thin gray dash-dot horizontal lines show the bifurcation points. The curves are trajectories, solid being converging and dashed being diverging.

$x^* = \pm\sqrt{r}$, while $x^* = 0$ becomes unstable. The bifurcation diagram resembles a pitchfork, with a stable symmetric branch splitting into two stable branches. See the third row of Figure 4.3 for a numerical illustration. □

Example 4.11 (Subcritical pitchfork bifurcation). The subcritical case is described by:

$$\frac{\mathrm{d}x}{\mathrm{d}t} = rx + x^3. \tag{4.41}$$

Here, for $r < 0$, $x^* = 0$ is stable, while $x^* = \pm\sqrt{-r}$ (unstable) exist. At $r = 0$, all three equilibria merge, and for $r > 0$, only $x^* = 0$ remains but is now unstable. See the fourth row of Figure 4.3 for a numerical illustration. □

4.2.4 Tipping points

Tipping points represent critical thresholds where small changes in external forcing can trigger significant and often irreversible shifts in the state of the Earth system [167, 166]. Examples include the collapse of ice sheets, shutdown of ocean circulation, or rapid dieback of ecosystems. These transitions are not gradual but abrupt, and once crossed, the system may not return to its original state even if the forcing is reduced. Understanding such responses is crucial for anticipating risks to both natural and human systems.

Bifurcation theory provides a mathematical framework to study tipping points in dynamical systems [257, 22]. In simple terms, a bifurcation occurs when the number or stability of equilibrium states changes as a control parameter varies. For example, when greenhouse gas concentration (treated as a model parameter) increases, the system may shift from multiple stable climate states to only one. Saddle-node bifurcations are especially relevant, as they capture the disappearance of a stable state at a critical threshold, which is precisely the mechanism behind many tipping phenomena in the Earth system. Through bifurcation diagrams, one can visualize where stable and unstable states exist and identify the conditions under which tipping points emerge.

It is worth noting that simple deterministic models often cannot perfectly describe real-world systems. These systems are subject to small-scale variability and unresolved processes that act like noise. Therefore, stochastic models (see Chapter 6) are often used to account for these fluctuations. Near a tipping point, random perturbations can push the system across stability boundaries earlier than predicted deterministically, leading to noise-induced tipping. Moreover, stochastic methods facilitate the identification of early-warning signals such as increasing variance and autocorrelation, which arise from "critical slowing down" as a system approaches a bifurcation. These tools provide both theoretical insight and practical diagnostics for detecting climate systems at risk of tipping before the transition occurs.

4.3 Chaotic systems

Chaotic systems are a class of nonlinear dynamical systems commonly observed in nature [176, 37]. One of their defining characteristics is sensitivity to initial conditions, meaning that even infinitesimal perturbations in the starting state can lead to drastically different trajectories over time. Such sensitivity implies that deterministic predictions are only reliable over short time horizons, as small uncertainties in the initial state amplify exponentially, rendering long-term path-wise forecasts unreliable.

In contrast to systems influenced by stochastic noise, where randomness explicitly drives unpredictability, chaotic systems exhibit intrinsic unpredictability arising purely from deterministic nonlinear dynamics. Nevertheless, as will be seen in subsequent chapters, stochastic models often serve as effective surrogates for characterizing the statistical behavior of deterministic chaotic systems.

4.3.1 Lyapunov exponents

A central mathematical concept underlying chaotic behavior is the Lyapunov exponent, which quantifies the exponential rate of divergence or convergence of nearby trajectories in dynamical systems [271, 163]. For a d-dimensional system, there exist d Lyapunov exponents $\{\lambda_1 \geq \lambda_2 \geq \cdots \geq \lambda_d\}$, forming the Lyapunov spectrum.

For a dynamical system $\dot{\mathbf{x}} = \mathbf{f}(\mathbf{x})$, consider a reference trajectory $\mathbf{x}(t)$ and an infinitesimal perturbation $\delta\mathbf{x}(0)$, which evolves into $\delta\mathbf{x}(t)$ along the flow. Different perturbation directions may grow or decay at different exponential rates. The maximal Lyapunov exponent corresponds to the largest possible exponential growth rate among all infinitesimal perturbations. It can be defined as

$$\lambda_1 = \lim_{t\to\infty} \lim_{\|\delta\mathbf{x}(0)\|\to 0} \frac{1}{t} \ln \frac{\|\delta\mathbf{x}(t)\|}{\|\delta\mathbf{x}(0)\|}. \tag{4.42}$$

Although this expression does not explicitly display a maximization over directions, the limit selects the dominant growth rate. Indeed, unless the initial perturbation lies exactly in a lower-dimensional invariant subspace, its component along the most unstable direction will eventually dominate, leading to exponential growth at rate λ_1.

A positive maximal Lyapunov exponent ($\lambda_1 > 0$) is a strong indicator of chaotic behavior, reflecting exponential divergence of nearby trajectories. If $\lambda_1 = 0$, the system may exhibit neutral dynamics, such as periodic or quasiperiodic motion. If the maximal exponent satisfies $\lambda_1 < 0$, then all nearby trajectories converge exponentially, indicating a stable equilibrium.

4.3.2 Intuitive explanation of the mechanism of chaotic systems

While a rigorous treatment and computation of Lyapunov exponents requires advanced mathematical tools beyond the scope of this discussion, an intuitive understanding can be gained through a simple example.

Example 4.12 (Exponential growth mechanism in a prototypical unstable system). Consider the linear ODE:

$$\frac{\mathrm{d}u}{\mathrm{d}t} = -\gamma u, \qquad \text{with initial condition} \qquad u(0). \tag{4.43}$$

Suppose the initial condition is perturbed by a small amount $\delta u(0)$, leading to a modified solution. If the parameter $\gamma = \gamma_- < 0$, the true solution and the perturbed solution evolve as $u(t) = u(0)\exp(-\gamma_- t)$ and $u^\delta(t) = (u(0) + \delta u(0))\exp(-\gamma_- t)$, respectively. The difference between these trajectories is:

$$\delta u(t) = \delta u(0)\exp(-\gamma_- t). \tag{4.44}$$

Since $\gamma_- < 0$, the term $\exp(-\gamma_- t)$ grows exponentially, causing the separation $\delta u(t)$ to increase rapidly. This illustrates exponential sensitivity to initial conditions, which is a key ingredient of chaotic dynamics. In genuinely chaotic systems, nonlinear interactions prevent unbounded growth by folding trajectories back into a bounded region of phase space, producing sustained complex behavior. □

A classic example of a chaotic system is the Lorenz 63 model, which was introduced by Edward Lorenz in 1963 as a simplified representation of atmospheric convection. The model consists of three coupled nonlinear differential equations [175]:

$$\begin{aligned}
\frac{\mathrm{d}x}{\mathrm{d}t} &= \sigma(y - x), \\
\frac{\mathrm{d}y}{\mathrm{d}t} &= x(\rho - z) - y, \\
\frac{\mathrm{d}z}{\mathrm{d}t} &= xy - \beta z,
\end{aligned} \tag{4.45}$$

where x represents the rate of convective overturning, y corresponds to horizontal temperature variations, and z describes vertical temperature variations. The parameters σ, ρ, and β can be derived from physical properties such as the Prandtl number, Rayleigh number, and geometric aspects of the fluid layer. With standard parameter values ($\sigma = 10$, $\rho = 28$, $\beta = 8/3$), the L63 model is known for its butterfly effect, where a small change in initial conditions leads to entirely different outcomes afterwards. Remarkably, when visualized in phase space, the solution of the Lorenz 63 model exhibits a double-lobed structure resembling a butterfly, which has become an iconic representation of chaotic attractors.

Example 4.13 (Chaotic behavior in the Lorenz 63 model). Figure 4.4 illustrates the behavior of the Lorenz 63 model with the standard parameters. Panel (a) compares two simulations (only the x component is shown) with nearly identical initial conditions:

$$\begin{aligned} \text{Original:} \quad & x(0) = 1, \quad & y(0) = 1, \quad & z(0) = 1, \\ \text{Perturbed:} \quad & x(0) = 1.001, \quad & y(0) = 1, \quad & z(0) = 1. \end{aligned} \tag{4.46}$$

Although the initial conditions differ by just 0.1%, the solutions diverge completely after about 5 time units. Panel (b) quantifies this divergence through the exponential growth of distance between trajectories on a logarithmic scale, which reveals the positive Lyapunov exponent that characterizes chaotic systems. Panel (c) displays the characteristic butterfly-shaped attractor in phase space, where both trajectories eventually trace the same qualitative pattern despite their divergence. This example shows the fundamental challenge of predicting chaotic systems: even infinitesimal uncertainties in initial conditions lead to exponentially growing errors that quickly make deterministic forecasts unreliable.

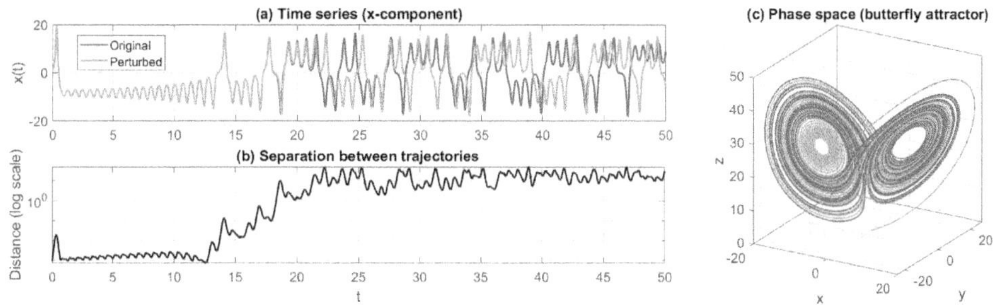

FIGURE 4.4: Simulation of the Lorenz 63 model. Panel (a): two trajectories with nearly identical initial conditions diverge rapidly due to chaotic dynamics. Panel (b): difference between the two trajectories as a function of time. Panel (c): phase-space plot reveals the butterfly-shaped attractor characteristic of the Lorenz 63 system.

□

For the Lorenz 63 model above, the Lyapunov exponent spectrum is given by $\lambda_1 \approx 0.906$, $\lambda_2 \approx 0$, $\lambda_3 \approx -14.572$. The positive first Lyapunov exponent λ_1 confirms chaos.

4.3.3 Deterministic and stochastic complexity

Chaotic behavior is classically defined in deterministic dynamical systems, where unpredictability arises purely from nonlinear dynamics without any randomness. These systems exhibit sensitivity to initial conditions, leading to exponential divergence of nearby trajectories, as quantified by positive Lyapunov exponents. In contrast, stochastic systems incorporate explicit randomness (e.g., noise or external forcing). Their unpredictability arises from both intrinsic dynamics and exogenous random perturbations. While randomness itself generates uncertainty, it can also interact with nonlinear dynamics to amplify variability and produce complex, irregular behavior resembling deterministic chaos. Although the underlying mechanisms differ, deterministic chaotic systems and stochastic systems often share statistical features, such as mixing behavior or ergodicity. Because of these similarities, stochastic models are frequently used as effective surrogates for capturing the statistical behavior of deterministic chaotic systems in practical applications.

The subsequent chapters will further explore these ideas, including stochastic modeling approaches, ensemble forecasting, internal predictability, and model error. Understanding these topics is essential for developing reliable prediction methods for chaotic and turbulent systems.

4.4 Basic numerical schemes for solving ODEs

Analytic solutions are available only for very simple ODEs. Most practical ODE systems, therefore, require numerical methods to obtain approximate solutions. In this section, we briefly discuss basic numerical schemes for solving ODEs and their associated stability analysis.

4.4.1 Forward Euler method

Consider the initial value problem for a one-dimensional ODE:

$$\frac{\mathrm{d}u}{\mathrm{d}t} = f(u,t), \quad u(0) = u_0. \tag{4.47}$$

Expressed in differential form, this becomes $\mathrm{d}u = f(u,t)\,\mathrm{d}t$. To derive the numerical scheme, we discretize time using a small step size Δt. Approximating the above differential equation over the interval $[t, t+\Delta t]$ yields [252]: $u(t+\Delta t) - u(t) \approx f(u(t),t)\Delta t$. This leads to the forward Euler update formula:

$$u(t+\Delta t) = u(t) + f(u(t),t)\Delta t. \tag{4.48}$$

The forward Euler method computes the solution sequentially, starting from $u(0)$ and progressing through $u(\Delta t)$, $u(2\Delta t)$, and so on, until the solution at the final time $u(T)$. The method is first-order accurate and explicit, with stability properties that depend on both the step size Δt and the problem characteristics.

4.4.2 Error convergence analysis of the forward Euler method

The local truncation error of the forward Euler method can be derived through Taylor expansion of the exact solution [94]:

$$u(t+\Delta t) = u(t) + \Delta t\frac{\mathrm{d}u}{\mathrm{d}t} + \frac{\Delta t^2}{2}\frac{\mathrm{d}^2u}{\mathrm{d}t^2} + \mathcal{O}(\Delta t^3). \tag{4.49}$$

Comparing with (4.48), we see the method matches the first two terms exactly. The dominant error term is $\frac{\Delta t^2}{2}\frac{\mathrm{d}^2u}{\mathrm{d}t^2}$. Over a fixed time interval $[0,T]$, the total number of time steps is $T/\Delta t$. Accumulating the local truncation error over $T/\Delta t$ steps leads to a global error of order $\mathcal{O}(\Delta t)$. Therefore, the forward Euler method is first-order accurate in time. The error decreases linearly with step size reduction, as shown by the convergence relation:

$$\|u_{\text{exact}}(T) - u_{\text{num}}(T)\| \leq C\Delta t, \tag{4.50}$$

where C depends on the final time T and the solution's derivatives.

4.4.3 Instability in stiff systems

The forward Euler method exhibits conditional stability that becomes particularly problematic for stiff systems [243], which contain fast-decaying components alongside much slower dynamics, thereby imposing severe time-step restrictions for stability. To understand this behavior, we first examine the linear test equation:

$$\frac{\mathrm{d}u}{\mathrm{d}t} = \lambda u, \quad u(0) = u_0, \quad \mathrm{Re}(\lambda) < 0, \tag{4.51}$$

whose exact solution $u(t) = u_0 e^{\lambda t}$ decays exponentially to zero. Applying the forward Euler scheme yields:

$$u_{n+1} = u_n + \lambda \Delta t u_n = (1 + \lambda \Delta t) u_n. \tag{4.52}$$

Applying (4.52) recursively over time, the numerical solution after n steps becomes:

$$u_n = (1 + \lambda \Delta t)^n u_0. \tag{4.53}$$

For numerical stability, we require $|1 + \lambda \Delta t| < 1$ so that perturbations decay as $n \to \infty$.

Example 4.14 (Numerical validation of a simple stiff system). Consider $\lambda = -1000$ in (4.51), representing a rapidly decaying system. The exact solution becomes negligible ($< 0.5\%$ of initial value) by $t = 0.005$. The stability condition requires:

$$\Delta t < \frac{2}{|\lambda|} = 0.002. \tag{4.54}$$

Figure 4.5 shows numerical solutions computed using different time-step sizes Δt, illustrating the stability behavior of this stiff system in accordance with (4.54). For $\Delta t = 0.0010$, we obtain $1 + \lambda \Delta t = 0$, so $u_n = 0^n u_0$. The numerical solution collapses to zero in a single step, representing excessive numerical damping. For $\Delta t = 0.0021$, which violates the stability condition, $u_n = (1 - 2.1)^n u_0 = (-1.1)^n u_0$. The solution becomes unstable and shows oscillatory divergence. Finally, for $\Delta t = 0.0020$, which corresponds to $|1 + \lambda \Delta t| = 1$, the solution neither decays nor grows in magnitude. Instead, the solution presents an oscillation pattern with the amplitude being preserved.

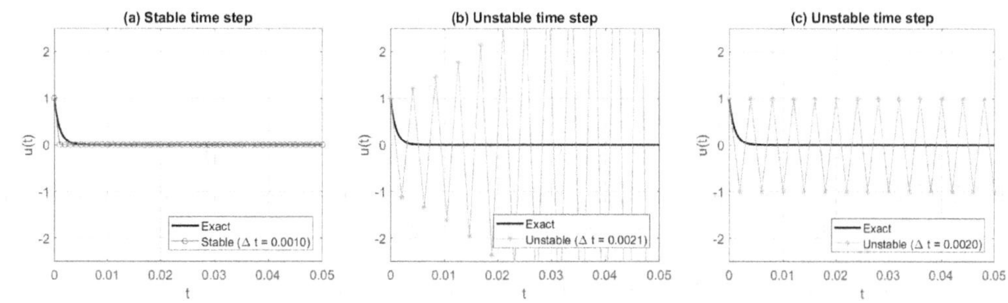

FIGURE 4.5: Numerical validation of the stiff system (4.51) with $\lambda = -1000$.

This example demonstrates how violating the stability condition leads to catastrophic growth of numerical error, despite the exact solution decaying rapidly. □

In multi-dimensional linear systems (4.20), stiffness arises when the constant coefficient matrix $\mathbf{A}$ has eigenvalues with widely separated magnitudes. Let $\lambda_1, \ldots, \lambda_n$ denote the eigenvalues of A, and suppose $|\lambda_{\max}| \gg |\lambda_{\min}|$. The stiffness ratio $S = |\lambda_{\max}|/|\lambda_{\min}|$ quantifies this separation. For stiff problems ($S \gg 1$), the fast modes associated with eigenvalues having large negative real parts decay very rapidly, while slow modes evolve over much longer time scales. Since forward Euler must satisfy

$$\Delta t < \frac{2}{|\lambda_{\max}|}, \tag{4.55}$$

the time step is determined by the fastest decaying mode, even if we are primarily interested in resolving the slower dynamics. This severe restriction makes explicit methods inefficient for stiff systems and motivates implicit methods such as backward Euler, which remain stable for any Δt when $\text{Re}(\lambda) < 0$.

Intuitively, stiffness arises because explicit methods must use a time step small enough to resolve the fastest-decaying components, even when these components become dynamically insignificant very quickly. In general, large time steps may fail to capture rapid transient behavior between successive updates. However, stiffness presents a stronger difficulty: even after the fast transients have vanished and no fine-scale details remain to be resolved, explicit methods must still use a time step small enough to maintain numerical stability. In the example above with $\lambda = -1000$, the solution effectively vanishes within a short transient interval. However, the forward Euler method must still satisfy the stability restriction $\Delta t < 2/|\lambda|$, forcing extremely small time steps long after the fast transient has decayed. Thus, stiffness is not primarily an accuracy issue but a stability issue: the numerical method is constrained by the fastest time scale in the system, even if the primary interest lies in the slower dynamics. A similar phenomenon occurs in damped oscillatory systems with widely separated time scales. If a system contains a rapidly damped oscillatory component alongside a slowly evolving mode, explicit methods must resolve the fast oscillation frequency to maintain stability, even though the oscillatory component may contribute little to the long-term dynamics.

Stiffness also arises when solving partial differential equations (PDEs) with high-frequency Fourier modes (large wavenumber k), since diffusion-type operators produce eigenvalues $\lambda_k \sim -k^2$. More technical details can be found in Section 4.5 and Chapter 5. The stability constraint $\Delta t < 2/|\lambda_k|$ then forces impractically small time steps to resolve rapidly decaying high-k modes, even when their energy contribution is negligible. Reduced-order models (ROMs) address this issue by projecting the system onto a low-dimensional subspace spanned by dominant modes (small k), thereby filtering stiff high-frequency components while preserving the essential dynamics. Techniques such as Proper Orthogonal Decomposition (POD) or Fourier mode truncation enable larger time steps and reduced computational cost without sacrificing accuracy at the resolved scales.

Geometrically, explicit methods advance the solution by evaluating the vector field explicitly at the current time level. When fast-decaying components are present, this explicit treatment can overreact to rapid variations and lead to numerical instability unless the time step is sufficiently small. Implicit methods, in contrast, determine the next state by treating the vector field implicitly at the future time level. This implicit coupling naturally damps fast components and stabilizes the numerical evolution. As a result, implicit schemes are particularly well suited for stiff systems. To illustrate this difference, consider the backward Euler method applied to (4.51):

$$u_{n+1} = u_n + \lambda \Delta t \, u_{n+1}. \tag{4.56}$$

Solving for u_{n+1} gives

$$u_{n+1} = \frac{1}{1 - \lambda \Delta t} u_n, \tag{4.57}$$

so that

$$u_n = \left(\frac{1}{1 - \lambda \Delta t} \right)^n u_0. \tag{4.58}$$

If $\mathrm{Re}(\lambda) < 0$, then $|1 - \lambda \Delta t| > 1$ for any $\Delta t > 0$, implying

$$\left| \frac{1}{1 - \lambda \Delta t} \right| < 1. \tag{4.59}$$

Thus, backward Euler remains stable for arbitrarily large time steps. This property is known as unconditional stability. In contrast to explicit methods, implicit schemes are not constrained by the fastest decaying mode in the system.

Although implicit methods offer superior stability properties, they introduce additional computational cost. At each time step, backward Euler requires solving an algebraic equation for u_{n+1}. For nonlinear systems, this typically involves iterative procedures such as Newton's method. In high-dimensional problems, implicit schemes require solving large linear (or nonlinear) systems at every time step. These solves can be computationally expensive and may dominate the overall cost of the simulation. Therefore, implicit methods trade increased stability for increased per-step computational effort. Whether an implicit or explicit method is preferable depends on the balance between stiffness, accuracy requirements, and available computational resources.

4.4.4 Higher-order numerical methods

The first-order accuracy of the forward Euler method, together with its conditional stability for stiff systems, is often insufficient for demanding applications. More advanced numerical schemes achieve higher convergence rates through improved temporal discretization strategies, thereby enhancing accuracy, efficiency, and stability.

Runge-Kutta methods

The Runge-Kutta (RK) family builds on Euler's approach by employing multiple function evaluations per time step [232, 162]. The second-order RK2 method (Midpoint Method) provides $\mathcal{O}(\Delta t^2)$ accuracy through intermediate evaluations:

$$\begin{aligned} k_1 &= f(u_n, t_n), \\ k_2 &= f(u_n + \tfrac{\Delta t}{2} k_1, t_n + \tfrac{\Delta t}{2}), \\ u_{n+1} &= u_n + \Delta t k_2. \end{aligned} \tag{4.60}$$

The classical RK4 method combines four weighted evaluations:

$$\begin{aligned} k_1 &= f(u_n, t_n), \\ k_2 &= f(u_n + \tfrac{\Delta t}{2} k_1, t_n + \tfrac{\Delta t}{2}), \\ k_3 &= f(u_n + \tfrac{\Delta t}{2} k_2, t_n + \tfrac{\Delta t}{2}), \\ k_4 &= f(u_n + \Delta t k_3, t_n + \Delta t), \\ u_{n+1} &= u_n + \tfrac{\Delta t}{6}(k_1 + 2k_2 + 2k_3 + k_4), \end{aligned} \tag{4.61}$$

which achieves fourth-order accuracy with only four function evaluations per step. A Taylor expansion of the four RK4 stages k_1, k_2, k_3, k_4 shows that the combined update reproduces exactly all terms up to Δt^4, implying a local truncation error of order five and hence fourth-order global accuracy.

The classical RK4 method is widely used because it achieves fourth-order accuracy using only function evaluations, without requiring higher derivatives or implicit solves. It offers an excellent balance between accuracy, stability, computational cost, and implementation simplicity, making it a standard choice for non-stiff dynamical systems.

Linear multistep methods

These schemes leverage historical solution data to enhance accuracy. The explicit Adams-Bashforth methods demonstrate this through polynomial extrapolation [28], with the second-order version:

$$u_{n+1} = u_n + \tfrac{\Delta t}{2}(3f_n - f_{n-1}), \tag{4.62}$$

where $f_n := f(u_n, t_n)$, and third-order variant:

$$u_{n+1} = u_n + \tfrac{\Delta t}{12}(23f_n - 16f_{n-1} + 5f_{n-2}). \tag{4.63}$$

Implicit methods such as Crank-Nicolson, which belongs to the family of the Adams-Moulton methods, achieve second-order accuracy ($\mathcal{O}(\Delta t^2)$) and offer unconditional stability for linear problems with $\mathrm{Re}(\lambda) < 0$ [81]:

$$u_{n+1} = u_n + \tfrac{\Delta t}{2}\left(f(u_{n+1}, t_{n+1}) + f(u_n, t_n)\right). \tag{4.64}$$

Runge-Kutta methods are single-step methods, relying only on information from the current time level. In contrast, linear multistep methods utilize solution values from previous steps to achieve higher accuracy.

Similar to (4.50), the convergence order p ensures the global error decreases as $|u(t_n) - u_n| \leq C\Delta t^p$, where $u(t_n)$ is the exact solution while u_n is the numerical solution. Higher-order methods demand greater computational effort per step but permit larger Δt for equivalent accuracy. The stability regions also vary significantly, with implicit methods typically offering superior stability properties at the cost of nonlinear equation solutions.

Example 4.15 (Comparison of accuracy for different numerical schemes). Figure 4.6 compares different numerical schemes in solving the following equation

$$\frac{\mathrm{d}u}{\mathrm{d}t} = -2u + \sin(t), \qquad u(0) = 1, \tag{4.65}$$

where the exact solution is given by

$$u(t) = \frac{1}{5}\left(2\sin(t) - \cos(t) + 6e^{-2t}\right). \tag{4.66}$$

The results confirm that forward Euler is first-order accurate, while both RK2 and Crank-Nicolson achieve second-order accuracy. Although RK2 and Crank-Nicolson have the same theoretical order of accuracy, Crank-Nicolson exhibits a smaller error constant in this example. RK4 is the most accurate among the four methods considered, consistent with its fourth-order convergence rate.

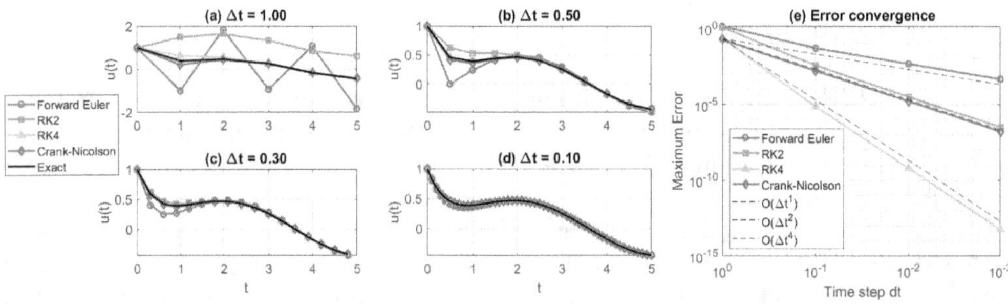

FIGURE 4.6: Comparison of the four methods in solving (4.65): forward Euler, RK2, Crank-Nicolson, and RK4. The exact solution is provided in (4.66). Panels (a)–(d): numerical trajectories with different time steps Δt. Panel (e): error convergence.

□

4.5 Partial differential equations

A partial differential equation (PDE) relates an unknown function of several variables to its partial derivatives [96, 248]. It characterizes how physical quantities evolve across space and time. Although analytic solutions are unavailable for most PDEs, certain fundamental equations admit explicit solutions under appropriate initial and boundary conditions. For unbounded domains $x \in (-\infty, \infty)$, one typically imposes decay conditions as $|x| \to \infty$, reducing the problem to finding solutions that match the initial profile $u(x, 0) = g(x)$.

4.5.1 Basic PDEs and their solutions

Below we illustrate two paradigmatic examples where $u(x, t)$ depends on time t and one spatial dimension x. These PDE systems reveal distinct physical mechanisms and mathematical behaviors.

Example 4.16 (Transport equation). The transport equation,

$$\frac{\partial u}{\partial t} + c\frac{\partial u}{\partial x} = 0 \tag{4.67}$$

models undisturbed propagation with constant velocity c. If the spatial domain is infinite, i.e., $x \in (-\infty, \infty)$, its general solution $u(x, t) = g(x - ct)$ shows that the initial profile $g(x)$ translates rigidly at speed c without distortion. Particularly, as will be seen in Chapter 5, high-frequency waves travel at the same phase velocity as low-frequency ones. The direction of propagation depends on the sign of c: positive values yield rightward motion (eastward in geophysical contexts), while negative values produce leftward propagation. The equation's linearity ensures superposition principles hold, allowing complex solutions to be built from simpler components. Remarkably, this PDE preserves all initial discontinuities indefinitely, making it fundamentally different from diffusive systems, such as the heat equation to be discussed shortly.

□

Example 4.17 (Heat equation). Another well-known PDE system is the heat equation, which governs thermal diffusion and random walks. Consider the heat equation on an infinite domain:

$$\frac{\partial u}{\partial t} = \nu\frac{\partial^2 u}{\partial x^2}, \quad x \in \mathbb{R}, \quad t > 0 \tag{4.68}$$

with initial condition $u(x, 0) = \delta(x)$, where $\delta(x)$ is the Dirac delta function. The delta function is zero everywhere except at $x = 0$, integrates to one, and represents an idealized point impulse. The solution of the heat equation in (4.68) is:

$$u(x, t) = \frac{1}{\sqrt{4\pi\nu t}} e^{-\frac{x^2}{4\nu t}}, \tag{4.69}$$

which has a Gaussian-like profile.

The width of the solution $\sigma(t) = \sqrt{2\nu t}$ grows diffusively, which reflects how thermal energy spreads. The parameter ν (diffusivity) controls how rapidly gradients dissipate. A dimensional analysis of (4.68) shows that for a system of characteristic length scale L, the diffusion time scale is L^2/ν. This explains why large systems take disproportionately longer to equilibrate. Despite the diffusion, the total heat $\int_{-\infty}^{\infty} u(x, t)\, dx = 1$ remains constant. Different from the transport equation, the heat equation exhibits infinite propagation speed, where any localized initial condition immediately affects all space, although with exponentially small amplitude far from the source. This equation also serves as a prototype for smoothing operators, as it immediately renders any initial condition infinitely differentiable for $t > 0$. The heat equation underlies the distribution of random motion of particles, where ν relates to the particle's mean free path and velocity. See Chapter 6 for its connection to Brownian motion. □

4.5.2 Wave mode analysis

Wave mode analysis is a fundamental tool for studying linear PDEs. For nonlinear PDEs, it is typically applied to the linearized system, which often captures the dominant local dynamics [179]. Consider a general linear PDE system

$$\mathcal{L}\mathbf{u} = 0, \tag{4.70}$$

where $\mathbf{u}(\mathbf{x}, t)$ denotes the unknown variables and $\mathcal{L}$ is a linear differential operator in space $\mathbf{x}$ and time t.

To understand the fundamental building blocks of the solution, we look for plane-wave solutions of the form

$$\mathbf{u}(\mathbf{x}, t) = \hat{\mathbf{u}}\, e^{i(\mathbf{k}\cdot\mathbf{x} - \omega t)}, \tag{4.71}$$

where $\mathbf{k}$ is the wave vector, ω is the frequency, and $\hat{\mathbf{u}}$ is a constant vector describing the relative amplitudes of the components of $\mathbf{u}$. Here we allow complex-valued exponentials for analytical convenience. Real-valued solutions are obtained by superposition of conjugate modes. The key reason this ansatz is useful is that differentiation becomes multiplication: $\partial_t \to -i\omega$ and $\nabla \to i\mathbf{k}$. Consequently, substituting (4.71) into the PDE converts the differential equation into a linear algebraic system

$$\mathbf{D}(\omega, \mathbf{k})\, \hat{\mathbf{u}} = 0, \tag{4.72}$$

where $\mathbf{D}(\omega, \mathbf{k})$ is a matrix depending on ω and $\mathbf{k}$. For nontrivial solutions $\hat{\mathbf{u}} \neq 0$ to exist, the matrix must be singular:

$$\det\left[\mathbf{D}(\omega, \mathbf{k})\right] = 0. \tag{4.73}$$

This condition is called the dispersion relation. It determines which frequencies ω are allowed for each spatial wave vector $\mathbf{k}$. Each root $\omega_j(\mathbf{k})$ corresponds to a distinct wave mode, and the associated eigenvector $\mathbf{p}_j(\mathbf{k})$ describes how different physical variables participate in that mode.

For an infinite domain, the general solution is obtained by superposing all admissible wave modes:

$$\mathbf{u}(\mathbf{x}, t) = \sum_j \int A_j(\mathbf{k})\, \mathbf{p}_j(\mathbf{k})\, e^{i(\mathbf{k}\cdot\mathbf{x} - \omega_j(\mathbf{k})t)}\, \mathrm{d}\mathbf{k}, \tag{4.74}$$

where the coefficients $A_j(\mathbf{k})$ are determined by the initial condition. In a periodic domain, the integral reduces to a discrete sum over admissible wave numbers.

Example 4.18 (Rotating shallow water equations on an f-plane). The shallow water equations (SWEs) are a set of partial differential equations that describe the flow of a fluid layer with a free surface under the influence of gravity, assuming the horizontal scale is much larger than the vertical scale [270]. These equations are derived from depth-integrating the Navier-Stokes equations and are widely used in various areas of fluids and geophysics. The SWEs conserve mass and momentum, accounting for key physical processes like wave propagation, advection, and bottom topography effects. Their simplicity and applicability to large-scale geophysical flows make them fundamental in hydrodynamics and environmental modeling.

We consider the linearized rotating shallow water equations on an f-plane in two spatial dimensions:

$$\begin{aligned} \frac{\partial u}{\partial t} - fv + g\frac{\partial h}{\partial x} &= 0, \\ \frac{\partial v}{\partial t} + fu + g\frac{\partial h}{\partial y} &= 0, \\ \frac{\partial h}{\partial t} + H\left(\frac{\partial u}{\partial x} + \frac{\partial v}{\partial y}\right) &= 0, \end{aligned} \tag{4.75}$$

where u is the zonal velocity (along the x direction), v the meridional velocity (along the y direction), h the height perturbation, H the mean depth, f the Coriolis parameter, and g gravity. Here, the f-plane approximation means that the Coriolis parameter f, which represents the effect of Earth's rotation, is treated as a constant. This approximation is appropriate for studying fluid motion over a limited geographical region, where variations in Earth's rotation with latitude can be neglected.

To analyze wave behavior, we assume solutions of the form $\hat{\mathbf{u}}\, e^{i(\mathbf{k}\cdot\mathbf{x}-\omega t)}$, corresponding to oscillatory waves with wavenumber $\mathbf{k}$ and frequency ω, where $\hat{\mathbf{u}} = (\hat{u}, \hat{v}, \hat{h})^{\mathrm{T}}$, $\mathbf{k} = (k_1, k_2)^{\mathrm{T}}$ and $\mathbf{x} = (x, y)^{\mathrm{T}}$. Substituting this form into the linear system leads to the dispersion relation:

$$\mathbf{D}(\omega, \mathbf{k}) \begin{pmatrix} \hat{u} \\ \hat{v} \\ \hat{h} \end{pmatrix} = \begin{pmatrix} -i\omega & -f & igk_1 \\ f & -i\omega & igk_2 \\ iHk_1 & iHk_2 & -i\omega \end{pmatrix} \begin{pmatrix} \hat{u} \\ \hat{v} \\ \hat{h} \end{pmatrix} = \mathbf{0}. \tag{4.76}$$

Nontrivial solutions require

$$\det\big[\mathbf{D}(\omega, \mathbf{k})\big] = 0. \tag{4.77}$$

This yields the characteristic polynomial:

$$\omega\big(\omega^2 - f^2 - c^2|\mathbf{k}|^2\big) = 0, \qquad c = \sqrt{gH}. \tag{4.78}$$

The roots give three characteristic wave modes:
(i) Geostrophic mode: $\omega = 0$. Setting $\omega = 0$ in (4.76) yields

$$\begin{cases} -f\hat{v} + igk_1\hat{h} = 0, \\ f\hat{u} + igk_2\hat{h} = 0, \\ iH(k_1\hat{u} + k_2\hat{v}) = 0, \end{cases} \quad\Longrightarrow\quad \hat{u} = -\frac{igk_2}{f}\hat{h}, \quad \hat{v} = \frac{igk_1}{f}\hat{h}. \tag{4.79}$$

This mode is balanced (Coriolis force balances the pressure-gradient force) and is non-propagating ($\omega = 0$), so it represents a steady large-scale adjustment rather than a traveling wave.
(ii) Inertia-gravity wave modes: $\omega = \pm\sqrt{f^2 + c^2|\mathbf{k}|^2}$. For $\omega \neq 0$, solving the first two equations in (4.76) for $(\hat{u}, \hat{v})$ in terms of $\hat{h}$ gives

$$\hat{u} = \frac{g}{\omega^2 - f^2}\big(\omega k_1 + ifk_2\big)\hat{h}, \qquad \hat{v} = \frac{g}{\omega^2 - f^2}\big(\omega k_2 - ifk_1\big)\hat{h}. \tag{4.80}$$

Substituting (4.80) into the third equation yields the dispersion relation

$$\omega^2 = f^2 + c^2|\mathbf{k}|^2, \qquad c = \sqrt{gH}, \tag{4.81}$$

which matches (4.78). These are propagating waves driven by gravity and modified by rotation. Their velocity components are generally out of phase (due to the if terms), producing elliptical particle motion.

Because the system (4.75) is linear, the full solution is obtained by superposing all admissible wave modes. For each wave vector $\mathbf{k}$, there exists one geostrophic mode and two inertia-gravity modes, each evolving independently in time. Thus, the general solution consists of a steady balanced component (the geostrophic mode) together with oscillatory propagating waves (the inertia-gravity modes) across all spatial scales.

Physically, this means that any initial disturbance in the fluid can be decomposed into a slowly varying balanced flow and rapidly oscillating wave motions. The balanced part represents large-scale adjustment governed by a near cancellation between pressure-gradient and Coriolis forces, while the inertia-gravity waves carry energy away through oscillations. Because the equations are linear, these components do not interact, and each evolves according to its own dispersion relation.

□

4.6 Code listings for figures and examples

- `Chapter_4_Cubic_Multiple_Equilibria.m`: generates Figure 4.1 in Example 4.6, which illustrates the multiple equilibria of the cubic scalar model.
- `Chapter_4_Predator_Prey_Model.m`: generates Figure 4.2 in Example 4.7, which shows the phase portrait of the predator–prey model together with the time evolution of the two variables.
- `Chapter_4_Different_Bifurcations.m`: generates Figure 4.3 in Examples 4.8–4.11, which presents representative examples of different bifurcation types.
- `Chapter_4_Lorenz_63.m`: generates Figure 4.4 in Example 4.13, which illustrates the sensitivity of trajectories to small initial perturbations in the Lorenz 63 model, a classical chaotic system.
- `Chapter_4_Stiff_ODE.m`: generates Figure 4.5 in Example 4.14, which compares solution behavior under different numerical integration time steps for a stiff system.
- `Chapter_4_Solver_Comparison.m`: generates Figure 4.6 in Example 4.15, which compares the accuracy of different numerical schemes.

5

Applied Analysis Toolkit

5.1 Fourier methods

Fourier analysis is among the most widely used tools in signal processing, physics, and engineering. For periodic functions, the Fourier series represents the signal as a discrete sum of harmonic modes, while for non-periodic functions, the Fourier transform extends this idea to a continuous spectrum of frequencies. Conceptually, both approaches decompose a signal into its constituent frequency components, which can then be analyzed separately.

5.1.1 Basic concept of Fourier series

Consider a one-dimensional time series $f(t)$ over a time window $t \in [0, L]$ with periodic boundary conditions $f(0) = f(L)$. The Fourier decomposition allows $f(t)$ to be represented as a summation of sine and cosine functions [261]:

$$f(t) = \sum_{k=0}^{\infty} \left[a_k \cos\left(\frac{2\pi kt}{L}\right) + b_k \sin\left(\frac{2\pi kt}{L}\right)\right]. \tag{5.1}$$

In many science and engineering applications, the right-hand side is usually written in complex form:

$$f(t) = \sum_{k=-\infty}^{\infty} c_k e^{i\frac{2\pi kt}{L}}. \tag{5.2}$$

There are two groups of parameters. The quantity $\tilde{k} = \frac{2\pi k}{L}$ is the angular frequency associated with the k-th harmonic, which represents the temporal frequencies. The parameter c_k (or a_k and b_k) is the Fourier coefficient, which is the amplitude of the k-th component, representing its strength in contributing to the entire signal. The coefficient c_k (and similarly for a_k and b_k) can be computed explicitly:

$$c_k = \frac{1}{L}\int_0^L f(t)e^{-i\frac{2\pi kt}{L}}\,\mathrm{d}t. \tag{5.3}$$

Example 5.1 (Fourier decomposition and its finite truncation for denoising). One of the most straightforward applications of Fourier decomposition is to denoise a signal. Let us assume the true signal is generated from the following function:

$$f(t) = 0.7\sin(2\pi\nu_1 t) + \sin(2\pi\nu_2 t) + 0.3\sin(2\pi\nu_3 t) + \epsilon(t). \tag{5.4}$$

where $\nu_1 = 50$, $\nu_2 = 120$, and $\nu_3 = 300$ with Hz as the unit. Recall that 1 Hz means one oscillation (one full period) per second. Throughout this example, the time variable t is measured in seconds. Therefore, a frequency of 50 Hz corresponds to 50 sinusoidal periods

DOI: 10.1201/9781003738459-5

per second, or equivalently a period of $1/50 = 0.02$ seconds for the first component. In (5.4), $\epsilon(t)$ is Gaussian white noise with zero mean and variance 0.25 at each time instant.

Figure 5.1 shows the true signal of $f(t)$ (Panel (a)) along with the three Fourier components (Panels (b)–(d)). The residual is given by the noise. Panel (e) shows the spectrum of the signal, which describes how the total variance (or energy) of the time series is distributed across different frequencies. The three peaks correspond to the three Fourier components, while the wiggles at other frequencies are due to the random noise.

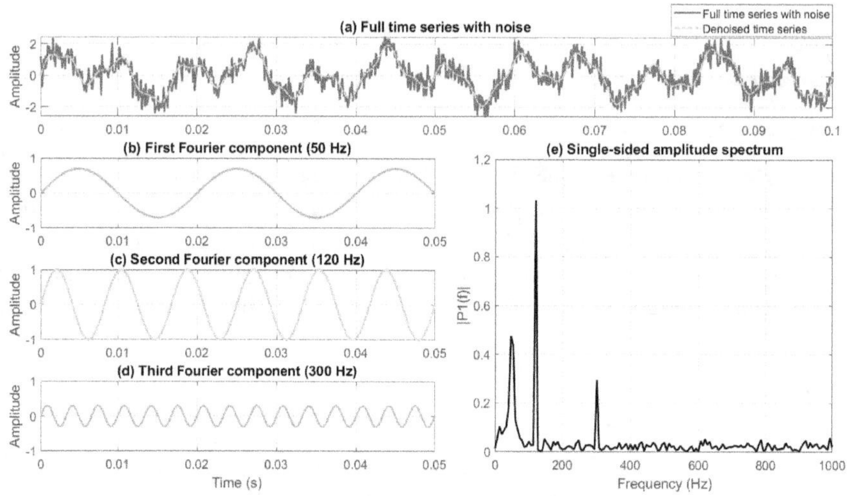

FIGURE 5.1: Fourier decomposition of the time series in (5.4). Panel (a): true signal and the denoised one. Panels (b)–(d): the three Fourier components. Panel (e): the spectrum of the true signal.

□

Recall in Chapter 3 that PCA has been used to extract the dominant features of data and reduce the noise, where PCA was applied to the scatter data. Nevertheless, PCA can also be used for time series data to denoise. One natural question to ask is the difference in the results using PCA and Fourier decomposition in denoising. Before showing a simple example, let us state how the PCA is used for time series denoising. The key intuition is that PCA denoising relies on separating structured patterns from unstructured random fluctuations. It is very different from explicitly decomposing the signal by frequency as in Fourier methods.

Consider a one-dimensional time series, $y = (y_1, y_2, \ldots, y_N)$. Construct the trajectory matrix (also known as the Hankel matrix) $\mathbf{X}$,

$$\mathbf{X} = \begin{pmatrix} y_1 & y_2 & \cdots & y_L \\ y_2 & y_3 & \cdots & y_{L+1} \\ \vdots & \vdots & \ddots & \vdots \\ y_K & y_{K+1} & \cdots & y_N \end{pmatrix}, \tag{5.5}$$

where L is the window length and $K = N - L + 1$. Then we follow the standard PCA algorithm described in Chapter 3. Recall that each row of $\mathbf{X}$ is a data point, while each column is a feature. In the above Hankel matrix, rows are time-delayed snapshots (also known as the sliding windows of the time series). They represent overlapping subsequences, which are like sample points in standard PCA, but now they are time-localized windows.

Columns are time lags, namely, delayed versions of the original series. They are analogs to features in standard PCA, but here they are temporal shifts. From an intuitive perspective, if the underlying signal contains coherent temporal patterns (such as oscillations or trends), these patterns will appear repeatedly across the sliding windows and thus produce large-variance directions captured by the leading principal components. In contrast, random noise varies irregularly from window to window and is spread across many directions, leading to small-variance components that can be truncated.

The Hankel embedding represents the time series in an L-dimensional delay-coordinate space, where each row corresponds to a state vector formed by time delays, while PCA extracts dominant temporal modes. Therefore, denoising is achieved by reconstructing the time series using only the leading principal components, which preserves the dominant temporal structures.

The fundamental distinction between these two methods is that Fourier decomposition is a projection onto a pre-defined harmonic basis, whereas PCA constructs an optimal low-rank approximation in the least-squares sense. Fourier modes are ordered by frequency, while PCA modes are ordered by variance. Consequently, Fourier is optimal when the signal is sparse in frequency space, whereas PCA is optimal when the signal admits a low-rank representation in delay-coordinate space.

Example 5.2 (Comparison between Fourier and PCA in denoising). Consider the time series generated from the following model:

$$f(t) = 2\sin(2\pi\nu_1 t) + 1.5\sin(2\pi\nu_2 t) + \sin(2\pi\nu_3 t) + \epsilon(t). \tag{5.6}$$

where $\nu_1 = 5$, $\nu_2 = 20$, and $\nu_3 = 50$. The residual term $\epsilon(t)$ represents the additional noise imposed on top of the original signal (containing only the three sinusoidal functions). It satisfies a Gaussian white noise with zero mean and unit variance at each time instant.

Figure 5.2 shows a comparison between using Fourier decomposition and PCA in denoising the one-dimensional time series generated from (5.6). As the true noise-free signal is the superposition of three Fourier modes, we expect that the reconstruction using the Fourier method, shown in Panel (b), will yield a nearly perfect result. This is because the noise is much smaller than the signal, as shown in Panel (c). In contrast, PCA does not completely recover the true signal when using the same number of modes (which is 3). See Panel (e). Note that, for oscillatory signals, each frequency component typically corresponds to a pair of principal components (forming a two-dimensional oscillatory subspace), which explains why more PCA modes are required to reconstruct the signal compared with the number of Fourier frequencies. Panel (f) indicates that the reconstruction with five modes yields a much more accurate result, comparable to that obtained with three Fourier modes. The variance explained in Panel (f) is computed relative to the total variance of the noisy signal. Therefore, a value close to 100% indicates that both the signal and the noise are included. The variance explained by the true signal is about 74%, which corresponds to 5 PCA modes.

It is worth highlighting that the comparison in this specific test does not mean Fourier consistently outperforms PCA, especially since the true signal here is built upon Fourier modes. The PCA-based time series denoising method does not require any assumption about signal periodicity. It has an adaptive basis and learns dominant patterns directly from data. Therefore, it has a wide application for non-sinusoidal signals and captures complex dynamics more effectively than the Fourier. □

Fourier decomposition can also be applied in space rather than time. In such a case, we usually rescale the domain to $[0, 2\pi]$. Consider a periodic signal $g(x)$, where $x \in [0, 2\pi]$. Its

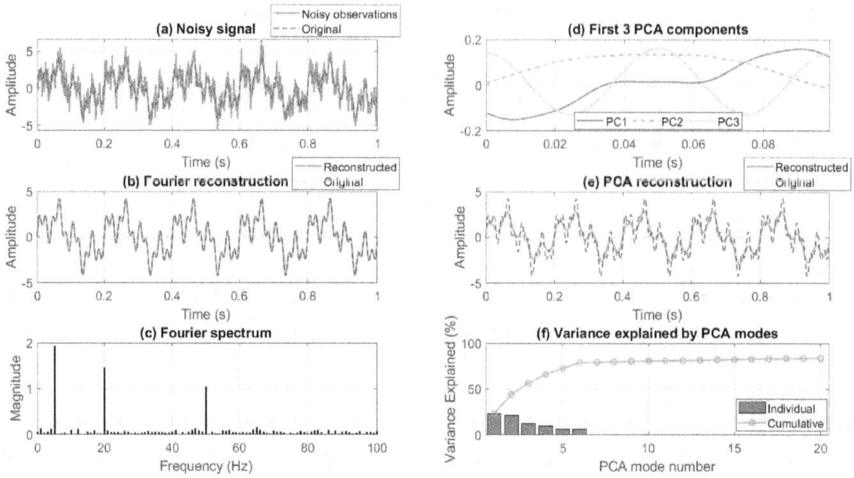

FIGURE 5.2: Comparison between using Fourier decomposition and PCA in denoising the one-dimensional time series generated from (5.6). Panel (a): the noisy time series and its original noise-free time series. Panel (b): the reconstructed time series from the Fourier method. Panel (c): the Fourier spectrum. Panel (d): the first 3 PCA modes. Panel (e): the reconstructed time series from PCA. Panel (f): variance explained by the accumulation of different number of PCA modes.

Fourier representation with respect to x is as follows:

$$g(x) = \sum_{k=-\infty}^{\infty} c_k e^{ikx}, \tag{5.7}$$

where the notations are simplified since the domain has been rescaled to $[0, 2\pi]$, the fundamental wavenumber is 1, and the admissible wavenumbers are the integers $k \in \mathbb{Z}$. Each term e^{ikx} corresponds to a spatial wave with wavelength $\lambda_k = \frac{2\pi}{|k|}$.

5.1.2 Fourier series applied to solving PDEs

In many real-world problems, especially in fluid and geophysics, the variable of interest depends on both time and space. In particular, the spatial coordinate is usually one, two, or three dimensions, representing the Cartesian coordinates.

Example 5.3 (Fourier series solution of a linear one-dimensional PDE). To illustrate how Fourier series are applied to PDEs, consider the one-dimensional heat equation on a periodic domain $x \in [0, 2\pi]$,

$$\frac{\partial u}{\partial t} = \kappa \frac{\partial^2 u}{\partial x^2}, \tag{5.8}$$

with periodic boundary conditions. Expand $u(x,t)$ into a Fourier series,

$$u(x,t) = \sum_{k\in\mathbb{Z}} \hat{u}_k(t) e^{ikx}. \tag{5.9}$$

Substituting this expansion into (5.8) gives

$$\sum_k \frac{\mathrm{d}\hat{u}_k}{\mathrm{d}t} e^{ikx} = \kappa \sum_k (-k^2)\hat{u}_k(t) e^{ikx}. \tag{5.10}$$

Since the Fourier modes e^{ikx} are linearly independent, the coefficients must satisfy

$$\frac{\mathrm{d}\hat{u}_k}{\mathrm{d}t} = -\kappa k^2 \hat{u}_k. \tag{5.11}$$

Thus, the PDE is reduced to a set of decoupled ordinary differential equations, each admitting the explicit solution

$$\hat{u}_k(t) = \hat{u}_k(0)\, e^{-\kappa k^2 t}. \tag{5.12}$$

Each Fourier mode evolves independently and decays at a rate proportional to k^2, with higher-wavenumber modes dissipating faster. □

Below, we focus on the two-dimensional case, which is the most widely seen scenario. Fourier representation can be applied to each snapshot of the variable, where a snapshot refers to the spatial field observed at a fixed time instant. Denote by $\mathbf{x} = (x, y)^{\mathrm{T}}$ the two-dimensional coordinates and $\mathbf{u}(\mathbf{x}, t) = (u(\mathbf{x}, t), v(\mathbf{x}, t))^{\mathrm{T}}$ the two-dimensional field, say the flow velocity, along the x and y directions. Correspondingly, $\mathbf{k} = (k_1, k_2)$ contains the two components of the wavenumbers. Assume periodic boundary conditions on $[0, 2\pi]^2$. The typical representation of the full field is as follows [61],

$$\mathbf{u}(\mathbf{x}, t) = \sum_{\mathbf{k}\in\mathbf{K}} \hat{u}_{\mathbf{k}}(t) e^{i\mathbf{k}\cdot\mathbf{x}} \mathbf{r}_{\mathbf{k}}, \tag{5.13}$$

where $\mathbf{K}$ is a certain set within the two-dimensional lattice grids for integers. In (5.13), the Fourier coefficient $\hat{u}_{\mathbf{k}}(t)$ is a function of time. Nevertheless, it is still a scalar. To be consistent with the dimension of the left-hand side, a vector $\mathbf{r}_{\mathbf{k}} = (r^x_{\mathbf{k}}, r^y_{\mathbf{k}})^{\mathrm{T}}$ is added. The role of this eigenvector is to include the relationship between the two components u and v as they are usually not independent.

Example 5.4 (Fourier representation of incompressible flows). For the flow field $\mathbf{u}(\mathbf{x}, t) = (u(\mathbf{x}, t), v(\mathbf{x}, t))^{\mathrm{T}}$, the incompressibility condition reads $\nabla \cdot \mathbf{u} = 0$, which means compressible of the flow in one direction will cause the expansion in the other direction such that the total area (in the two-dimensional case) is conserved. In the component-wise form, the incompressibility condition leads to [233]

$$\frac{\partial u}{\partial x} + \frac{\partial v}{\partial y} = 0. \tag{5.14}$$

Plugging (5.13) into (5.14) yields

$$\sum_{\mathbf{k}\in\mathbf{K}} \hat{u}_{\mathbf{k}}(t) ik_1 e^{i\mathbf{k}\cdot\mathbf{x}} r^x_{\mathbf{k}} + \sum_{\mathbf{k}\in\mathbf{K}} \hat{u}_{\mathbf{k}}(t) ik_2 e^{i\mathbf{k}\cdot\mathbf{x}} r^y_{\mathbf{k}} = 0. \tag{5.15}$$

Since different Fourier bases are orthogonal to each other, we can handle each wavenumber independently. This means for each fixed $\mathbf{k} \neq \mathbf{0}$, we have

$$k_1 r^x_{\mathbf{k}} + k_2 r^y_{\mathbf{k}} = 0, \tag{5.16}$$

which means

$$\mathbf{r}_{\mathbf{k}} = \frac{1}{\sqrt{k_1^2 + k_2^2}} \begin{pmatrix} -k_2 \\ k_1 \end{pmatrix}, \tag{5.17}$$

since the direction vector is normalized to unit length. The zero mode corresponds to spatial mean flow. It must be treated separately. □

5.1.3 Energy spectrum represented by Fourier modes

Fourier analysis is widely used to decompose signals into contributions from different spatial scales. In fluid dynamics and geophysics, it is commonly used to compute the energy spectrum, which describes how the total kinetic energy of a flow is distributed across different wavenumbers.

Consider the two-dimensional velocity field in (5.13) defined on a periodic domain. The total kinetic energy at time t is given by $\frac{1}{2}\int |\mathbf{u}(\mathbf{x},t)|^2\,\mathrm{d}\mathbf{x}$. Parseval's identity states that the L^2 norm of a function in physical space is equal to the sum of the squared magnitudes of its Fourier coefficients in spectral space. Using this identity, the kinetic energy can be expressed equivalently in Fourier space as $\frac{1}{2}\sum_{\mathbf{k}} |\hat{\mathbf{u}}_{\mathbf{k}}(t)|^2$, where $\hat{\mathbf{u}}_{\mathbf{k}}(t)$ denotes the Fourier coefficient of the velocity field corresponding to wavevector $\mathbf{k}$. To characterize how energy is distributed across spatial scales, one groups together all wavevectors with the same magnitude $k = |\mathbf{k}|$. The isotropic energy spectrum is then defined as

$$E(k,t) = \frac{1}{2}\sum_{|\mathbf{k}|\in S(k)} |\hat{\mathbf{u}}_{\mathbf{k}}(t)|^2, \tag{5.18}$$

where $S(k)$ denotes a thin shell in Fourier space consisting of all wavevectors whose magnitudes lie in a small interval around k. The time-averaged energy spectrum is often used in practice. The energy spectrum reveals important structural properties of the flow.

Example 5.5 (Energy spectrum: laminar and turbulent signals). Figure 5.3 shows two incompressible flow fields constructed from a superposition of Fourier modes with wavevectors $\mathbf{k} \in [-64, 64]^2$.

The first row corresponds to a laminar flow field, which exhibits regular spatial patterns. In this case, the energy spectrum is sharply peaked at a small number of low wavenumbers, with a dominant contribution from $\mathbf{k} = 4$. This dominant mode corresponds to a spatial pattern that repeats four times across the domain in each periodic direction, explaining the regularly spaced structures visible in the flow field.

In contrast, the second row corresponds to a turbulent flow field, characterized by irregular and multiscale spatial structures. Here, energy is distributed across a broad range of wavenumbers rather than being concentrated at a single scale. The corresponding energy spectrum is much flatter and spreads over many modes, indicating the presence of structures at multiple spatial scales. Such multiscale energy distribution makes turbulent flows significantly more difficult to predict than laminar ones. □

5.1.4 Fourier transform

It has been shown in the previous subsections that the Fourier series is designed for periodic signals. It represents such a signal as a sum of sine and cosine waves whose frequencies are fixed multiples of a fundamental frequency.

The Fourier transform extends this idea to non-periodic signals, which occur in many real-world signals. Different from the Fourier series, which sums a finite (or countably infinite) set of periodic functions, the Fourier transform describes a signal using a continuous range of frequencies. This is because when a signal has periodic patterns, a finite (or countably infinite) number of discrete frequencies is often sufficient to describe it. In contrast, for non-periodic signals, the frequency spectrum becomes continuous rather than discrete.

Given a sufficiently regular function $f(x)$ (e.g., integrable or square-integrable), its Fourier transform is defined as [215]:

$$\hat{f}(k) = \mathcal{F}\{f(x)\} = \int_{-\infty}^{\infty} e^{-ikx} f(x)\,\mathrm{d}x, \tag{5.19}$$

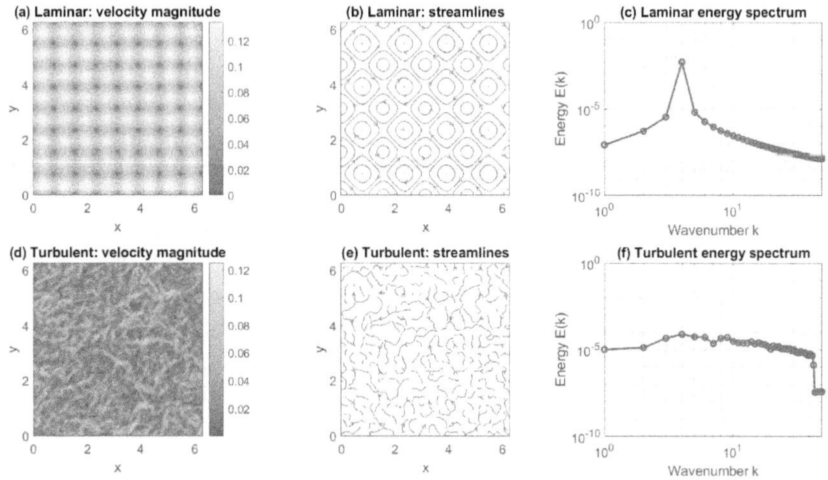

FIGURE 5.3: Comparison of laminar and turbulent flow fields. Panel (a): the amplitude of the velocity field. Panel (b): streamlines, which are curves tangent to the velocity field. Panel (c): energy spectrum shown in a log-log plot for the laminar case. Panels (d)–(f): corresponding plots for the turbulent case.

where k is the wavenumber (or frequency) variable. The inverse Fourier transform, which recovers the original function, is given by:

$$f(x) = \mathcal{F}^{-1}\{\hat{f}(k)\} = \frac{1}{2\pi}\int_{-\infty}^{\infty} e^{ikx}\hat{f}(k)\,\mathrm{d}k. \tag{5.20}$$

Similar to the discrete Fourier series, direct calculation using the definition in (5.19) shows the following basic properties of the Fourier transform, provided suitable decay conditions hold.

First, the Fourier transform of the n-th derivative of a function equals the multiplication of n-th power of ik by the function, namely

$$\mathcal{F}\left(\frac{\mathrm{d}^n f}{\mathrm{d}x^n}\right) = (ik)^n \hat{f}(k). \tag{5.21}$$

Second, the Fourier transform satisfies the linear property:

$$\mathcal{F}\{af(x) + bg(x)\} = a\mathcal{F}\{f(x)\} + b\mathcal{F}\{g(x)\}. \tag{5.22}$$

Third, the Fourier transform has the so-called shift property:

$$\mathcal{F}\{f(x-a)\} = e^{-ika}\mathcal{F}\{f(x)\}. \tag{5.23}$$

Lastly, the inverse Fourier transform of a product of two functions in the frequency domain equals the convolution of their individual inverse transforms:

$$\mathcal{F}^{-1}\{\hat{f}(k)\hat{g}(k)\}(x) = (f * g)(x) = \int_{-\infty}^{\infty} f(y)g(x-y)\,\mathrm{d}y. \tag{5.24}$$

Conversely, the Fourier transform of a convolution is the product of their Fourier transforms:

$$\mathcal{F}\{(f * g)(x)\}(k) = \hat{f}(k)\hat{g}(k). \tag{5.25}$$

Note that although the formulae above are written in terms of the spatial variable x, this x can be regarded as a generic variable. In particular, when x represents time, convolution describes how the value of a signal at a given time depends on its nearby values.

Multiplying two functions in the frequency domain means adjusting the strength of each frequency component, where some frequencies are weakened, while others are preserved or amplified. In contrast, convolution in the space (or time) domain describes how the value of a signal at one location (or time) is influenced by its nearby values. Roughly speaking, frequency multiplication is equivalent to original (space- or time-domain) convolution. If $\hat{h}(k) = \hat{f}(k)\hat{g}(k)$, then $h(x)$ is obtained by sliding the function $g(x)$ along $f(x)$ and taking a weighted average of their overlap at each location x. In this sense, convolution combines information from a neighborhood around each point rather than using only the value at that point. For example, if $\hat{g}(k)$ suppresses high frequencies (a low-pass filter), then multiplying $\hat{f}(k)$ by $\hat{g}(k)$ removes rapid oscillations in the frequency domain. In the time domain, this operation corresponds to convolving $f(x)$ with $g(x)$, which replaces sharp variations in $f(x)$ by local averages, resulting in a smoother signal.

Example 5.6 (Fourier transform for a Gaussian distribution). Let $f(x)$ be a Gaussian function centered at zero with standard deviation σ:

$$f(x) = \frac{1}{\sqrt{2\pi\sigma^2}} e^{-\frac{x^2}{2\sigma^2}}. \tag{5.26}$$

The Fourier transform $\hat{f}(k)$ is:

$$\hat{f}(k) = \mathcal{F}\{f(x)\} = \frac{1}{\sqrt{2\pi\sigma^2}} \int_{-\infty}^{\infty} e^{-\frac{x^2}{2\sigma^2}} e^{-ikx}\, dx \tag{5.27}$$

Complete the square in the exponent:

$$-\frac{x^2}{2\sigma^2} - ikx = -\frac{1}{2\sigma^2}\left(x^2 + 2i\sigma^2 kx\right) = -\frac{1}{2\sigma^2}\left((x + i\sigma^2 k)^2 - \sigma^4 k^2\right). \tag{5.28}$$

Thus:

$$\hat{f}(k) = \frac{1}{\sqrt{2\pi\sigma^2}} e^{-\frac{\sigma^2 k^2}{2}} \int_{-\infty}^{\infty} e^{-\frac{(x+i\sigma^2 k)^2}{2\sigma^2}}\, dx. \tag{5.29}$$

The integral evaluates to $\sqrt{2\pi\sigma^2}$, so the Fourier transform leads to

$$\hat{f}(k) = e^{-\frac{\sigma^2 k^2}{2}}, \tag{5.30}$$

which is also a Gaussian-like function. Note that if $f(x) = \frac{1}{\sqrt{2\pi\sigma^2}} e^{-\frac{(x-\mu)^2}{2\sigma^2}}$ contains a nonzero mean, then the Fourier transform is:

$$\hat{f}(k) = e^{-ik\mu} e^{-\frac{\sigma^2 k^2}{2}}, \tag{5.31}$$

which can be computed by using the shift property $\mathcal{F}\{f(x-\mu)\} = e^{-ik\mu}\mathcal{F}\{f(x)\}$. Similarly, the inverse Fourier transform of a Gaussian function will be a Gaussian as well.

The fact that a Gaussian remains a Gaussian under the Fourier transform is extremely useful in applications because it greatly simplifies analysis in both physical space and frequency space. First, many physical processes naturally produce Gaussian profiles. For example, solutions of the heat equation and diffusion processes have Gaussian shapes. Since the Fourier transform of a Gaussian is again a Gaussian, such problems can be analyzed very cleanly using Fourier methods. Second, Gaussian functions are widely used in signal and image processing as smoothing filters. Because their Fourier transforms are also Gaussian, applying a Gaussian filter in physical space corresponds to multiplying by a Gaussian in frequency space. This provides smooth attenuation of high frequencies without introducing oscillations or artifacts. Third, in probability and statistics, the normal distribution plays a central role. The fact that its Fourier transform (characteristic function) has a simple closed form makes many theoretical calculations tractable and explains why Gaussian models are so mathematically convenient. □

Example 5.7 (Solving a first-order linear ODE using the Fourier transform). Consider the differential equation:

$$\frac{\mathrm{d}y}{\mathrm{d}t} + ay(t) = f(t), \tag{5.32}$$

where a is a constant and $f(t)$ is a known function. The initial condition is assumed to be $y(0) = 0$. We further assume $y(t) = 0$ for $t < 0$, so the Fourier transform can be applied in the distributional sense. Our goal is to find $y(t)$. The Fourier transform of a function $g(t)$ is defined as:

$$\mathcal{F}\{g(t)\} = G(k) = \int_{-\infty}^{\infty} g(t)e^{-ikt}\,\mathrm{d}t. \tag{5.33}$$

Applying the Fourier transform to both sides of the ODE and using the derivative and linear properties:

$$ikY(k) + aY(k) = F(k), \tag{5.34}$$

where $Y(k) = \mathcal{F}\{y(t)\}$ and $F(k) = \mathcal{F}\{f(t)\}$. Solving for $Y(k)$ in the frequency domain gives the solution of the Fourier transform:

$$Y(k) = \frac{F(k)}{ik + a}. \tag{5.35}$$

Finally, applying inverse Fourier transform to find $y(t)$ in the original time domain yields

$$y(t) = \mathcal{F}^{-1}\{Y(k)\} = \mathcal{F}^{-1}\left\{\frac{F(k)}{ik + a}\right\}. \tag{5.36}$$

Let us now consider a special case, where $f(t)$ is given by an impulse response $f(t) = \delta(t)$, which means the impulse is imposed at $t = 0$. Plugging this into (5.19), we have $F(k) = 1$. Therefore,

$$Y(k) = \frac{1}{ik + a}. \tag{5.37}$$

The inverse Fourier transform gives:

$$y(t) = e^{-at}u(t), \tag{5.38}$$

where $u(t)$ is the Heaviside step function that equals to 0 for $t < 0$ and 1 for $t > 0$. The solution $y(t)$ stays zero until the impulse response occurs at $t = 0$, and then it decays exponentially. □

Fourier transform is also a powerful tool to solve PDEs.

Example 5.8 (Solving the transport equation using the Fourier transform and the Fourier series). Recall the transport equation:

$$\frac{\partial u}{\partial t} + c\frac{\partial u}{\partial x} = 0. \tag{5.39}$$

If the spatial domain is infinite, i.e., $x \in (-\infty, \infty)$, we can solve this using Fourier transforms. Recall the Fourier transform of $u(x,t)$:

$$\hat{u}(k,t) = \mathcal{F}\{u(x,t)\} = \int_{-\infty}^{\infty} u(x,t)e^{-ikx}\,\mathrm{d}x. \tag{5.40}$$

Taking the Fourier transform of both sides of (5.39):

$$\mathcal{F}\left\{\frac{\partial u}{\partial t}\right\} + c\mathcal{F}\left\{\frac{\partial u}{\partial x}\right\} = 0. \tag{5.41}$$

Since the spatial derivative becomes the multiplication factor under the Fourier transform, (5.41) yields

$$\frac{\mathrm{d}\hat{u}}{\mathrm{d}t} + ick\hat{u} = 0, \tag{5.42}$$

where as the only derivative is in the temporal direction we have replaced $\partial\hat{u}/\,\mathrm{d}t$ by $\mathrm{d}\hat{u}/\,\mathrm{d}t$. This simple ODE has the solution:

$$\hat{u}(k,t) = \hat{u}(k,0)e^{-ickt}, \tag{5.43}$$

where $\hat{u}(k,0)$ is the Fourier transform of the initial condition $u(x,0) = g(x)$.

To find the solution in physical space, we apply the inverse Fourier transform:

$$u(x,t) = \mathcal{F}^{-1}\{\hat{u}(k,t)\} = \frac{1}{2\pi}\int_{-\infty}^{\infty} \hat{u}(k,0)e^{i(kx-ckt)}\,\mathrm{d}k. \tag{5.44}$$

Recognizing that $\hat{u}(k,0) = \hat{g}(k)$, we can write:

$$u(x,t) = \frac{1}{2\pi}\int_{-\infty}^{\infty} \hat{g}(k)e^{ik(x-ct)}\,\mathrm{d}k = g(x-ct). \tag{5.45}$$

As a remark, if we consider a finite domain with periodic boundary conditions, we can use Fourier series. Without loss of generality, we can assume $x \in [0, 2\pi]$ and $u(0,t) = u(2\pi, t)$. We express the solution as a Fourier series:

$$u(x,t) = \sum_{n=-\infty}^{\infty} \hat{u}_n(t)e^{ik_nx}, \quad k_n = n \quad (\text{since } L = 2\pi). \tag{5.46}$$

Substituting this Fourier ansatz into the transport equation:

$$\frac{\partial}{\partial t}\left(\sum_{n=-\infty}^{\infty} \hat{u}_n(t)e^{inx}\right) + c\frac{\partial}{\partial x}\left(\sum_{n=-\infty}^{\infty} \hat{u}_n(t)e^{inx}\right) = 0$$

$$\sum_{n=-\infty}^{\infty}\left(\frac{\mathrm{d}\hat{u}_n}{\mathrm{d}t} + icn\hat{u}_n\right)e^{inx} = 0.$$

For each Fourier mode, we obtain the following simple ODE:

$$\frac{\mathrm{d}\hat{u}_n}{dt} + icn\hat{u}_n = 0, \tag{5.47}$$

the solution of which is given by

$$\hat{u}_n(t) = \hat{u}_n(0)e^{-icnt}. \tag{5.48}$$

Let the initial condition be $u(x,0) = g(x)$. Its Fourier coefficients are:

$$\hat{u}_n(0) = \frac{1}{2\pi} \int_0^{2\pi} g(x) e^{-inx} \, dx. \tag{5.49}$$

The final solution is:

$$u(x,t) = \sum_{n=-\infty}^{\infty} \hat{u}_n(0) e^{in(x-ct)}. \tag{5.50}$$

Note that each Fourier mode propagates with phase velocity c. This is called a non-dispersive wave, which contrasts with a dispersive wave, in which different frequency components propagate at different speeds, causing wave packets to spread over time. In addition, since there is no damping and forcing, all modes maintain their amplitude. The initial profile translates rigidly while maintaining periodicity.

□

5.2 Dimensional analysis

Dimensional analysis is a powerful tool that simplifies complex problems by studying the units of physical quantities. Its conceptual roots can be traced back to ancient Greek ideas of similarity and proportion in geometry. It became a formal mathematical method in the early 20th century [41].

The underlying idea of dimensional analysis is that meaningful physical laws must be dimensionally consistent and independent of the particular system of units used. For example, Newton's law $F = ma$ holds whether force is measured in newtons or pounds, provided the units are consistent. This principle allows us to identify dimensionless combinations of variables, called Π groups. These Π groups are products of the original variables raised to suitable powers such that all physical dimensions cancel. They often reveal fundamental scaling relationships and characteristic time or length scales that clarify the underlying physics.

The Buckingham Π-theorem formalizes this idea [277]. Suppose a problem involves n dimensional variables whose dimensions span a space of rank m (that is, m independent fundamental dimensions are required to express them). Then the problem can be equivalently described using $n-m$ independent dimensionless Π groups. The rank m is determined by the variables appearing in the problem, not necessarily by the full list of SI base units. The Π-theorem uses these dimensions as building blocks to derive dimensionless forms of physical laws.

Example 5.9 (Constructing a dimensionless Π group). Assume a physical problem involves the variables

$$L \ \text{(length)}, \qquad T \ \text{(time)}, \qquad v \ \text{(velocity)}. \tag{5.51}$$

Here $n = 3$ variables. Their dimensions are

$$[L] = L, \qquad [T] = T, \qquad [v] = LT^{-1}. \tag{5.52}$$

These quantities involve $m = 2$ independent fundamental dimensions: L (length) and T (time). The Buckingham Π-theorem states that the system can therefore be described using $n - m = 1$ dimensionless group. One such group is $\Pi = \frac{vT}{L}$. Thus the relationship among the variables can be written in the form $f(\Pi) = 0$. □

Dimensional analysis also has practical significance. During World War II, G. I. Taylor famously used it to estimate the energy released by the first atomic bomb using only photographs of its shockwave. Today, it is essential for experimental design, model scaling (such as wind tunnel testing), and consistency checks of governing equations. For physicists and engineers, dimensional analysis is often the first step toward understanding a problem before detailed calculations are performed.

5.2.1 Similarity

We begin with geometric similarity, in which systems share the same shape but differ in overall size. We then introduce dynamical similarity, in which systems may differ in scale yet exhibit equivalent governing physics when expressed in terms of appropriate dimensionless parameters. In science and engineering applications, dynamical similarity often plays the more fundamental role.

Example 5.10 (Geometric similarity: scaling a triangle). Consider a right triangle with legs a and b, and hypotenuse c. If we scale the triangle by a factor k, how do the new dimensions relate to the original ones? Since all sides (a, b, c) have dimensions of length $[L]$, multiplying each by the dimensionless factor k yields

$$a' = ka, \quad b' = kb, \quad c' = kc. \tag{5.53}$$

Because all lengths scale uniformly, the angles remain unchanged. The two triangles are, therefore, geometrically similar. □

Example 5.11 (Dynamical similarity: simple pendulum). Consider a pendulum of length l and mass m swinging under gravity g. We seek to determine how the period T depends on these variables. The dimensional variables are:

1. period T with dimension $[T]$,
2. length l with dimension $[L]$,
3. mass m with dimension $[M]$, and
4. gravitational acceleration g with dimension $[LT^{-2}]$.

The dimensions involved span $[M]$, $[L]$, and $[T]$, so the rank is $m = 3$. Since there are $n = 4$ variables, the Buckingham Π-theorem implies that there is $n - m = 1$ independent dimensionless group. We seek a dimensionless combination of the form $\Pi = T^a l^b m^c g^d$. Requiring $[\Pi] = 1$ gives

$$[T]^a [L]^b [M]^c [LT^{-2}]^d = 1, \tag{5.54}$$

which yields the system

$$\begin{cases} \text{Mass } (M) & : c = 0, \\ \text{Length } (L) & : b + d = 0, \\ \text{Time } (T) & : a - 2d = 0. \end{cases} \tag{5.55}$$

Choosing $d = 1$ gives $b = -1$ and $a = 2$, so $\Pi = \frac{T^2 g}{l}$. Since Π is dimensionless, dimensional analysis implies

$$T = C\sqrt{\frac{l}{g}}, \tag{5.56}$$

where C is a dimensionless constant. More generally, the period may also depend on additional dimensionless parameters, such as the swing amplitude θ_0. In that case, $T/\sqrt{l/g} =$

$f(\theta_0)$, for some dimensionless function f. In the small-angle limit, $f(\theta_0) \approx 2\pi$, giving

$$T \approx 2\pi\sqrt{\frac{l}{g}}. \tag{5.57}$$

Importantly, the mass m does not appear in the final relation. Therefore, pendulums with the same l and g (and comparable amplitude) exhibit dynamically similar motion. □

5.2.2 Non-dimensional numbers and laboratory experiments

Dimensional analysis is widely used in laboratory experiments to ensure that small-scale models faithfully represent real-world systems. If we aim to test the aerodynamic performance or structural resilience of an aircraft, conducting experiments on the full-scale airplane is often impractical due to cost and risk. Instead, scaled models are used. However, it is not sufficient to match geometric similarity alone; dynamical similarity must also be achieved.

A non-dimensional number is a quantity formed by combining physical variables so that all units cancel, producing a pure number. Such numbers describe the relative importance of competing physical effects. Non-dimensional numbers often emerge from dimensional analysis (e.g., Buckingham Π-theorem) and help organize complex physical systems into universal regimes.

Example 5.12 (Reynolds number and modeling aircraft in the laboratory). Suppose we wish to construct a laboratory model that captures the aerodynamic behavior of a real aircraft. For the real aircraft, let the typical chord length be $L_{\text{real}} = 2\,\text{m}$, the characteristic velocity be $U_{\text{real}} = 200\,\text{m/s}$, and the air density be $\rho = 1.2\,\text{kg/m}^3$. The laboratory model has a reduced chord length $L_{\text{model}} = 0.2\,\text{m}$, that is, a $10 : 1$ geometric scaling.

Matching geometric similarity is straightforward: all linear dimensions are scaled by the same factor. However, to reproduce the correct aerodynamic behavior, we must also match the relevant non-dimensional parameters governing the flow. One of the most important such parameters in fluid mechanics is the Reynolds number, which measures the ratio of inertial forces to viscous forces:

$$\text{Re} = \frac{\rho U L}{\mu}, \tag{5.58}$$

where ρ is the fluid density, U is the characteristic velocity, L is the characteristic length, and μ is the dynamic viscosity. Their units are kg/m^3, m/s, m, and $\text{kg/(m\,s)}$, respectively. A direct dimensional check confirms that Re is dimensionless.

Assuming both experiments use air under similar conditions so that ρ and μ remain unchanged, achieving dynamical similarity requires matching Reynolds numbers: $\text{Re}_{\text{real}} = \text{Re}_{\text{model}}$. Thus, $\rho U_{\text{real}} L_{\text{real}} = \rho U_{\text{model}} L_{\text{model}}$, which simplifies to $U_{\text{model}} L_{\text{model}} = U_{\text{real}} L_{\text{real}}$. Since $L_{\text{real}} = 10\,L_{\text{model}}$, we obtain $U_{\text{model}} = 10\,U_{\text{real}} = 2000\,\text{m/s}$. Therefore, to maintain Reynolds number similarity using air, the model would need to move ten times faster than the real aircraft. This reflects the fact that as length decreases, viscous effects become relatively more significant. Increasing the velocity restores the same balance between inertial and viscous forces.

However, $U_{\text{model}} = 2000\,\text{m/s}$ is hypersonic and physically unrealistic for a laboratory wind tunnel. In practice, alternative strategies are used. One approach is to increase the air density using a pressurized wind tunnel, which allows Reynolds number matching at lower speeds. Another approach is to use a different fluid with higher viscosity or density to adjust the Reynolds number without requiring extreme velocities.

□

5.2.3 Deriving approximate models

Dimensional analysis also helps simplify equations and guide numerical solvers.

Example 5.13 (Using non-dimensional numbers to simplify equations). Consider the temperature transport equation describing heat transport in a moving fluid:

$$\frac{\partial T}{\partial t} + u\frac{\partial T}{\partial x} = \alpha\frac{\partial^2 T}{\partial x^2}, \tag{5.59}$$

which is a one-dimensional advection-diffusion equation. Here T is temperature (unit: K), u is velocity (unit: m/s), α is thermal diffusivity (unit: $\mathrm{m^2/s}$), and x is the spatial coordinate (unit: m). The advection term represents heat transport by fluid motion, while the diffusion term describes heat conduction. Introduce the following dimensionless variables:

$$\tilde{T} = \frac{T}{T_1}, \qquad \tilde{x} = \frac{x}{L}, \qquad \tilde{t} = \frac{t}{L/u}, \tag{5.60}$$

where T_1 is a characteristic temperature scale and L/u is the advective time scale. Substituting these variables into (5.59) yields

$$\frac{\partial \tilde{T}}{\partial \tilde{t}} + \frac{\partial \tilde{T}}{\partial \tilde{x}} = \frac{1}{\mathrm{Pe}}\frac{\partial^2 \tilde{T}}{\partial \tilde{x}^2}, \tag{5.61}$$

where the Peclet number is

$$\mathrm{Pe} = \frac{uL}{\alpha}, \tag{5.62}$$

which measures the ratio between advective and diffusive effects. Equation (5.61) clearly reveals the relative importance of the two mechanisms. If $\mathrm{Pe} \gg 1$, advection dominates. If $\mathrm{Pe} \ll 1$, diffusion dominates. For example, consider cooling a highly viscous oil with $u = 10^{-6}$ m/s and layer thickness $L = 0.01$ m. With thermal diffusivity $\alpha = 10^{-7}\ \mathrm{m^2/s}$,

$$\mathrm{Pe} = \frac{uL}{\alpha} = \frac{10^{-6} \times 0.01}{10^{-7}} = 0.1 \ll 1. \tag{5.63}$$

Thus, diffusion is the dominant mechanism and the advection term can be neglected at leading order, yielding a simplified diffusion equation. Directly comparing the numerical values of u and α in the dimensional equation would be misleading, because the coefficients carry different physical units. Only after non-dimensionalization does their relative importance become meaningful. Even when both terms are retained, the Peclet number indicates whether an advection-dominated or diffusion-dominated numerical scheme is appropriate. □

5.2.4 Broad applications of dimensional and scaling analysis

Dimensional and scaling analysis has much wider applications. We will further provide two interesting examples before closing this section.

Example 5.14 (A scaling argument for the Pythagorean theorem). Consider a right triangle $\triangle ABC$ with right angle at C. Let $AB = c$ be the hypotenuse, $BC = a$, and $AC = b$. Denote by $\alpha = \angle B$ one of the acute angles. Let D be the foot of the perpendicular from C onto AB (see Figure 5.4).

The area of a triangle has physical dimension $[L]^2$. For any right triangle, if we choose its hypotenuse as the characteristic length and one acute angle as the dimensionless parameter, dimensional reasoning implies that its area must take the form

$$S = (\text{hypotenuse})^2\, \Phi(\text{acute angle}), \tag{5.64}$$

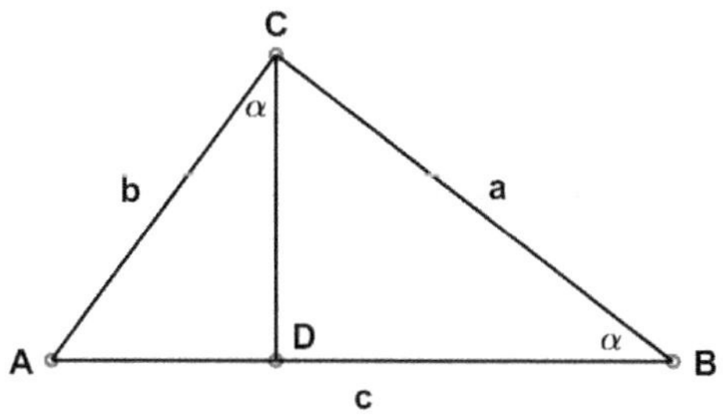

FIGURE 5.4: A right triangle used for the dimensional analysis proof of the Pythagorean theorem.

where Φ is a dimensionless function depending only on the angle. Applying this principle to the large triangle $\triangle ABC$, we obtain

$$S_{ABC} = c^2\Phi(\alpha). \tag{5.65}$$

Now consider the two smaller right triangles $\triangle ACD$ and $\triangle BCD$. By similarity of right triangles (the altitude construction preserves angles), both smaller triangles are also right triangles and are similar to the original one. For $\triangle ACD$, the hypotenuse is $AC = b$, and the acute angle $\angle ACD = \alpha$. Thus,

$$S_{ACD} = b^2\Phi(\alpha). \tag{5.66}$$

For $\triangle BCD$, the right angle is at D, so its hypotenuse is $BC = a$. The acute angle $\angle BCD = \alpha$. Therefore, using the hypotenuse as the characteristic length and choosing the acute angle equal to α as the dimensionless parameter, we obtain

$$S_{BCD} = a^2\Phi(\alpha). \tag{5.67}$$

Because the area of the large triangle equals the sum of the areas of the two smaller ones, $S_{ABC} = S_{ACD} + S_{BCD}$, we combine (5.65), (5.66), and (5.67) to obtain

$$c^2\Phi(\alpha) = b^2\Phi(\alpha) + a^2\Phi(\alpha). \tag{5.68}$$

Dividing both sides by the common nonzero factor $\Phi(\alpha)$ gives

$$c^2 = a^2 + b^2. \tag{5.69}$$

This completes the proof of the Pythagorean theorem via scaling and similarity arguments. □

Dimensional analysis can also be a powerful tool in the area of functional analysis, in particular to determine the powers in various inequalities.

Example 5.15 (Dimensional analysis in proving the Sobolev inequality). The Sobolev embedding theorem is a fundamental result in PDE and functional analysis [4]. Here, we will show how to use dimensional analysis to determine the power p^* in the following Sobolev inequality. Proficiency in applying such inequalities and determining their exponents is invaluable for analysis.

For $f \in C_c^1(\mathbb{R}^n)$ and $1 \le p < n$,

$$\|f\|_{L^{p^*}(\mathbb{R}^n)} \le m_{n,p}\|\nabla f\|_{L^p(\mathbb{R}^n)}, \tag{5.70}$$

where $C_c^1(\mathbb{R}^n)$ is the continuously differentiable function space with compact support within the n-dimensional real space and $m_{n,p}$ is a constant depending on n and p. The norm $\|\nabla f\|_{L^p(\mathbb{R}^n)}$ is defined as:

$$\|\nabla f\|_{L^p(\mathbb{R}^n)} = \left(\int_{\mathbb{R}^n} |\nabla f|^p \, \mathrm{d}\mathbf{x}\right)^{\frac{1}{p}}. \tag{5.71}$$

Let us determine the critical Sobolev exponent p^* using dimensional analysis. We denote $[f]$ by the dimension of f and L the dimension of length. So, the dimension of ∇f is $[f]/[L]$. The dimension of $\|\nabla f\|_{L^p(\mathbb{R}^n)}$ on the right-hand side of (5.70) is

$$\dim\left(\|\nabla f\|_{L^p(\mathbb{R}^n)}\right) = \dim\left(\int_{\mathbb{R}^n} |\nabla f|^p \, \mathrm{d}\mathbf{x}\right)^{\frac{1}{p}} = \left(\left(\frac{[f]}{[L]}\right)^p [L]^n\right)^{\frac{1}{p}} = [f][L]^{\frac{n-p}{p}}, \tag{5.72}$$

where $[L]^n$ comes from the integration over the n-dimensional space. Similarly, the left-hand side is

$$\dim\left(\|f\|_{L^{p^*}(\mathbb{R}^n)}\right) = \dim\left(\int_{\mathbb{R}^n} |f|^{p^*} \, \mathrm{d}\mathbf{x}\right)^{\frac{1}{p^*}} = \left([f]^{p^*}[L]^n\right)^{\frac{1}{p^*}} = [f][L]^{\frac{n}{p^*}}. \tag{5.73}$$

Since the powers of L in (5.72) and (5.73) must be the same, we have

$$\frac{n-p}{p} = \frac{n}{p^*}, \tag{5.74}$$

which leads to $p^* = \frac{pn}{n-p}$.

This dimensional analysis method helps determine the power in any inequality. The direction of the inequality ($\geq$ or $\leq$) typically reflects that higher-order derivatives (here, ∇f) control lower-order terms (here, f). □

5.3 Asymptotic analysis and perturbation theory

Asymptotic analysis is a wide class of methods to understand the limiting behavior of a system when one of the parameters goes to zero or infinity. It provides approximations (often via asymptotic expansions) when the parameter approaches the limit. Perturbation theory is a specific application of asymptotic analysis, in which the solution to a complicated problem is approximated by expanding around the exact solution of a related simpler problem. It heavily relies on asymptotic expansions.

5.3.1 Regular perturbations

Regular perturbation theory is used to approximate solutions to problems that are "close" to an exactly solvable problem. The key idea is to expand the solution in a power series of a small parameter ϵ, where the unperturbed problem ($\epsilon = 0$) is exactly solvable while the perturbed problem ($\epsilon \neq 0$) introduces a small correction that can be systematically approximated [137].

Regular perturbation applies when the perturbation does not change the fundamental nature of the solution and when the resulting asymptotic expansion remains uniformly valid in the domain of interest. In the next subsection, we will see the cases where the solution has sudden boundary layers or vanishing highest-order derivatives, which have to be dealt with by singular perturbation methods.

Example 5.16 (Finding an approximate solution to a quadratic algebraic equation with small linear coefficient). Let us find an approximate solution to:

$$x^2 + \epsilon x - 1 = 0, \tag{5.75}$$

for $|\epsilon| \ll 1$. An asymptotic expansion in powers of ϵ provides an approximate solution. We expand x as

$$x = x_0 + \epsilon x_1 + \epsilon^2 x_2 + \mathcal{O}(\epsilon^3). \tag{5.76}$$

Substituting (5.76) into (5.75) yields

$$(x_0 + \epsilon x_1 + \epsilon^2 x_2)^2 + \epsilon(x_0 + \epsilon x_1) - 1 = 0. \tag{5.77}$$

Reordering terms with respect to ϵ leads to

$$x_0^2 - 1 + \epsilon(2x_0x_1 + x_0) + \epsilon^2(x_1^2 + 2x_0x_2 + x_1) + \mathcal{O}(\epsilon^3) = 0. \tag{5.78}$$

Then we can solve the equation corresponding to each order of ϵ

$$\begin{aligned}
\mathcal{O}(1): &\quad x_0^2 - 1 = 0 &\implies&\quad x_0 = 1 \quad \text{(choosing the positive root)},\\
\mathcal{O}(\epsilon): &\quad 2x_0x_1 + x_0 = 0 &\implies&\quad x_1 = -\frac{1}{2},\\
\mathcal{O}(\epsilon^2): &\quad x_1^2 + 2x_0x_2 + x_1 = 0 &\implies&\quad x_2 = \frac{1}{8}.
\end{aligned} \tag{5.79}$$

Therefore, the approximate solution is given by

$$x \approx 1 - \frac{\epsilon}{2} + \frac{\epsilon^2}{8}. \tag{5.80}$$

For this simple problem, we can easily write down the exact solution:

$$x = \frac{-\epsilon + \sqrt{\epsilon^2 + 4}}{2} \approx 1 - \frac{\epsilon}{2} + \frac{\epsilon^2}{8} + \ldots \tag{5.81}$$

which serves as a useful validation of the perturbation result above. □

5.3.2 Singular perturbation and boundary layer problem

A singular perturbation problem differs from a regular perturbation problem in that setting the small parameter equal to zero changes the order or type of the equation, leading to a qualitatively different solution structure. More precisely, the solution cannot be uniformly approximated by an asymptotic expansion [275].

Differential equations that contain a small parameter as the coefficient of the highest order term fall into the singular perturbation category. These problems typically exhibit boundary layers, that is, thin regions near boundaries or interfaces where the solution changes rapidly over a short spatial (or temporal) scale, while varying smoothly elsewhere. As a result, the solution evolves on two different scales.

Example 5.17 (Boundary layer problem). We are given the second-order ODE:

$$\epsilon y'' + 2y' + y = 0 \tag{5.82}$$

with boundary conditions $y(0) = 0$ and $y(1) = 1$, where $|\epsilon| \ll 1$.

We may start with the regular perturbation method, which keeps the leading order term and ignores those containing ϵ. This gives

$$2y' + y = 0, \tag{5.83}$$

which is a first-order ODE. Its solution is:

$$y = Ae^{-x/2}. \tag{5.84}$$

Now the issue appears. Since we have ignored the highest-order term (the second derivative here), the solution of the resulting first-order ODE contains only one constant to be determined. However, we have two boundary conditions, and unfortunately, they will give different answers. Let us apply the boundary condition $y(1) = 1$, which gives

$$Ae^{-1/2} = 1 \implies A = e^{1/2}. \tag{5.85}$$

We call this solution the outer solution:

$$y_{\text{outer}}(x) = e^{(1-x)/2}. \tag{5.86}$$

Note that this solution does not satisfy the boundary condition $y(0) = 0$. Therefore, additional work needs to be done to arrive at the complete solution.

Let us introduce a stretched coordinate near $x = 0$: $\xi = \frac{x}{\epsilon^\alpha}$. Rewrite the ODE in terms of ξ:

$$\epsilon^{1-2\alpha}\frac{\mathrm{d}^2 y}{\mathrm{d}\xi^2} + \epsilon^{-\alpha}2\frac{\mathrm{d}y}{\mathrm{d}\xi} + y = 0. \tag{5.87}$$

Since we have to keep the highest derivative term (otherwise the solution degenerates to the outer solution), we wish to balance at least two of the three terms in (5.87). To this end, we have two choices to determine this α. First, we can match the order of the coefficients of the second-order and zeroth-order terms, assuming the first-order derivative term is negligible. To satisfy the former, we will have $\alpha = 0.5$. However, this means the coefficient of the first-order derivative term will be $\epsilon^{-1/2}$, which will dominate the other two terms. The second alternative is to match the order of the coefficients of the second-order and the first-order terms, which leads to $\alpha = 1$. In such a case, the first two terms dominate the third term. The leading-order inner solution equation is given by

$$\frac{\mathrm{d}^2 y}{\mathrm{d}\xi^2} + 2\frac{\mathrm{d}y}{\mathrm{d}\xi} = 0. \tag{5.88}$$

The characteristic equation of (5.88) is $r^2 + 2r = 0$, giving $r = 0$ and -2. Thus, the solution of (5.88), named as the inner solution and denoted by $y_{\text{inner}}(\xi)$, is:

$$y_{\text{inner}}(\xi) = C + De^{-2\xi}. \tag{5.89}$$

Applying $y(0) = 0$ gives $C + D = 0$ or $D = -C$, which leads to the inner solution

$$y_{\text{inner}}(\xi) = C(1 - e^{-2\xi}). \tag{5.90}$$

The remaining constant C can be determined by matching the inner and outer solutions. The inner solution as $\xi \to \infty$ leads to $y_{\text{inner}} \approx C$. The outer solution as $x \to 0$ gives $y_{\text{outer}} \approx e^{1/2}$. Matching them determines C:

$$C = e^{1/2}. \tag{5.91}$$

This matching is required because the limits $\xi \to \infty$ and $x \to 0$ describe the same intermediate physical region, where $\xi \to \infty$ corresponds to moving away from the boundary layer in the inner coordinates while $x \to 0$ corresponds to approaching the boundary layer from the outer region. Requiring the two solutions to agree in this overlap region ensures a smooth and consistent composite solution across scales. Thus, the inner solution is:

$$y_{\text{inner}}(\xi) = e^{1/2}(1 - e^{-2\xi}). \tag{5.92}$$

Combining the inner and outer solutions and subtracting the common part $e^{1/2}$ yields the full solution:

$$y_{\text{full}}(x) = y_{\text{outer}}(x) + y_{\text{inner}}(\xi) - e^{1/2} = e^{(1-x)/2} - e^{1/2}e^{-2x/\epsilon}. \tag{5.93}$$

Figure 5.5 shows the inner, outer, and full solutions of the equation (5.82). With $\epsilon \to 0$, the area of the boundary layer shrinks and the full solution becomes more consistent with the outer solution, while the sudden change in the boundary layer near $x = 0$ is more dramatic.

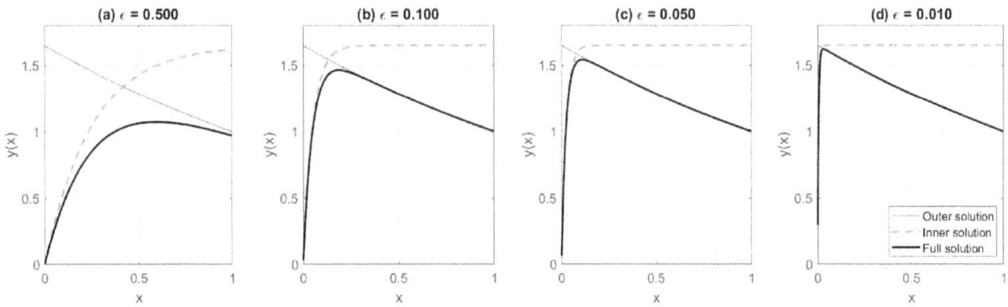

FIGURE 5.5: Inner, outer, and composite solutions of (5.82) illustrating the boundary layer near $x = 0$. Different panels show the solution with different ϵ values.

□

5.3.3 Multiscale expansion: Eliminating secular terms and capturing frequency shifts

When nonlinear terms are present, the behavior of a system can become much more complicated. Even weak nonlinearities may produce qualitative effects such as amplitude-dependent frequencies, energy transfer between modes, and long-time modulation of oscillations. Standard regular perturbation expansions often fail to capture these features, as they may generate secular terms that grow in time and eventually invalidate the approximation. To construct uniformly valid approximations over long time intervals, more refined perturbation techniques are required. In what follows, we introduce the method of multiple time scales [91] to illustrate how slow modulation of amplitude or phase can be incorporated systematically. Other related approaches, such as the Poincaré-Lindstedt method [218, 267] and averaging techniques, follow similar principles and are briefly discussed later.

The Duffing oscillator with weak nonlinearity is used to illustrate the strengths and weaknesses of different methods. The Duffing oscillator reads [92]:

$$\frac{\mathrm{d}^2 y}{\mathrm{d}t^2} + y + \epsilon y^3 = 0, \quad y(0) = 1, \quad \frac{\mathrm{d}y}{\mathrm{d}t}(0) = 0, \tag{5.94}$$

where $\epsilon \ll 1$ is a small parameter.

Example 5.18 (Solving the Duffing oscillator via regular perturbation). Let us start with the regular perturbation discussed in Section 5.3.1. Assume an asymptotic expansion

$$y(t) = y_0(t) + \epsilon y_1(t) + \epsilon^2 y_2(t) + \cdots . \tag{5.95}$$

Substituting into (5.94) and matching powers of ϵ yields

$$\begin{aligned} \mathcal{O}(1): \quad & y_0'' + y_0 = 0, \quad y_0(0) = 1, \; y_0'(0) = 0 \;\Rightarrow\; y_0(t) = \cos t, \\ \mathcal{O}(\epsilon): \quad & y_1'' + y_1 = -y_0^3 = -\cos^3 t = -\frac{3}{4}\cos t - \frac{1}{4}\cos 3t, \quad y_1(0) = 0, \; y_1'(0) = 0. \end{aligned} \tag{5.96}$$

A solution satisfying the $\mathcal{O}(\epsilon)$ initial conditions is

$$y_1(t) = -\frac{3}{8} t \sin t + \frac{1}{32}\big(\cos 3t - \cos t\big). \tag{5.97}$$

Therefore,

$$y(t) = \cos t + \epsilon \left[-\frac{3}{8} t \sin t + \frac{1}{32}\big(\cos 3t - \cos t\big)\right] + \mathcal{O}(\epsilon^2). \tag{5.98}$$

The term $-\frac{3}{8} t \sin t$ is a secular term: it grows with t and causes the regular perturbation approximation to lose validity on long time intervals (typically $t \sim \mathcal{O}(1/\epsilon)$), even though the exact Duffing solution remains bounded. See Figure 5.6. □

Example 5.19 (Solving the Duffing oscillator via the multiscale expansion). To avoid the appearance of secular terms in the regular perturbation expansion, we apply the method of multiple time scales. Introduce two time variables:

$$T_0 = t, \qquad T_1 = \epsilon t, \tag{5.99}$$

where T_0 captures the fast oscillations on an $\mathcal{O}(1)$ time scale, while T_1 describes slow modulation occurring over long times $t = \mathcal{O}(1/\epsilon)$. The variable $T_1 = \epsilon t$ is referred to as a slow time scale because an $\mathcal{O}(1)$ variation in T_1 corresponds to a much longer physical time interval $t = \mathcal{O}(1/\epsilon)$. Since $\epsilon \ll 1$, changes in T_1 accumulate gradually compared with those in $T_0 = t$, which vary on the fast $\mathcal{O}(1)$ oscillatory time scale. Equivalently, derivatives with respect to T_1 are multiplied by ϵ in the chain rule,

$$\frac{\mathrm{d}}{\mathrm{d}t} = \frac{\partial}{\partial T_0} + \epsilon \frac{\partial}{\partial T_1}, \qquad \frac{\mathrm{d}^2}{\mathrm{d}t^2} = \frac{\partial^2}{\partial T_0^2} + 2\epsilon \frac{\partial T_0}{\partial T_1} + \epsilon^2 \frac{\partial^2}{\partial T_1^2}, \tag{5.100}$$

showing that the influence of T_1 enters the dynamics only at higher order.

Assume an asymptotic expansion of the form

$$y(t) = y_0(T_0, T_1) + \epsilon y_1(T_0, T_1) + \mathcal{O}(\epsilon^2). \tag{5.101}$$

Substituting into the Duffing equation

$$\frac{\mathrm{d}^2 y}{\mathrm{d}t^2} + y + \epsilon y^3 = 0 \tag{5.102}$$

and collecting powers of ϵ gives:

Let us start by the governing equation for leading order $\mathcal{O}(1)$:

$$\frac{\partial^2 y_0}{\partial T_0^2} + y_0 = 0. \tag{5.103}$$

The general solution is

$$y_0(T_0, T_1) = A(T_1)\cos\theta, \qquad \theta = T_0 + \phi(T_1), \tag{5.104}$$

where $A(T_1)$ and $\phi(T_1)$ represent slowly varying amplitude and phase.

Next, the governing equation of the first order $\mathcal{O}(\epsilon)$ is given by:

$$\frac{\partial^2 y_1}{\partial T_0^2} + y_1 = -2\frac{\partial^2 y_0}{\partial T_0 \partial T_1} - y_0^3. \tag{5.105}$$

Compute the required derivatives $\frac{\partial y_0}{\partial T_0} = -A\sin\theta$, $\frac{\partial y_0}{\partial T_1} = A'\cos\theta - A\phi'\sin\theta$, and $\frac{\partial^2 y_0}{\partial T_0 \partial T_1} = -A'\sin\theta - A\phi'\cos\theta$. Therefore,

$$\frac{\partial^2 y_1}{\partial T_0^2} + y_1 = 2A'\sin\theta + 2A\phi'\cos\theta - A^3\cos^3\theta. \tag{5.106}$$

Using $\cos^3\theta = \frac{3}{4}\cos\theta + \frac{1}{4}\cos 3\theta$, we obtain

$$\frac{\partial^2 y_1}{\partial T_0^2} + y_1 = \underbrace{2A'\sin\theta + \left(2A\phi' - \frac{3}{4}A^3\right)\cos\theta}_{\text{resonant terms}} - \frac{1}{4}A^3\cos 3\theta. \tag{5.107}$$

To eliminate secular growth, the coefficients of the resonant terms $\sin\theta$ and $\cos\theta$ must vanish, which gives $2A' = 0$ and $2A\phi' - \frac{3}{4}A^3 = 0$. Hence, $A(T_1) = A_0$ and $\phi'(T_1) = \frac{3}{8}A_0^2$. With the initial conditions $y(0) = 1$ and $y'(0) = 0$, we obtain $A_0 = 1$ and $\phi(0) = 0$, so

$$\phi(T_1) = \frac{3}{8}T_1 = \frac{3}{8}\epsilon t. \tag{5.108}$$

The remaining equation becomes

$$\frac{\partial^2 y_1}{\partial T_0^2} + y_1 = -\frac{1}{4}A_0^3\cos 3\theta, \tag{5.109}$$

with particular solution

$$y_1(T_0, T_1) = \frac{A_0^3}{32}\cos 3\theta. \tag{5.110}$$

Substituting back gives the final solution,

$$y(t) = \cos\left(t + \frac{3}{8}\epsilon t\right) + \epsilon\frac{1}{32}\cos\left(3\left(t + \frac{3}{8}\epsilon t\right)\right) + \mathcal{O}(\epsilon^2). \tag{5.111}$$

This approximation remains uniformly valid on long time intervals $t = \mathcal{O}(1/\epsilon)$ and captures the amplitude-dependent frequency shift:

$$\omega = 1 + \frac{3}{8}\epsilon. \tag{5.112}$$

The secular growth present in the regular perturbation expansion is eliminated by allowing the phase to evolve on the slow time scale. See Figure 5.6.

□

In addition to the multiscale expansion, several alternative perturbation methods have been developed to address the limitations of regular expansions. The Poincaré-Lindstedt method removes secular growth by rescaling time and renormalizing the oscillation frequency, which is particularly effective for periodic systems. The method of averaging provides a systematic way to derive slow evolution equations for amplitudes and phases in weakly nonlinear oscillators. More generally, renormalization group methods and matched asymptotic expansions extend these ideas to systems with multiple scales, boundary layers, or singular perturbations. Each technique is tailored to a specific type of structure in the governing equations.

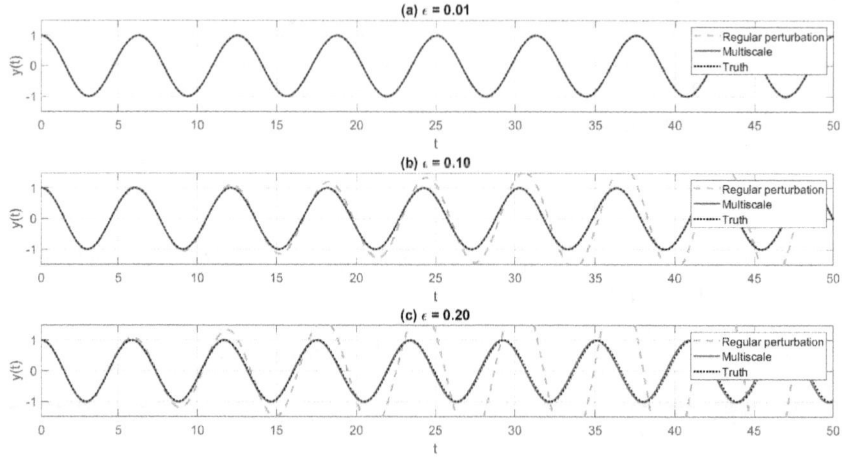

FIGURE 5.6: Comparison of the three perturbation methods to the Duffing equation with different ϵ values. The truth and the solution from the multiscale expansion almost overlap with each other while the solution from the regular perturbation shows significant error as time increases due to the secular term, especially with large ϵ values.

5.3.4 Asymptotics of integrals (Laplace method)

Asymptotic expansion has broader applications. One typical scenario is to approximate the integrals [255]. While different approximate methods exist, we will focus on the Laplace method, which approximates integrals of the form [202]:

$$I(\lambda) = \int_a^b e^{-\lambda f(x)} g(x)\,\mathrm{d}x \quad \text{as} \quad \lambda \to \infty. \tag{5.113}$$

To begin with, assume we can find the dominant point $x^* \in [a, b]$, which corresponds to the local minimum of $f(x)$ (i.e., $f'(x^*) = 0,\ f''(x^*) > 0$). Then we apply a Taylor expansion around x^*,

$$f(x) \approx f(x^*) + \frac{1}{2} f''(x^*)(x - x^*)^2. \tag{5.114}$$

With this, we approximate $g(x) \approx g(x^*)$ near x^* (if $g(x)$ varies slowly). Finally, we extend limits and evaluate:

$$I(\lambda) \approx g(x^*) e^{-\lambda f(x^*)} \int_{-\infty}^{\infty} e^{-\frac{\lambda}{2} f''(x^*)(x-x^*)^2}\,\mathrm{d}x = g(x^*) e^{-\lambda f(x^*)} \sqrt{\frac{2\pi}{\lambda f''(x^*)}}. \tag{5.115}$$

The reason we can extend the integration limits is that when $\lambda \gg 1$, the integrand is concentrated in an $O(\lambda^{-1/2})$ neighborhood of the interior minimizer x^*; outside this neighborhood it is exponentially small. Extending $[a, b]$ to $(-\infty, \infty)$ therefore introduces an error that is exponentially small in λ. With the extended limits, the integral reduces to a Gaussian integral after an appropriate change of variables.

Note that the equality in (5.115) can be derived by performing a change of variables

$$y = \sqrt{\frac{\lambda f''(x^*)}{2}}\,(x - x^*), \qquad \mathrm{d}x = \sqrt{\frac{2}{\lambda f''(x^*)}}\,\mathrm{d}y. \tag{5.116}$$

such that the integral in (5.115) becomes

$$\int_{-\infty}^{\infty} e^{-\frac{\lambda}{2}f''(x^*)(x-x^*)^2}\, dx = \sqrt{\frac{2}{\lambda f''(x^*)}} \int_{-\infty}^{\infty} e^{-y^2}\, dy. \tag{5.117}$$

Then, using the standard Gaussian integral $\int_{-\infty}^{\infty} e^{-y^2}\, dy = \sqrt{\pi}$, we obtain (5.115)

Example 5.20 (Gamma function and Stirling's approximation). A classical application of the Laplace method is the asymptotic approximation of the Gamma function. For a given complex number n with $\mathrm{Re}(n) > -1$, the Gamma function $\Gamma(n+1)$ is defined by

$$\Gamma(n+1) = \int_0^{\infty} e^{-x} x^n\, dx. \tag{5.118}$$

For a positive integer n, this definition yields $\Gamma(n+1) = n!$, which will be the case we focus on below. When n is large, the integrand $e^{-x}x^n$ becomes highly peaked, meaning that the main contribution to the integral comes from a small neighborhood around its maximum. To make this structure explicit, we rewrite the integrand in exponential form:

$$n! = \int_0^{\infty} \exp(-x + n\log x)\, dx = \int_0^{\infty} e^{-f(x)}\, dx, \tag{5.119}$$

where

$$f(x) = x - n\log x. \tag{5.120}$$

Below, let us show the step-by-step application of the Laplace method in approximating the Gamma function.

Step 1: Identify the dominant point.

The Laplace method tells us that the dominant contribution comes from the point x^* where $f(x)$ attains its minimum. This point balances the exponential decay of e^{-x} with the rapid growth of x^n. Differentiating the function $f(x)$ leads to

$$f'(x) = 1 - \frac{n}{x}. \tag{5.121}$$

Setting $f'(x) = 0$ yields $x^* = n$, which confirms the intuitive expectation that the integral is concentrated near $x = n$ when n is large.

Step 2: Quadratic approximation near the minimum.

To quantify the shape of the integrand near x^*, we compute the second derivative:

$$f''(x) = \frac{n}{x^2}, \qquad f''(n) = \frac{1}{n}. \tag{5.122}$$

Near x^*, we approximate $f(x)$ by its second-order Taylor expansion,

$$f(x) \approx f(n) + \frac{1}{2}f''(n)(x-n)^2, \tag{5.123}$$

so that the integrand becomes approximately Gaussian.

Step 3: Evaluate the Gaussian integral.

Applying the Laplace approximation gives

$$n! \approx e^{-f(n)}\sqrt{\frac{2\pi}{f''(n)}}. \tag{5.124}$$

Since $f(n) = n - n\log n$ and $f''(n) = \frac{1}{n}$, we obtain

$$n! \approx e^{-n+n\log n}\sqrt{2\pi n} = \sqrt{2\pi n}\left(\frac{n}{e}\right)^n. \tag{5.125}$$

This is Stirling's approximation, which reveals that the factorial grows essentially like $(n/e)^n$, with a subleading prefactor $\sqrt{2\pi n}$ arising from Gaussian fluctuations around the dominant point.

Panel (a) of Figure 5.7 includes a numerical validation of using Stirling's approximation for the Gamma function as a function of n. The difference between the approximation and the truth is almost invisible. Panels (b) and (c) show the Gaussian approximation of the integrand in computing the Gamma function (5.118). The Gaussian approximation becomes more accurate with a larger n. □

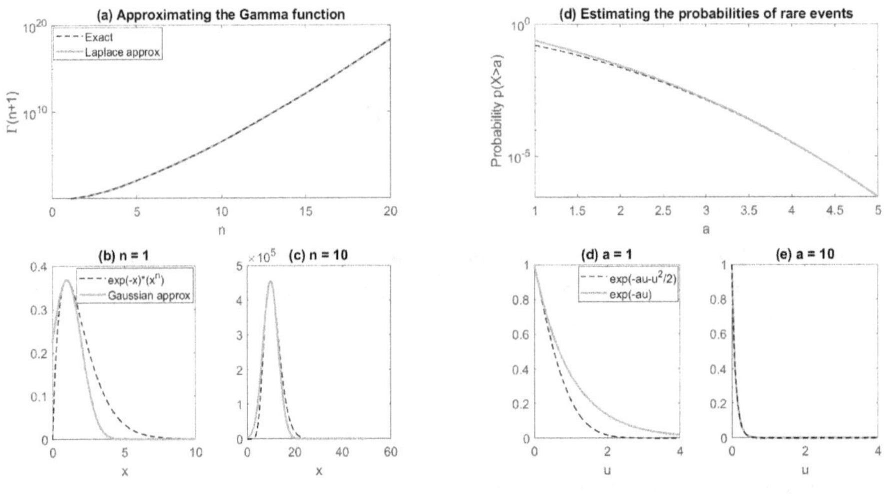

FIGURE 5.7: Numerical validation of the Laplace method. Panel (a): using Stirling's approximation for the Gamma function as a function of n. Panels (b)–(c): comparison of the exact integrand $e^{-x}x^n$ in computing the Gamma function (5.118) and its Gaussian approximation (5.123) based on the Laplace method with $n = 1$ and $n = 10$, respectively. Panel (d): estimating the probabilities of rare events as a function of a using (5.130). Panels (e)–(f): the difference between the leading-order term e^{-au} in the integrand $e^{-au-\frac{u^2}{2}}$ in (5.128) with $a = 1$ and $a = 10$, respectively.

Example 5.21 (Estimating the probabilities of rare events). For a standard Gaussian random variable X with PDF $p(x) = \frac{1}{\sqrt{2\pi}}e^{-x^2/2}$, we wish to estimate:

$$P(X > a) = \int_a^\infty p(x)\,\mathrm{d}x \quad \text{for large } a \tag{5.126}$$

Let $f(x) = -\log p(x)$, such that

$$P(X > a) = \int_a^\infty e^{-f(x)}\,\mathrm{d}x. \tag{5.127}$$

Since $p(x) = \frac{1}{\sqrt{2\pi}}e^{-x^2/2}$, we have $f(x) = \frac{u^2}{2} + \text{const.}$. However, $f'(x) = x$ and its minimal occurs at $x^* = a$. When $a \gg 0$, $f'(a) \neq 0$. So, standard Laplace method does not directly work. Nevertheless, we can still apply asymptotic expansion to solve this problem. Let us implement a change of variable $u = x - a$ such that

$$P(X > a) = \frac{1}{\sqrt{2\pi}}\int_a^\infty e^{-\frac{x^2}{2}}\,\mathrm{d}x = \frac{1}{\sqrt{2\pi}}e^{-\frac{a^2}{2}}\int_0^\infty e^{-au-\frac{u^2}{2}}\,du \tag{5.128}$$

for $a \gg 0$, the term e^{-au} dominates over $e^{-u^2/2}$. Clearly, when $a \gg 0$, only values of u near zero play a role in the above integral as e^{-au} decays very fast when u is away from zero. This means $-\frac{u^2}{2}$ is small such that we can approximate $e^{-u^2/2} \approx 1 - \frac{u^2}{2}$. Therefore,

$$P(X > a) \approx \frac{1}{\sqrt{2\pi}}e^{-\frac{a^2}{2}}\int_0^\infty e^{-au}\left(1 - \frac{u^2}{2}\right)\,du = \frac{1}{\sqrt{2\pi}}e^{-\frac{a^2}{2}}\left(\frac{1}{a} - \frac{1}{a^3}\right). \tag{5.129}$$

Therefore, the leading-order approximation is

$$P(X > a) \approx \frac{e^{-\frac{a^2}{2}}}{a\sqrt{2\pi}}. \tag{5.130}$$

Panel (d) in Figure 5.7 includes the numerical validations of estimating the probabilities of rare events as a function of a using (5.130). As is expected, the approximation in (5.130) becomes more accurate with a larger value of a. Panels (e)–(f) shows the difference between the leading-order term e^{-au} in the integrand $e^{-au-\frac{u^2}{2}}$ in (5.128). Again, the leading-order term becomes more dominant as a increases. □

Example 5.22 (Connection to importance sampling in estimating the probabilities of rare events). The asymptotic analysis in Example 5.21 is closely related to the idea of importance sampling for estimating rare-event probabilities discussed in Example 2.14. Recall that

$$P(X > a) = \mathbb{E}\left[\mathbf{1}_{\{X>a\}}\right] = \mathbb{E}_q\left[\mathbf{1}_{\{X>a\}}\frac{p(X)}{q(X)}\right], \tag{5.131}$$

where q is the proposal density function used for sampling. A common choice in rare-event simulation for Gaussian tails is a shifted Gaussian distribution $q(x) = \mathcal{N}(\theta, 1)$. The parameter θ is chosen so that samples are concentrated near the rare-event region.

For the Gaussian tail probability considered here, one natural choice is $\theta = a$, which shifts the mean of the sampling distribution to the boundary point $x = a$, where the rare event $\{X > a\}$ begins to occur. Therefore, $q(x)$ is given by

$$q(x) = \frac{1}{\sqrt{2\pi}}e^{-(x-a)^2/2} = \frac{1}{\sqrt{2\pi}}e^{-x^2/2+ax-a^2/2}. \tag{5.132}$$

Using importance sampling,

$$P(X > a) = \mathbb{E}_q\left[\mathbf{1}_{\{X>a\}}\frac{p(X)}{q(X)}\right], \qquad \frac{p(x)}{q(x)} = e^{-ax+a^2/2}. \tag{5.133}$$

Writing $X = a + U$, where $U \sim \mathcal{N}(0,1)$ under q, so that $\mathbf{1}_{\{X>a\}} = \mathbf{1}_{\{U>0\}}$, we obtain

$$\begin{aligned} P(X > a) &= \mathbb{E}\left[\mathbf{1}_{\{X>a\}} e^{a^2/2 - aX}\right] = \mathbb{E}\left[\mathbf{1}_{\{U>0\}} e^{a^2/2 - a(a+U)}\right] \\ &= \mathbb{E}\left[\mathbf{1}_{\{U>0\}} e^{-a^2/2 - aU}\right] = e^{-a^2/2}\,\mathbb{E}\left[\mathbf{1}_{\{U>0\}} e^{-aU}\right] \\ &= \frac{1}{\sqrt{2\pi}} e^{-a^2/2} \int_0^\infty e^{-au - \frac{u^2}{2}}\, du, \end{aligned} \tag{5.134}$$

where the factor $e^{-a^2/2}$ captures the dominant exponential decay of the Gaussian tail, while the remaining integral accounts for fluctuations around the boundary point $x = a$. This recovers the result in (5.128). □

5.4 Mean-fluctuation decomposition

5.4.1 Definition

The mean-fluctuation decomposition, also known as the Reynolds decomposition, is a fundamental technique developed initially in fluid mechanics in studying turbulent flows, which have irregular and random features that are difficult to predict precisely. Nowadays, mean-fluctuation decomposition has wide applications. The key idea is to separate a variable in a system, which is turbulent or random, into its mean and fluctuating components to allow for the statistical studies of the system.

Consider a variable $u(x,t)$ that depends on both time and space. In the content of fluid, u can be a spatiotemporal variable describing velocity, pressure, temperature, etc. The variable u is assumed to be turbulent or stochastic. The mean-fluctuation decomposition splits it into [5]:

$$u = \overline{u} + u', \tag{5.135}$$

where the mean of the fluctuation term is zero by design, namely, $\overline{u'} = 0$. Notably, the mean of the product satisfies

$$\overline{u \cdot v} = \overline{u} \cdot \overline{v} + \overline{u'v'}, \tag{5.136}$$

where $\overline{u'v'}$ is so-called flux term. The relationship in (5.136) highlights that the mean of the product does not equal the product of the two individual means. Although the mean of the two individual fluctuation terms is zero, the flux term $\overline{u'v'}$ represents the covariance between u and v and is generally nonzero unless the fluctuations are uncorrelated. Such a flux term will play a significant role in adjusting the mean of the product. See Example 5.24 below for a concrete illustration.

5.4.2 Types of averaging in mean-fluctuation decomposition

Three primary averaging methods exist for decomposing a variable $u(x,t)$ into mean and fluctuating components, each suited for different physical scenarios.

1. Ensemble average:

$$u(x,t) = \overline{u}(x,t) + u'(x,t), \tag{5.137}$$

where $\overline{u}(x,t)$ is the ensemble average taking at each fixed location x and t,

$$\overline{u}(x,t) = \lim_{N\to\infty} \frac{1}{N} \sum_{n=1}^{N} u_n(x,t) \tag{5.138}$$

with N being the number of ensemble members (e.g., from the Monte Carlo simulations) and u_n being the n-th realization.

2. Time average:

$$u(x,t) = \overline{u}(x) + u'(x,t), \tag{5.139}$$

where $\overline{u}(x)$ is the time-averaged (mean) component at each fixed location x,

$$\overline{u}(x) = \lim_{T\to\infty} \frac{1}{T} \int_{t_0}^{t_0+T} u(x,t)\, \mathrm{d}t. \tag{5.140}$$

with T being the temporal range.

3. Spatial average:

$$u(x,t) = \overline{u}(t) + u'(x,t), \tag{5.141}$$

where $\overline{u}(t)$ is the spatial average at each time instant, giving a global mean field that assumes statistical homogeneity,

$$\overline{u}(t) = \frac{1}{V} \iiint_V u(x,t)\, \mathrm{d}V, \tag{5.142}$$

with V being the total spatial domain.

This strict spatial averaging is independent of space (x) by definition. For reduced-order modeling (such as LES subgrid modeling), $\overline{u}$ is typically a filtered field (not strictly an average):

$$\overline{u}(x,t) = \int G(x-x')u(x',t)\, \mathrm{d}x', \tag{5.143}$$

where G is a filter kernel (e.g., box or Gaussian). In numerical implementations, $\overline{u}(x,t)$ in (5.143) is frequently simplified by downsampling to a coarse grid. For example, when the original field has 128^3 grid points and the coarse solution has 16^3 grid points, each coarse-grid value can be calculated by averaging the original u values over the corresponding eight fine-grid points. This operation acts as a low-pass filter that attenuates high-wavenumber components while retaining the large-scale structures.

These different mean-fluctuation decompositions have their own uses, where concrete examples will be shown in the remainder of this chapter and the subsequent chapters. The temporal average applies to the situation where the variable (often the flow field) is statistically steady and, therefore, it has no long-term temporal trends. When the flow is statistically stationary, the time-averaged field satisfies $\frac{\partial \overline{u}}{\partial t} = 0$. It results in a system that is only spatially varying, which is much simpler. However, the temporally-averaged dynamics often fail for describing transient behavior. The ensemble average is based on repeatable experiments or realizations. It captures the mean dynamics of the transient effects (as $\frac{\partial \overline{u}}{\partial t} \neq 0$). Although repeating experiments is challenging in practice, theoretical analysis can consistently be implemented when knowing the basic statistical properties of the variables. Finally, the spatial average is a valid approximation to describe the statistical homogeneity in space (e.g., isotropic turbulence). It widely appears in engineering applications such as the large eddy simulation (LES) subgrid modeling. Different mean-fluctuation decompositions can also be combined. For example, the ensemble average can be applied to the LES solution to describe the mean behavior of the spatially averaged flow field.

These averages lead to different coarse-grained equations. On the numerical side, when spatial filtering is accompanied by grid coarsening (as in LES), the number of resolved degrees of freedom is reduced. Time averaging removes explicit time dependence in statistically stationary flows. Ensemble averaging yields deterministic equations for mean quantities but does not reduce the underlying state dimension. In some reduced or averaged models, the removal of fast fluctuations can allow larger effective time steps, although numerical stability ultimately depends on the spectral properties of the resulting system. In all these cases, the computational efficiency can be improved significantly. On the theoretical side, the mean equation can reveal the large-scale features that may otherwise be polluted by small-scale fluctuations. The mean equations are less complicated than the original equation and are often easier for theoretical analysis.

Example 5.23 (Decomposing a simple 1D velocity signal with time average). Assume we measure velocity $u(t)$ at a point in a turbulent flow:

$$u(t) = 5 + 2\sin(t). \tag{5.144}$$

Applying the time average, the mean component is $\overline{u} = 5$ (constant) while the fluctuating component is $u'(t) = 2\sin(t)$. □

Example 5.24 (Mean-fluctuation decomposition of a Gaussian random variable). Let u be a Gaussian random variable with mean μ and variance σ^2. Since there is no space and time, the mean-fluctuation decomposition is implemented in the sense of ensemble average. The mean component is $\overline{u} = \mu$, which is a constant, and the fluctuation component is $u' = u - \mu$, where $\overline{u'} = 0$.

By definition, the variance is

$$\sigma^2 = \mathbb{E}[(u-\mu)^2] = \mathbb{E}[u^2 - 2u\mu + \mu^2] = \mathbb{E}[u^2] - 2\mu\mathbb{E}[u] + \mu^2 = \overline{u^2} - \mu^2 = \overline{u'^2} \tag{5.145}$$

according to (5.136) (by setting $v = u$ and $v' = u'$). The well-known relationship in (5.145), putting into the content of the mean-fluctuation decomposition, indicates two crucial facts. First, the flux term, which is the average of the second-order fluctuations, represents the variance. Second, the mean of the square of the random variable $\overline{u^2}$ does not equal the square of the mean of the random variable μ^2. Their difference is precisely the variance of the random variable $\overline{u'^2}$. □

The simple example 5.24 illustrates a fundamental property in the mean dynamics of a turbulent or nonlinear stochastic system. Due to the nonlinearity, the mean equation will contain flux terms that characterize the statistics of small-scale features (in fluctuation), which introduce the interactions between large and small scales. This example also shows that if the statistical property of the random variable (for example the mean and the variance in a Gaussian distribution) is known, then the ensemble average of a function of such a random variable can sometimes be computed theoretically (with possibly approximations) without the need of carrying out repeated experiments or running Monte Carlo simulations.

5.4.3 Reynolds-averaged Navier-Stokes (RANS) equations

One of the original applications of the mean-fluctuation decomposition (Reynolds decomposition) is to simplify the Navier-Stokes equations by deriving the so-called Reynolds-averaged Navier-Stokes (RANS) equations [226]. It is useful to show different types of the decomposition in providing RANS from different perspectives.

The Navier–Stokes equations are a set of mathematical equations that describe how fluids move. They express a fundamental principle: the motion of a fluid at any point is determined by a balance between forces, including pressure differences, internal friction (viscosity), and possibly external influences such as gravity. The incompressible Navier-Stokes equations are:

$$\frac{\partial \mathbf{u}}{\partial t} + (\mathbf{u}\cdot\nabla)\mathbf{u} = -\frac{1}{\rho}\nabla p + \nu\nabla^2\mathbf{u}, \quad \nabla\cdot\mathbf{u} = 0, \tag{5.146}$$

where $\mathbf{u}$ is the velocity field, p is the pressure, ρ is the density, and ν is the viscosity.
1. The RANS equations under time averaging read:

$$(\overline{\mathbf{u}} \cdot \nabla)\overline{\mathbf{u}} = -\frac{1}{\rho}\nabla\overline{p} + \nu\nabla^2\overline{\mathbf{u}} - \nabla \cdot \overline{\mathbf{u}' \otimes \mathbf{u}'}, \qquad \nabla \cdot \overline{\mathbf{u}} = 0. \tag{5.147}$$

Due to the time average, the time derivative term disappears in the RANS, resulting in a steady flow field. However, due to the nonlinear advection, one additional term involving the average of the quadratic interaction of the fluctuation $\nabla \cdot \overline{\mathbf{u}' \otimes \mathbf{u}'}$ appears. The quadratic term $\overline{\mathbf{u}' \otimes \mathbf{u}'}$ is called Reynolds stress. The symbol $\otimes$ denotes the outer product of two vectors, producing a second-order tensor with components $(\mathbf{u}' \otimes \mathbf{u}')_{ij} = u'_i u'_j$, whose average defines the Reynolds stress. Such RANS equations are skillful in characterizing the statistically steady flows, for example, the fully developed channel flow. However, the limitation is also obvious since it fails for describing the transient behavior due to the lack of the time derivative term $\partial\overline{\mathbf{u}}/\partial t$.
2. The RANS equations under ensemble averaging read:

$$\frac{\partial\overline{\mathbf{u}}}{\partial t} + (\overline{\mathbf{u}} \cdot \nabla)\overline{\mathbf{u}} = -\frac{1}{\rho}\nabla\overline{p} + \nu\nabla^2\overline{\mathbf{u}} - \nabla \cdot \overline{\mathbf{u}' \otimes \mathbf{u}'}, \qquad \nabla \cdot \overline{\mathbf{u}} = 0. \tag{5.148}$$

which can capture the averaged transient behavior. It is computationally more expensive than the time-averaged RANS equations (5.147), but is cheaper than the original Navier-Stokes equations.
3. The RANS equations under spatial averaging (in the sense of a spatially filtered field) read:

$$\frac{\partial\overline{\mathbf{u}}}{\partial t} + (\overline{\mathbf{u}} \cdot \nabla)\overline{\mathbf{u}} = -\frac{1}{\rho}\nabla\overline{p} + \nu\nabla^2\overline{\mathbf{u}} - \nabla \cdot \overline{\mathbf{u}' \otimes \mathbf{u}'}, \qquad \nabla \cdot \overline{\mathbf{u}} = 0. \tag{5.149}$$

While the RANS equations with spatial averaging in (5.149) have the same form as those with ensemble averaging in (5.148), they operate on fundamentally different scales: spatial averaging coarsens the resolution significantly compared to ensemble averaging. This approach describes homogeneous turbulence well and forms the basis for LES subgrid modeling. However, spatially averaged RANS equations are invalid for analyzing inhomogeneous flows or features that vary significantly in space, such as boundary layers.

It is worth highlighting that all RANS methods require modeling $\overline{\mathbf{u}' \otimes \mathbf{u}'}$, which usually do not have explicit and straightforward expressions. Therefore, approximations are needed. This is called the closure problem. When $\overline{\mathbf{u}' \otimes \mathbf{u}'}$ is approximated by functions of the mean variable $\mathbf{u}$, the system is closed.

5.5 Code listings for figures and examples

- `Chapter_5_Fourier_Denoising.m`: generates Figure 5.1 in Example 5.1, which demonstrates the Fourier decomposition of a one-dimensional time series and its denoising.
- `Chapter_5_Fourier_PCA_Comparison.m`: generates Figure 5.2 in Example 5.2, which compares Fourier decomposition and PCA for denoising a one-dimensional time series.
- `Chapter_5_Fourier_Spectra.m`: generates Figure 5.3 in Example 5.5, which analyzes laminar and turbulent flow fields using Fourier decomposition.
- `Chapter_5_Boundary_Layer.m`: generates Figure 5.5 in Example 5.17, which presents the solution of a boundary value problem.

- `Chapter_5_Comparison_Perturbation_Methods.m`: generates Figure 5.6 in Examples 5.18–5.19, which compares different perturbation methods for the Duffing equation.
- `Chapter_5_Laplace_Method.m`: generates Figure 5.7 in Examples 5.21–5.22, which applies the Laplace method to approximate the Gamma function and compute rare-event probabilities.

6

Stochastic Toolkit

6.1 Stochastic processes

A stochastic process is a collection of random variables $\{X_t\}_{t\in\mathbb{R}_+}$, representing the evolution of a random quantity over time. Roughly speaking, $X_t(\omega)$ is a "random" function of time: for each fixed time t^*, $X_{t^*}(\omega)$ is a random variable, and for each fixed sample point ω^*, $X_t(\omega^*)$ is a (deterministic) function of time, called a sample path [110].

6.1.1 Wiener process

The Wiener process $W(t)$, also known as the standard Brownian motion, is a commonly used stochastic process [89]. A real-valued Wiener process satisfies the following basic properties:

1. $W(0) = 0$ almost surely (a.s.).
2. $W(t)$ has continuous sample paths almost surely.
3. $W(t)$ has independent, stationary Gaussian increments: for $0 \leq s < t$,

$$W(t) - W(s) \sim \mathcal{N}(0, t-s), \tag{6.1}$$

   and increments over disjoint time intervals are independent.

Thus, the sample paths of a Wiener process are continuous and almost surely nowhere differentiable. For each time step $\Delta t_i = t_i - t_{i-1} \geq 0$, the increment $\Delta W_i \equiv W(t_i) - W(t_{i-1})$ satisfies

$$\mathbb{E}[\Delta W_i] = 0 \qquad \text{and} \qquad \mathbb{E}[(\Delta W_i)^2] = \Delta t_i. \tag{6.2}$$

A complex Wiener process has independent real and imaginary parts, both being real Wiener processes. A factor of $1/\sqrt{2}$ is often introduced so that $\mathbb{E}\big[|W_c(t)|^2\big] = t$, matching the real case:

$$W_c(t) = \frac{1}{\sqrt{2}}\big(W_R(t) + iW_I(t)\big). \tag{6.3}$$

In addition, if $W(t)$ is a Wiener process, then $-W(t)$ is also a Wiener process.

Example 6.1 (Variance of the Wiener process). The variance of $W(t)$ grows linearly with time:

$$\mathrm{Var}(W(t)) = t. \tag{6.4}$$

To see this, note that for any $t \geq 0$, $W(t) = W(t) - W(0) \sim \mathcal{N}(0, t)$, since $W(0) = 0$ a.s. and by the Gaussian increment property. Therefore,

$$\mathrm{Var}(W(t)) = \mathbb{E}[W(t)^2] - \mathbb{E}[W(t)]^2 = t - 0 = t. \tag{6.5}$$

DOI: 10.1201/9781003738459-6

Figure 6.1 illustrates sample trajectories of $W(t)$, its mean and variance over $t \in [0, 1]$, and the PDFs at select times. To generate trajectories numerically, we discretize $[0, 1]$ with $\Delta t = 0.001$ and compute

$$W(t_i) = W(t_{i-1}) + \Delta W_i, \qquad \Delta W_i \sim \mathcal{N}(0, \Delta t) \tag{6.6}$$

for $i = 1, \ldots, 1000$. These numerical results confirm (6.4). The irregular wiggles in the trajectories are also consistent with the fact that Wiener sample paths are almost surely nowhere differentiable. □

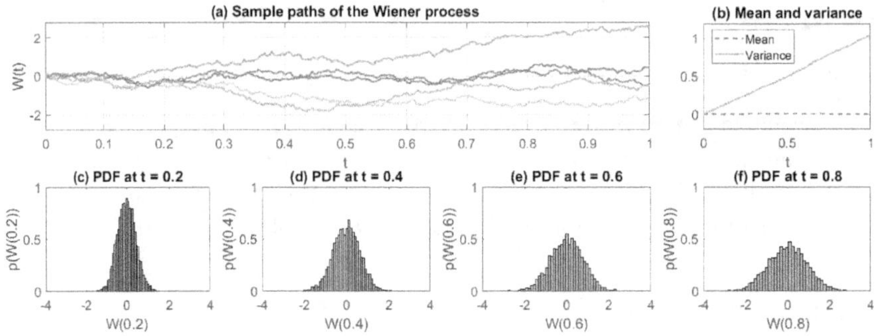

FIGURE 6.1: Wiener process. Panel (a): sample trajectories. Panel (b): sample mean and sample variance of $W(t)$ from $t = 0$ to $t = 1$. Panels (c)–(f): PDFs of $W(t)$ at different time instants, estimated using 5,000 samples.

Example 6.2 (First passage time). What is the probability that $W(t)$ reaches the value $a > 0$ before $t = T$? In other words, given $a > 0$, we seek $P(\tau_a \leq T)$, where τ_a is the so-called first passage time, defined as:

$$\tau_a = \inf\{t \geq 0 : W(t) = a\}. \tag{6.7}$$

The problem can be solved using the following idea. If $W(t)$ hits a at $\tau_a \leq T$, then for $t \geq \tau_a$, the path can be "reflected" around a. By symmetry, paths that end above a at $t = T$ are balanced by reflected paths ending below a. This "reflection" property leads to the factor 2 in the following calculation:

$$P(\tau_a \leq T) = 2P(W(T) \geq a) = 2\left(1 - \Phi\left(\frac{a}{\sqrt{T}}\right)\right), \tag{6.8}$$

where Φ is the standard normal cumulative distribution function (CDF). The result in (6.8) can be derived using the fact that the random variable $W(T)$ is Gaussian because a Wiener process has independent, normally distributed increments. In particular, $W(T) \sim \mathcal{N}(0, T)$. The appearance of $\sqrt{T}$ comes from the fact that the variance of $W(T)$ grows linearly in time, so its standard deviation is $\sqrt{T}$. Writing $a/\sqrt{T}$ simply rescales $W(T)$ to a standard normal random variable, which allows us to express the probability using the standard normal CDF Φ.

As a quick illustration, if $a = 1$ and $T = 1$, then

$$P(\tau_1 \leq 1) = 2(1 - \Phi(1)) \approx 0.3174. \tag{6.9}$$

Figure 6.2 shows a numerical validation of the theoretic result. These trajectories again confirm the continuous but nowhere differentiable properties of the Wiener process. Panel

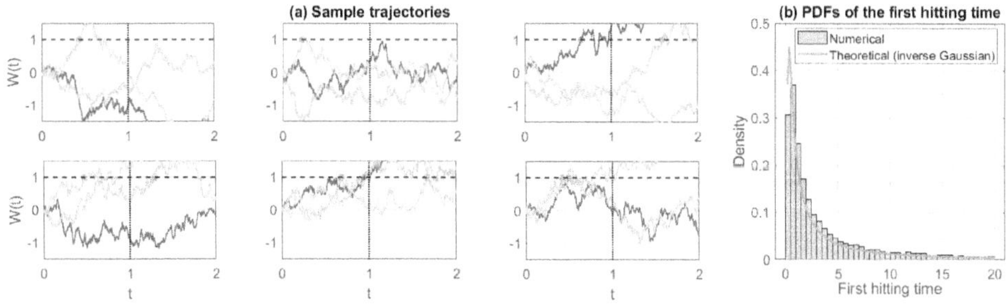

FIGURE 6.2: Wiener processes and first passage time. Panel (a): 18 random trajectories of independent Wiener processes. If a trajectory hits $W(t) = 1$ before $t = 1$, then the cross section is marked by a dot. Panel (b): PDFs of the first hitting time.

(a) includes 18 random trajectories of independent Wiener processes. Panel (b) shows the distribution of the first hitting time that $W(t)$ hits $W(t) = 1$, which satisfies an inverse Gaussian distribution. This inverse Gaussian distribution can be obtained by differentiating the CDF (6.8) with respect to T:

$$p_{\tau_a}(T) = \frac{\mathrm{d}}{\mathrm{d}T} P(\tau_a \leq T) = \frac{a}{\sqrt{2\pi T^3}} \exp\left(-\frac{a^2}{2T}\right), \qquad T > 0. \tag{6.10}$$

Here, for any fixed $T > 0$, the quantity $P(\tau_a \leq T)$ gives the probability that the process reaches level a by time T. Differentiating this probability with respect to T yields the probability density function of τ_a. The numerical results based on 10,000 independent trajectories fit well with the theoretical conclusion. □

6.1.2 Markov process

An important class of stochastic processes is the Markov process, in which the future state is independent of the past, given the present state [195]. Consider a stochastic process $\mathbf{X}$ sampled at discrete times $t_0 < t_1 < t_2 < \ldots$, with a joint probability density $p(\mathbf{x}_0, \mathbf{x}_1, \mathbf{x}_2, \ldots, \mathbf{x}_n, \ldots)$. The process satisfies the Markov property if, for all n,

$$p(\mathbf{x}_n \mid \mathbf{x}_{n-1}, \ldots, \mathbf{x}_0) = p(\mathbf{x}_n \mid \mathbf{x}_{n-1}), \tag{6.11}$$

meaning that the future state depends only on the present state and not on the earlier history. This "memoryless" property is central to the analysis of Markov processes.

The same idea extends to continuous time, where the conditional distribution of the future state depends only on the current state and not on the past trajectory.

Example 6.3 (Wiener process: A Markov process with Gaussian transitions). The Wiener process $W(t)$ is a canonical example of a continuous-time Markov process. Its Markov property stems from its independent increments. Its transition density, which describes the probability density of transitioning from one state to another in a given time interval, is given by

$$p(y, t+s \mid x, t) = \frac{1}{\sqrt{2\pi s}} e^{-(y-x)^2/(2s)}, \tag{6.12}$$

which is a Gaussian distribution centered at the previous step x with variance s. □

Example 6.4 (Deterministic ODEs as degenerate Markov processes). A deterministic ODE system $\dot{X}(t) = f(X(t))$ is also Markovian, but in a degenerate sense as it has no randomness. According to

$$X(t+s) = X(t) + \int_t^{t+s} f(X(u))\,\mathrm{d}u, \tag{6.13}$$

it is memoryless since $X(t+s)$ depends solely on $X(t)$, not on earlier history. □

6.2 Itô calculus

6.2.1 Itô integral

Let us start with a concept called non-anticipating. A function $G(x,t)$, where x may be a stochastic process, is called non-anticipating or adapted if its value at time t depends only on information available up to that time.

Example 6.5 (Anticipating and non-anticipating functions). The function $G(W(t),t) = W(t)$ is non-anticipating while $G(W(t),t) = W(t+1)$ is anticipating since it uses future information at any time t. □

The non-anticipating property ensures that each increment $W(t_j) - W(t_{j-1})$ is independent of $G(x(t_{j-1}), t_{j-1})$, which is crucial for Itô calculus.

Let $W(t)$ be a Wiener process and $G(x,t)$ a non-anticipating function. The Itô stochastic integral is defined as [111]:

$$\int_{t_0}^{t} G(x(s),s)\,\mathrm{d}W(s) := \lim_{n\to\infty} \sum_{j=1}^{n} G(x(t_{j-1}), t_{j-1})\big(W(t_j) - W(t_{j-1})\big), \tag{6.14}$$

where the limit is taken in the mean-square sense, which means that although individual realizations of the stochastic sum may fluctuate, the expected squared error between the approximation and the limiting integral vanishes as the partition is refined. Note that G is evaluated at the left endpoint t_{j-1} of each subinterval $[t_{j-1}, t_j]$.

The choice of evaluation point is fundamental in stochastic integration. Itô integral uses the left endpoint (t_{j-1}). In contrast, Stratonovich integral uses the midpoint $(\frac{t_{j-1}+t_j}{2})$. These yield different results because $W(t)$ has unbounded variation, which differs fundamentally from the Lebesgue integral for smooth processes, where the limit is independent of the evaluation point within $[t_{j-1}, t_j]$. However, they can be related through the Itô-Stratonovich correction formula. Unless otherwise specified, we will use the Itô integral throughout.

Example 6.6 (Difference between Itô and Stratonovich integrals). Consider the stochastic integral of $W(t)$ with respect to itself over $[0,T]$. Using a partition $0 = t_0 < t_1 < \cdots < t_n = T$, the Itô and the Stratonovich sums are given by

$$\begin{aligned} S_n^{\text{Ito}} &= \sum_{j=1}^{n} W(t_{j-1})\big(W(t_j) - W(t_{j-1})\big), \\ S_n^{\text{Strat}} &= \sum_{j=1}^{n} W\left(\frac{t_{j-1}+t_j}{2}\right)\big(W(t_j) - W(t_{j-1})\big), \end{aligned} \tag{6.15}$$

respectively.

Using $W(t_j) = W(t_{j-1}) + \Delta W_j$, we expand $W(t_j)^2 = \big(W(t_{j-1}) + \Delta W_j\big)^2 = W(t_{j-1})^2 + 2W(t_{j-1})\Delta W_j + (\Delta W_j)^2$. Rearranging terms yields $W(t_{j-1})\,\Delta W_j = \frac{1}{2}\Big(W(t_j)^2 - W(t_{j-1})^2 - (\Delta W_j)^2\Big)$. Therefore,

$$S_n^{\text{Ito}} = \frac{1}{2}\big(W(T)^2 - W(0)^2\big) - \frac{1}{2}\sum_{j=1}^{n}(\Delta W_j)^2. \tag{6.16}$$

Since $\sum_{j=1}^{n}(\Delta W_j)^2 \to T$ in probability as the partition is refined, we obtain

$$\int_0^T W(t)\,\mathrm{d}W(t) = \frac{1}{2}\big(W(T)^2 - T\big) \qquad \text{(Itô integral)}. \tag{6.17}$$

In contrast, in the Stratonovich integral, the stochastic Riemann sums may be written in the symmetric form

$$\sum_{j=1}^{n} \frac{W(t_{j-1}) + W(t_j)}{2}\,\big(W(t_j) - W(t_{j-1})\big), \tag{6.18}$$

which is equivalent in the limit to midpoint evaluation. This leads to:

$$S_n^{\text{Strat}} = \frac{1}{2}\sum_{j=1}^{n}\big(W(t_j)^2 - W(t_{j-1})^2\big) \longrightarrow \frac{1}{2}W(T)^2. \tag{6.19}$$

Thus

$$\int_0^T W(t)\circ\mathrm{d}W(t) = \frac{1}{2}W(T)^2 \qquad \text{(Stratonovich integral)}. \tag{6.20}$$

where, the symbol $\circ$ denotes a Stratonovich stochastic integral, in which the midpoint rule is used in the defining Riemann sums.

This example shows that different evaluation points lead to different limits. This does not occur for smooth functions but arises here because the Wiener process has unbounded variation. □

6.2.2 Properties of Itô integral

Let $G(t)$ and $H(t)$ be non-anticipating processes adapted to the Wiener process $W(t)$, and let $f(W)$ be a smooth function. The following examples illustrate fundamental properties of Itô calculus. They capture the essential features of stochastic integration that distinguish Itô calculus from classical calculus while showing some connections between these two categories of integrations. We will present these results through formal derivations rather than rigorous proofs to facilitate the understanding of a general audience.

Example 6.7 (Quadratic variation of the Wiener process). A key feature that distinguishes the Wiener process from smooth functions is its quadratic variation. Informally, while increments of a smooth function are of order Δt, increments of a Wiener process satisfy $(\Delta W)^2 \sim \Delta t$. As a result, squared increments accumulate to a finite limit.

This leads to the fundamental Itô identity

$$\int_{t_0}^{t} G(t')\,[\mathrm{d}W(t')]^2 = \int_{t_0}^{t} G(t')\,\mathrm{d}t, \tag{6.21}$$

which should be understood as a limit of discrete sums.

To see this, consider a partition $t_0 < t_1 < \cdots < t_n = t$ with $\Delta t_j = t_j - t_{j-1}$. For the Wiener process,

$$\sum_{j=1}^{n} \left(W(t_j) - W(t_{j-1})\right)^2 \longrightarrow t - t_0 \quad \text{in the mean-square sense.} \tag{6.22}$$

For equally spaced partitions ($\Delta t_j = \Delta t = (t - t_0)/n$), this follows from

$$\mathbb{E}\left[\sum_{j=1}^{n}(\Delta W_j)^2\right] = n\,\Delta t = t - t_0, \qquad \mathrm{Var}\left[\sum_{j=1}^{n}(\Delta W_j)^2\right] = 2n(\Delta t)^2 \to 0, \tag{6.23}$$

as $n \to \infty$. The expectation identifies the limiting value, while the variance going to zero ensures that random fluctuations vanish. The convergence result (6.22) plays a fundamental role in what follows. It shows that, although individual Brownian increments ΔW_j are random, their squared increments accumulate deterministically and behave like time increments in the continuum limit. This observation provides the basic building block for treating more general expressions: once $(\Delta W_j)^2$ is understood to contribute Δt_j in the limit, one can introduce a non-anticipating weight $G(t_{j-1})$ and analyze the resulting weighted sum using the same quadratic variation mechanism.

Now let $G(t)$ be a non-anticipating process and define the weighted sum

$$S_n = \sum_{j=1}^{n} G(t_{j-1})(\Delta W_j)^2. \tag{6.24}$$

We compare this with the deterministic sum $\sum_{j=1}^{n} G(t_{j-1})\Delta t_j$. To analyze the difference, define $M_j := (\Delta W_j)^2 - \Delta t_j$. Since Brownian increments have mean zero and variance Δt_j, we have $\mathbb{E}[M_j \mid \text{information up to } t_{j-1}] = 0$ and $\mathbb{E}[M_j^2] = 2(\Delta t_j)^2$. Then

$$S_n - \sum_{j=1}^{n} G(t_{j-1})\Delta t_j = \sum_{j=1}^{n} G(t_{j-1})M_j. \tag{6.25}$$

Because $G(t_{j-1})$ depends only on past information, it is independent of the future increment structure generating M_j, and the random variables M_j have zero conditional mean. Expanding the second moment gives

$$\mathbb{E}\left[\left(S_n - \sum_{j=1}^{n} G(t_{j-1})\Delta t_j\right)^2\right] = \sum_{j=1}^{n} \mathbb{E}\left[G^2(t_{j-1})\right] \mathbb{E}[M_j^2] = 2\sum_{j=1}^{n} \mathbb{E}\left[G^2(t_{j-1})\right] (\Delta t_j)^2, \tag{6.26}$$

since cross terms vanish by independence and zero mean. As the partition is refined (i.e., $\max_j \Delta t_j \to 0$), the right-hand side converges to zero provided $\mathbb{E}\int_{t_0}^{t} G^2(t')\,dt' < \infty$. Therefore, S_n converges in the mean-square sense to $\int_{t_0}^{t} G(t')\,dt'$, which justifies the identity (6.21).

□

Example 6.8 (Itô isometry). The Itô isometry states that for any non-anticipating process $G(t)$,

$$\mathbb{E}\left[\left(\int_{t_0}^{t} G(t')\,\mathrm{d}W(t')\right)^2\right] = \int_{t_0}^{t} \mathbb{E}[G^2(t')]\,\mathrm{d}t'. \tag{6.27}$$

To show this, we expand the squared sum:

$$\mathbb{E}\left[\left(\sum_{j=1}^{n} G(t_{j-1})\Delta W_j\right)^2\right] = \underbrace{\sum_{j=1}^{n} \mathbb{E}[G^2(t_{j-1})(\Delta W_j)^2]}_{\text{Diagonal terms}} + \underbrace{2\sum_{i<j} \mathbb{E}[G(t_{i-1})G(t_{j-1})\Delta W_i \Delta W_j]}_{\text{Cross terms}}. \tag{6.28}$$

For the diagonal terms, we have

$$\mathbb{E}[G^2(t_{j-1})(\Delta W_j)^2] = \mathbb{E}[G^2(t_{j-1})]\Delta t_j. \tag{6.29}$$

For the cross terms, we have

$$\mathbb{E}[G(t_{i-1})G(t_{j-1})\Delta W_i \Delta W_j] = \mathbb{E}[G(t_{i-1})G(t_{j-1})\Delta W_i]\mathbb{E}[\Delta W_j] = 0. \tag{6.30}$$

To understand why the cross terms vanish, recall two key properties of Brownian motion. First, Brownian motion has independent increments, meaning that the random displacement over a future time interval is independent of all motion that occurred earlier. Second, the process $G(t)$ is non-anticipating, so its value at time t_{j-1} can only depend on information available up to that time and cannot depend on future increments of the Wiener process.

When $i < j$, the increment $\Delta W_j = W(t_j) - W(t_{j-1})$ represents the random motion that occurs after time t_{j-1}. In contrast, the quantities $G(t_{i-1})$, $G(t_{j-1})$, and ΔW_i are all determined using information from earlier times. Because future Brownian motion is independent of the past, the product $G(t_{i-1})G(t_{j-1})\Delta W_i$ is independent of ΔW_j. This allows the expectation to factor as

$$\mathbb{E}[G(t_{i-1})G(t_{j-1})\Delta W_i \Delta W_j] = \mathbb{E}[G(t_{i-1})G(t_{j-1})\Delta W_i]\,\mathbb{E}[\Delta W_j]. \tag{6.31}$$

Since each Brownian increment has zero mean, $\mathbb{E}[\Delta W_j] = 0$, all cross terms vanish.

Combining these results and taking the limit yields $\sum_{j=1}^{n} \mathbb{E}[G^2(t_{j-1})]\Delta t_j \to \int_{t_0}^{t} \mathbb{E}[G^2(t')]\,\mathrm{d}t'$. □

6.2.3 Itô's stochastic differential equations (SDEs)

The Itô stochastic differential equation (SDE) can be formulated using the Itô stochastic integral. Given an initial condition $x(t_0) = x_0$ and for $t > t_0 \geq 0$, the SDE has two commonly used representations.

1. Physics formulation (formal notation):

$$\frac{\mathrm{d}x(t)}{\mathrm{d}t} = a(x(t), t) + b(x(t), t)\dot{W}(t), \tag{6.32}$$

where $\dot{W}(t)$ is interpreted as Gaussian white noise, though formally $\dot{W}(t) = \mathrm{d}W(t)/\,\mathrm{d}t$ does not exist since $W(t)$ is almost surely nowhere differentiable. Note that the SDE in (6.32) reduces to an ODE when $b \equiv 0$.

2. Mathematical formulation (rigorous notation):

$$\mathrm{d}x(t) = a(x(t), t)\,\mathrm{d}t + b(x(t), t)\,\mathrm{d}W(t), \tag{6.33}$$

which is standard in probability theory and numerical analysis.

The solution to (6.33) is given by:

$$x(t) = x(t_0) + \int_{t_0}^{t} a(x(s), s)\,\mathrm{d}s + \int_{t_0}^{t} b(x(s), s)\,\mathrm{d}W(s), \tag{6.34}$$

where $\int_{t_0}^{t} b(x(s), s)\,\mathrm{d}W(s)$ is the Itô integral in (6.14). For notation simplicity, $x(t)$, $W(t)$ and other functions, e.g., $a(x(t), t)$, are sometimes denoted as x_t, W_t and a_t, respectively, in SDEs. This formulation defines what it means for a process $x(t)$ to solve the SDE.

6.2.4 Itô's formula

Itô's formula (or Itô's lemma) provides a rule for computing the differential of a function of a stochastic process. This fundamental result serves as the stochastic calculus counterpart to the chain rule in classical calculus. The formula is a crucial tool for determining pathwise solutions to SDEs, deriving moment equations for stochastic processes, and constructing reduced-order models of complex systems.

Consider a scalar stochastic process x_t governed by the SDE

$$\mathrm{d}x_t = a_t\,\mathrm{d}t + b_t\,\mathrm{d}W_t, \tag{6.35}$$

where $a_t := a(x_t, t)$ and $b_t := b(x_t, t)$ are sufficiently regular coefficients. Let $f \in C^2(\mathbb{R})$. Then Itô's formula states that

$$\begin{aligned} \mathrm{d}f(x_t) &= f'(x_t)\,\mathrm{d}x_t + \frac{1}{2}f''(x_t)(\,\mathrm{d}x_t)^2 && \text{(6.36a)}\\ &= f'(x_t)(a_t\,\mathrm{d}t + b_t\,\mathrm{d}W_t) + \frac{1}{2}f''(x_t)(a_t\,\mathrm{d}t + b_t\,\mathrm{d}W_t)^2 && \text{(6.36b)}\\ &= f'(x_t)(a_t\,\mathrm{d}t + b_t\,\mathrm{d}W_t) + \frac{1}{2}f''(x_t)b_t^2\,\mathrm{d}t && \text{(6.36c)}\\ &= \left(f'(x_t)a_t + \frac{1}{2}f''(x_t)b_t^2\right)\,\mathrm{d}t + f'(x_t)b_t\,\mathrm{d}W_t. && \text{(6.36d)} \end{aligned}$$

Unlike classical calculus, second-order terms must be retained because Brownian motion has nonzero quadratic variation.

To interpret the symbolic identity $(\,\mathrm{d}W_t)^2 = \mathrm{d}t$, fix a small time interval $[t, t+\Delta t]$ and refine it into n subintervals $t = t_0 < t_1 < \cdots < t_n = t + \Delta t$, $\delta t := \frac{\Delta t}{n}$. Denote the Brownian increments by $\Delta W_i := W_{t_{i+1}} - W_{t_i}$. Each ΔW_i is random with $\Delta W_i \sim \mathcal{N}(0, \delta t)$, so $(\Delta W_i)^2 \neq \delta t$ pathwise. However, when we sum the squared increments over the refined partition, the accumulated quantity converges: $\sum_{i=0}^{n-1}(\Delta W_i)^2 \to \Delta t$, almost surely as $n \to \infty$. This limiting behavior is the quadratic variation of Brownian motion over $[t, t+\Delta t]$. In differential notation, it is encoded by the Itô multiplication rules $(\,\mathrm{d}W_t)^2 = \mathrm{d}t$, $\mathrm{d}t\,\mathrm{d}W_t = 0$, and $(\,\mathrm{d}t)^2 = 0$, which are not pointwise equalities but shorthand rules capturing the above limit under partition refinement. Using the scaling $\mathrm{d}W_t = \mathcal{O}(\sqrt{\mathrm{d}t})$, we have $(\,\mathrm{d}W_t)^2 = \mathcal{O}(\,\mathrm{d}t)$, $\mathrm{d}t\,\mathrm{d}W_t = \mathcal{O}(\,\mathrm{d}t^{3/2})$, and $(\,\mathrm{d}t)^2 = \mathcal{O}(\,\mathrm{d}t^2)$, so only the term involving $(\,\mathrm{d}W_t)^2$ contributes at order $\mathrm{d}t$. This explains the retention of the $\frac{1}{2}f''(x_t)b_t^2\,\mathrm{d}t$ term in (6.36c), which has no analogue in the deterministic chain rule. The final form (6.36d) separates the drift and diffusion contributions to $\mathrm{d}f(x_t)$.

Example 6.9 (Geometric Brownian motion). The geometric Brownian motion, also called exponential Brownian motion, is a well-known continuous-time stochastic process [228], where the logarithm of the random quantity follows a Brownian motion with drift. The geometric Brownian motion serves as a canonical example in stochastic differential equations and finds particular application in mathematical finance, notably for modeling stock prices in the Black-Scholes framework [55]. The geometric Brownian motion is described by:

$$\mathrm{d}x_t = \mu x_t\,\mathrm{d}t + \sigma x_t\,\mathrm{d}W_t, \tag{6.37}$$

where μ and σ are given constants. Let $f(x_t) = x_t^2$. Applying Itô's formula yields:

$$\begin{aligned} \mathrm{d}f(x_t) &= f'(x_t)\,\mathrm{d}x_t + \frac{1}{2}f''(x_t)(\,\mathrm{d}x_t)^2 \\ &= 2x_t(\mu x_t\,\mathrm{d}t + \sigma x_t\,\mathrm{d}W_t) + (\mu x_t\,\mathrm{d}t + \sigma x_t\,\mathrm{d}W_t)^2 \\ &= 2\mu x_t^2\,\mathrm{d}t + 2\sigma x_t^2\,\mathrm{d}W_t + \sigma^2 x_t^2\,\mathrm{d}t \\ &= (2\mu + \sigma^2)x_t^2\,\mathrm{d}t + 2\sigma x_t^2\,\mathrm{d}W_t. \end{aligned}$$

The appearance of σ^2 in the drift term reveals how noise intensity affects the deterministic part of the dynamics of $f(x_t)$. This term directly relates to the variance of x_t under statistical averaging.

□

Example 6.10 (Itô's product rule). For two Itô processes $x(t)$ and $y(t)$, the product rule states:

$$\mathrm{d}(x(t)y(t)) = x(t)\,\mathrm{d}y(t) + y(t)\,\mathrm{d}x(t) + \,\mathrm{d}x(t)\,\mathrm{d}y(t). \tag{6.38}$$

This result follows from applying Itô's lemma to $f(x, y) = xy$:

$$\mathrm{d}f = \frac{\partial f}{\partial x}\,\mathrm{d}x + \frac{\partial f}{\partial y}\,\mathrm{d}y + \frac{1}{2}\left(\frac{\partial^2 f}{\partial x^2}(\,\mathrm{d}x)^2 + 2\frac{\partial^2 f}{\partial x \partial y}\,\mathrm{d}x\,\mathrm{d}y + \frac{\partial^2 f}{\partial y^2}(\,\mathrm{d}y)^2\right). \tag{6.39}$$

Substituting the partial derivatives ($f_x = y$, $f_y = x$, $f_{xx} = f_{yy} = 0$, $f_{xy} = 1$) yields:

$$\mathrm{d}(xy) = y\,\mathrm{d}x + x\,\mathrm{d}y + \,\mathrm{d}x\,\mathrm{d}y. \tag{6.40}$$

As a concrete example, consider processes with:

$$\mathrm{d}x = \mu_x\,\mathrm{d}t + \sigma_x\,\mathrm{d}W \quad \text{and} \quad \mathrm{d}y = \mu_y\,\mathrm{d}t + \sigma_y\,\mathrm{d}W. \tag{6.41}$$

where both processes are driven by the same Wiener process $W(t)$. Their product then satisfies $\mathrm{d}x\,\mathrm{d}y = \sigma_x\sigma_y\,\mathrm{d}t$ using $\mathrm{d}W^2 = \mathrm{d}t$. □

6.2.5 Additive and multiplicative noises

In the SDE (6.32), if the noise coefficient $B(X(t), t)$ is independent of the state variable $X(t)$, then the noise is called additive noise. In this case, the noise affects the system uniformly, regardless of its state. For linear SDEs with additive Gaussian noise and Gaussian initial conditions, the solution remains Gaussian at all times. This is because the solution can be written as a linear combination of the initial state and Brownian increments, and linear combinations of Gaussian variables remain Gaussian.

In contrast, if $B(X(t), t)$ depends on $X(t)$, then the noise is called multiplicative, and the resulting SDE becomes more complex. Multiplicative noise can induce phenomena such as noise-induced transitions or stochastic bifurcations. Because the noise amplitude depends on the state, the random forcing becomes state-dependent. This interaction between the state and the noise amplitude typically leads to non-Gaussian distributions.

Example 6.11 (Comparison of linear scalar models with additive and multiplicative noises). Consider two linear SDEs. The first one has additive noise, which is called the Ornstein-Uhlenbeck (OU) process [111]:

$$\frac{\mathrm{d}x_t}{\mathrm{d}t} = (-ax_t + f) + \sigma\dot{W}_t, \tag{6.42}$$

with $a = 5$, $f = 0$, and $\sigma = 0.5$, while the second one has multiplicative noise and is given by the geometric Brownian motion (6.37) with $\sigma = 0.5$ and $\mu = 0.1$. Figure 6.3 shows the sample trajectories and their PDFs at $t = 1$, where both models start from deterministic initial conditions with $x(0) = 0$ for the OU process and $x(0) = 1$ for the geometric Brownian motion. The OU process has a Gaussian distribution while the geometric Brownian motion has a strongly non-Gaussian distribution with a one-sided fat tail. □

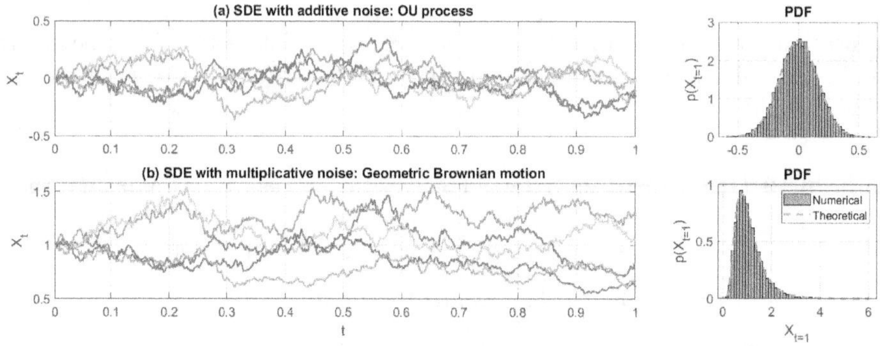

FIGURE 6.3: Sample trajectories and PDFs of the OU process (6.42) and the geometric Brownian motion (6.37).

6.3 Fokker-Planck equation and moment equations

6.3.1 Fokker-Planck equation

The Fokker-Planck equation is a PDE that governs the time evolution of the PDF for solutions of stochastic differential equations [105, 216]. The Fokker-Planck equation and its solutions play significant roles in data assimilation, ensemble prediction, and uncertainty quantification in providing statistical descriptions of dynamical systems, which will be discussed in the subsequent chapters.

Let $\langle\cdot\rangle$ denote the statistical average (equivalent to the overline notation $\bar{\cdot}$ used earlier, but the bracket notation is more conventional in this context). For a scalar random variable x with PDF $p(x)$ and any smooth function f, the expectation of $f(x)$ with respect to the density $p(x)$ is defined as $\langle f(x)\rangle := \int_{-\infty}^{\infty} f(x)p(x)\,\mathrm{d}x$. When $x(t)$ evolves according to the SDE (6.35), the time-dependent PDF $p(x,t)$ (vanishing at infinity) yields the ensemble average at each instant:

$$\langle f(x(t))\rangle := \int_{-\infty}^{\infty} f(x)p(x,t)\,\mathrm{d}x. \tag{6.43}$$

Applying Itô's formula (6.36) to $f(x(t))$, taking expectation, and using the fact that the Itô stochastic integral has zero mean, we obtain

$$\frac{\mathrm{d}}{\mathrm{d}t}\langle f(x(t))\rangle = \left\langle a(x(t),t)\frac{\partial f}{\partial x} + \frac{1}{2}b^2(x(t),t)\frac{\partial^2 f}{\partial x^2}\right\rangle \tag{6.44a}$$

$$= \int_{-\infty}^{\infty}\left(a\frac{\partial f}{\partial x} + \frac{1}{2}b^2\frac{\partial^2 f}{\partial x^2}\right)p(x,t)\,\mathrm{d}x \tag{6.44b}$$

$$= \int_{-\infty}^{\infty} f(x)\left(-\frac{\partial}{\partial x}(ap) + \frac{1}{2}\frac{\partial^2}{\partial x^2}(b^2 p)\right)\mathrm{d}x, \tag{6.44c}$$

where (6.44c) follows via integration by parts using the boundary conditions. On the other hand, direct differentiation of (6.43) gives:

$$\frac{\mathrm{d}}{\mathrm{d}t}\langle f(x(t))\rangle = \int_{-\infty}^{\infty} f(x)\frac{\partial p}{\partial t}\,\mathrm{d}x. \tag{6.45}$$

Note that in (6.45), although the random state $x(t)$ evolves in time, the function $f(x)$ appearing inside the integral is a fixed function of its argument x, meaning that the mathematical rule defining f does not change with time. For example, once $f(x) = x$, x^2, or $\sin x$ is chosen, the same function is evaluated at all times. The time dependence of the expectation $\langle f(x(t))\rangle$ therefore arises entirely from the probability density $p(x,t)$, which describes how the likelihood of different values of $x(t)$ changes as the system evolves. In this formulation, time affects how often each value of x occurs through $p(x,t)$, not how the function f assigns a value to x.

Equating (6.44) and (6.45) yields the Fokker-Planck equation:

$$\frac{\partial p}{\partial t} = -\frac{\partial}{\partial x}(ap) + \frac{1}{2}\frac{\partial^2}{\partial x^2}(b^2 p). \tag{6.46}$$

Despite possible nonlinear dependence of $a(x,t)$ and $b(x,t)$ on the state variable x, the Fokker-Planck equation is linear with respect to the density $p(x,t)$.

The Fokker-Planck equation is an advection-diffusion equation that includes a drift term (transport due to deterministic forcing $a(x,t)$) and a diffusion term (spreading due to Brownian motion via $b(x,t)$). The drift term typically induces convergence effects, causing the probability distribution to concentrate (e.g., through damping or mean-reverting dynamics). In contrast, the diffusion term causes spreading, counteracting the concentration gradient through random fluctuations. In steady-state systems (e.g., $\partial p/\partial t = 0$), these competing effects often reach equilibrium, resulting in a stable probability distribution where the net change vanishes.

Example 6.12 (Fokker-Planck equations for simple and special SDEs). Consider two special cases of the underlying SDE (6.35) with initial condition $p(x,0) = \delta(x)$.

- Purely deterministic case. When the SDE (6.35) reduces to

$$\mathrm{d}x = \mathrm{d}t, \tag{6.47}$$

  the Fokker-Planck equation contains only the drift term, which is also known as the Liouville equation. Its solution is given by

$$\frac{\partial p(x,t)}{\partial t} = -\frac{\partial p(x,t)}{\partial x} \quad \Rightarrow \quad p(x,t) = \delta(x-t). \tag{6.48}$$

  This solution propagates deterministically along the characteristic line $x = t$, matching the direct solution of (6.47).

- Purely stochastic case. When the SDE (6.35) contains only noise:

$$\mathrm{d}x = \mathrm{d}W_t, \tag{6.49}$$

  the Fokker-Planck equation retains only the diffusion term, yielding

$$\frac{\partial p(x,t)}{\partial t} = \frac{1}{2}\frac{\partial^2 p(x,t)}{\partial x^2} \quad \Rightarrow \quad p(x,t) = \frac{1}{\sqrt{2\pi t}}e^{-x^2/(2t)}. \tag{6.50}$$

  Here, the Fokker-Planck equation becomes identical to the heat equation. For multiple realizations of (6.49), the PDF exhibits diffusive spreading identical to thermal diffusion, with variance growing linearly in time.

□

The Fokker-Planck equation can also be defined for multi-dimensional SDEs, which is omitted here.

6.3.2 Moment equations from the Fokker-Planck equation

The time evolution of the moments of the random variable $x(t)$ can be obtained directly from the Fokker-Planck equation. Recall from Chapter 2 the n-th moment defined as [111]:

$$\langle x^n(t)\rangle = \int_{-\infty}^{\infty} x^n p(x,t)\,\mathrm{d}x. \tag{6.51}$$

Taking the time derivative and substituting the Fokker-Planck equation (6.46), we obtain

$$\frac{\mathrm{d}}{\mathrm{d}t}\langle x^n\rangle = \int_{-\infty}^{\infty} x^n \left[-\frac{\partial}{\partial x}(ap) + \frac{1}{2}\frac{\partial^2}{\partial x^2}(b^2 p)\right]\mathrm{d}x. \tag{6.52}$$

Applying integration by parts once to the drift term and twice to the diffusion term, and assuming that $p(x,t)$ and its derivatives decay sufficiently fast at infinity so that all boundary terms vanish, yields

$$\frac{\mathrm{d}}{\mathrm{d}t}\langle x^n\rangle = n\langle a(x,t)x^{n-1}\rangle + \frac{n(n-1)}{2}\langle b^2(x,t)x^{n-2}\rangle. \tag{6.53}$$

For the special cases of $n = 1$ and $n = 2$, we obtain:

$$\frac{\mathrm{d}}{\mathrm{d}t}\langle x\rangle = \langle a(x,t)\rangle, \tag{6.54a}$$

$$\frac{\mathrm{d}}{\mathrm{d}t}\langle x^2\rangle = 2\langle xa(x,t)\rangle + \langle b^2(x,t)\rangle, \tag{6.54b}$$

which shows that the diffusion term does not contribute directly to the evolution of the first moment, but it influences the second moment and thus the variance. In nonlinear systems, the diffusion may still affect the mean indirectly through coupling with higher moments. When $a(x,t)$ and $b(x,t)$ are linear functions of x, the moment equations form a closed system that can be solved exactly. For nonlinear coefficients, the moment equations typically lead to an infinite hierarchy, requiring appropriate closure schemes for practical applications.

6.4 Numerical schemes for SDEs

As with ODEs and PDEs, most SDEs rely on numerical methods to find solutions. In this section, we demonstrate some basic numerical schemes for solving SDEs. Recall the SDE in (6.33)

$$\mathrm{d}x(t) = a(x(t),t)\,\mathrm{d}t + b(x(t),t)\,\mathrm{d}W(t), \tag{6.55}$$

the exact solution of which from time t_n to t_{n+1} is given by

$$x(t_{n+1}) = x(t_n) + \int_{t_n}^{t_{n+1}} a(x(s),s)\,\mathrm{d}s + \int_{t_n}^{t_{n+1}} b(x(s),s)\,\mathrm{d}W(s). \tag{6.56}$$

6.4.1 Euler-Maruyama scheme

Euler-Maruyama method is one of the most widely used numerical schemes for solving SDEs. Denote by x_{n+1} the numerical solution of the truth $x(t_{n+1})$. Let $\Delta t = t_{n+1} - t_n$. The Euler-Maruyama scheme reads [151]:

$$x_{n+1} = x_n + a(x_n,t_n)\Delta t + b(x_n,t_n)\Delta W_n, \tag{6.57}$$

where the left endpoint is utilized in $b(x_n)$ as in Itô's calculus. The first two terms on the right-hand side of (6.57) are precisely the Euler scheme for solving the ODE. The last term is the numerical approximation of the stochastic part. Here, ΔW_n is $\sqrt{\Delta t} \cdot Z$, where $Z \sim \mathcal{N}(0, 1)$ is drawn from a standard Gaussian random variable.

6.4.2 Strong and weak convergence

The next task is to explore the convergence of the Euler-Maruyama and other numerical schemes. Different from ODEs, where convergence means that a single numerical trajectory approaches the exact solution pointwise as $\Delta t \to 0$, solutions of stochastic differential equations are themselves random. As a result, convergence must be understood in a statistical sense, making it essential to clarify in what way the numerical error $|x_n - x(t_n)|$ vanishes as the time step decreases.

Among different definitions of convergence for sequences of random variables, the two most common and useful concepts in numerical SDEs are [152]:
(a). The strong convergence, which measures the pathwise accuracy. The scheme has a strong order of convergence p if there exists a constant $C > 0$ such that

$$\left(\mathbb{E}\left[|x_n - x(t_n)|^2\right]\right)^{1/2} \leq C(\Delta t)^p. \quad \text{as} \quad \Delta t \to 0, \tag{6.58}$$

where x_n is the numerical solution and $x(t_n)$ is the exact solution at time t_n. The strong convergence measures how closely individual sample paths of the numerical solution approximate those of the true solution. The order p characterizes how fast this pathwise error decreases as the time step Δt is refined. For example, $p = \frac{1}{2}$ indicates that the typical error scales like $\sqrt{\Delta t}$. For the general SDE in (6.55), the Euler-Maruyama scheme attains strong order $p = \frac{1}{2}$, reflecting the intrinsic approximation error introduced by discretizing the stochastic (noise) term.
(b). The weak convergence, which measures the distributional accuracy. The scheme has a weak order of convergence p if for all sufficiently smooth test functions ϕ,

$$|\mathbb{E}[\phi(x_n)] - \mathbb{E}[\phi(x(t_n))]| \leq C(\Delta t)^p \quad \text{as} \quad \Delta t \to 0. \tag{6.59}$$

The weak convergence focuses on matching statistical properties (e.g., moments, probabilities) rather than exact paths. For $p = 1$, the error scales linearly with Δt. The Euler-Maruyama scheme often achieves weak order $p = 1$ because, although individual trajectories may be inaccurate, it correctly reproduces the average drift and diffusion effects of the SDE at each time step, leading to accurate evolution of the probability distribution to first order in Δt.

Strong convergence often leads to weak convergence. But for certain SDEs, the strong order can be enhanced (see next subsection).

6.4.3 Milstein Scheme

In the Euler-Maruyama scheme (6.57), the deterministic increment is of order Δt while the stochastic increment ΔW_n is of order $\sqrt{\Delta t}$. As a result, the stochastic approximation becomes the dominant source of strong error. To improve the strong accuracy, we can include higher-order terms from the Itô expansion, which leads to the Milstein scheme [196] for scalar SDEs:

$$x_{n+1} = x_n + a(x_n)\Delta t + b(x_n)\Delta W_n + \frac{1}{2}b(x_n)b'(x_n)(\Delta W_n^2 - \Delta t), \tag{6.60}$$

where b' denotes the derivative of the diffusion coefficient b with respect to x.

Rather than presenting the full derivation, the following example illustrates the key idea behind the Milstein correction term.

Example 6.13 (Milstein scheme for geometric Brownian motion). Consider the geometric Brownian motion SDE (6.37), where μ and σ are constants. Applying Itô's formula to $\ln x(t)$ gives:

$$\mathrm{d}\ln x(t) = \left(\mu - \frac{1}{2}\sigma^2\right)\,\mathrm{d}t + \sigma\,\mathrm{d}W(t). \tag{6.61}$$

Thus the exact solution satisfies

$$x_{n+1} = x_n \exp\left(\left(\mu - \frac{1}{2}\sigma^2\right)\Delta t + \sigma \Delta W_n\right). \tag{6.62}$$

Expanding the exponential and retaining terms up to order Δt (with $\Delta W_n^2 \sim \Delta t$), we obtain

$$\begin{aligned} x_{n+1} &\approx x_n\left(1 + \mu\Delta t + \sigma\Delta W_n + \frac{1}{2}\sigma^2(\Delta W_n^2 - \Delta t)\right) \\ &= x_n + a(x_n)\Delta t + b(x_n)\Delta W_n + \frac{1}{2}b(x_n)b'(x_n)(\Delta W_n^2 - \Delta t), \end{aligned} \tag{6.63}$$

where $a(x) = \mu x$ and $b(x) = \sigma x$ as in (6.55). □

The Euler-Maruyama method has weak order $p = 1$ and, in general, strong order $p = \frac{1}{2}$ for SDEs with sufficiently regular coefficients. In contrast, the Milstein method attains strong order $p = 1$ (and typically weak order $p = 1$) for scalar SDEs. When the diffusion coefficient is constant (additive noise), the Milstein correction term vanishes because $b'(x_n) = 0$, so the Milstein and Euler-Maruyama updates coincide; in this case, Milstein offers no advantage over Euler–Maruyama. In practice, Euler-Maruyama is widely used for SDEs with additive noise due to its simplicity. For multiplicative noise cases ($b = b(x)$), the Milstein method provides improved strong accuracy at slightly higher computational cost. Higher-order stochastic schemes are rarely used in practice because of their algebraic complexity, especially in multidimensional systems.

It is also worth emphasizing that high-order deterministic integrators (such as the classical fourth-order Runge-Kutta method) may still be required when resolving stiff or strongly nonlinear deterministic dynamics. However, when the primary goal is to capture low-order statistics (weak accuracy), lower-order schemes such as Euler-Maruyama are often sufficient. For path-dependent quantities (e.g., hitting times or extremes) or when trajectory-wise accuracy is important, higher strong-order schemes such as Milstein become significantly more relevant.

6.5 Some commonly used stochastic processes

6.5.1 Linear Gaussian SDEs

A linear Gaussian SDE is one of the simplest and most important classes of stochastic differential equations. Different from a linear ODE, which only requires linear dependence on the state variable in the deterministic dynamics, a linear Gaussian SDE must satisfy the following three conditions to preserve the Gaussianity of the solution. First, the initial condition must follow a Gaussian distribution. Second, the drift term must be linear in the state variable, so that the deterministic part of the evolution corresponds to a linear transformation, which maps Gaussian distributions to Gaussian distributions. Third, the stochastic forcing must be Gaussian and enter the system additively. In this chapter, we use linear Gaussian SDE to mean: Gaussian initial condition, linear drift, and additive Gaussian noise. Under these assumptions the solution remains Gaussian.

Let us consider the following real-valued scalar linear Gaussian SDE, which is also known as the OU process, which has been shown in Example 6.11,

$$\frac{\mathrm{d}x_t}{\mathrm{d}t} = (-ax_t + f_t) + \sigma \dot{W}_t, \tag{6.64}$$

where $\dot{W}_t$ is the standard Gaussian white noise (W_t is the Wiener process) and the subscript t means the variable is a function of time. In (6.64), a and σ are both constants with $a > 0$ representing the damping. The forcing f_t can either be a constant or a time-dependent deterministic function.

This simple linear Gaussian SDE provides an analytically solvable illustration of the time evolution of the statistics. Since a one-dimensional Gaussian distribution is entirely determined by its mean and variance, the information carried by the moment equations for the first two moments can fully characterize the Fokker-Planck equation, which directly describes the time evolution of the PDF. In this simple case, the moment equations provide a much simpler way to solve the time evolution of the PDF. Note that, although analytic solutions can be obtained by directly solving the Fokker-Planck equation for this linear Gaussian SDE, the derivation, involving the Fourier transform and the method of characteristics, is much more complicated. For general nonlinear SDEs, the analytic solution of the Fokker-Planck equation is not available, and numerically solving such a PDE can also be computationally expensive. In contrast, solving the ODEs for the leading few moments can provide effective approximations for the time evolution of the PDF.

Let us assume $f_t \equiv f$ (a constant) for the simplicity of presentation.

Example 6.14 (Solutions of the moment equations for the scalar linear Gaussian SDE). To derive the moment equations, we rely on the mean-fluctuation decomposition introduced in Section 5.4. Recall that x_t at each fixed time instant t is a random variable. The Reynolds decomposition separates its mean from the fluctuation part,

$$x_t = \overline{x}_t + x'_t, \tag{6.65}$$

where the mean is in the sense of ensemble average. Recall from Example 5.24 in Section 5.4 that the mean of x_t is precisely $\overline{x}_t$ while the variance is $\overline{(x'_t)^2}$.

To derive the mean equation from the SDE (6.64), we simply need to take the ensemble average over the entire system, which implies that the ensemble average is imposed on each individual term in the equation. Since a and f are constants (namely deterministic), they remain unchanged. The random variable x_t at each fixed t becomes $\overline{x}_t$. The Itô integral has zero mean, so the noise term does not contribute to the mean equation. Therefore, the mean equation yields the following deterministic ODE:

$$\frac{\mathrm{d}\overline{x}_t}{\mathrm{d}t} = -a\overline{x}_t + f. \tag{6.66}$$

Denote by $\overline{x}_0$ the average of the initial ensembles. The solution of (6.66) is given by

$$\overline{x}_t = \overline{x}_0 e^{-at} + \frac{f}{a}(1 - e^{-at}). \tag{6.67}$$

While finding the equation characterizing the time evolution of the variance $\overline{(x'_t)^2}$ seems not to be straightforward, we can introduce two intermediate steps by deriving the equation of the fluctuation x_t' and $(x_t')^2$ sequentially. Subtracting (6.66) from (6.64) allows an equation of x_t',

$$\frac{\mathrm{d}x_t'}{\mathrm{d}t} = -ax_t' + \sigma\dot{W}_t. \tag{6.68}$$

Itô's formula has to be applied to obtain the correct expression of the time evolution of $x_t'^2$ from x_t,

$$\mathrm{d}x_t'^2 = 2x_t'\,\mathrm{d}x_t' + (\,\mathrm{d}x_t')^2. \tag{6.69}$$

Inserting (6.68) into the two terms on the right-hand side of (6.69) yields

$$\frac{\mathrm{d}x_t'^2}{\mathrm{d}t} = -2ax_t'^2 + \sigma^2 + 2\sigma x_t'\dot{W}_t. \tag{6.70}$$

With (6.70) in hand, we take the ensemble average to arrive at

$$\frac{\mathrm{d}\overline{(x_t')^2}}{\mathrm{d}t} = -2a\overline{(x_t')^2} + \sigma^2. \tag{6.71}$$

Denote by $\overline{(x_0')^2}$ the variance of the initial condition, the solution of (6.71) is

$$\overline{(x_t')^2} = \overline{(x_0')^2}e^{-2at} + \frac{\sigma^2}{2a}(1 - e^{-2at}). \tag{6.72}$$

Thus, (6.67) and (6.72) provide the solutions of the time evolutions of the mean and variance for this linear Gaussian SDE. □

Example 6.15 (Solutions of the Fokker-Planck equation for the scalar linear Gaussian SDE). Let us start by defining $y = x - \frac{f}{a}$ to center the equilibrium and rewrite the Fokker-Planck equation:

$$\frac{\partial p}{\partial t} = a\frac{\partial}{\partial y}(yp) + \frac{\sigma^2}{2}\frac{\partial^2 p}{\partial y^2}. \tag{6.73}$$

The initial condition under this new coordinate becomes:

$$p(y,0) = \frac{1}{\sqrt{2\pi\overline{(x_0')^2}}}\exp\left(-\frac{(y-\overline{y}_0)^2}{2\overline{(x_0')^2}}\right), \quad \text{with} \quad \overline{y}_0 = \overline{x}_0 - \frac{f}{a}. \tag{6.74}$$

Recall the Fourier transform introduced in Chapter 5, $\hat{p}(k,t) = \int_{-\infty}^{\infty} e^{iky}p(y,t)\,\mathrm{d}y$. Applying it to the initial Gaussian distribution yields another Gaussian distribution (See Example 5.6),

$$\hat{p}(k,0) = \exp\left(ik\overline{y}_0 - \frac{k^2\overline{(x_0')^2}}{2}\right). \tag{6.75}$$

Similarly, applying it to the Fokker-Planck equation leads to the following equation:

$$\frac{\partial \hat{p}}{\partial t} + ak\frac{\partial \hat{p}}{\partial k} = -\frac{\sigma^2 k^2}{2}\hat{p}. \tag{6.76}$$

The method of characteristics allows us to solve the above system with the following two steps. Introduce a parameter s along characteristic curves:

$$\frac{\mathrm{d}\hat{p}}{\mathrm{d}s} = \frac{\partial \hat{p}}{\partial t}\frac{\mathrm{d}t}{\mathrm{d}s} + \frac{\partial \hat{p}}{\partial k}\frac{\mathrm{d}k}{\mathrm{d}s}. \tag{6.77}$$

Comparing with (6.76), we obtain the characteristic equations:

$$\frac{\mathrm{d}t}{\mathrm{d}s} = 1, \tag{6.78}$$

$$\frac{\mathrm{d}k}{\mathrm{d}s} = ak, \tag{6.79}$$

$$\frac{\mathrm{d}\hat{p}}{\mathrm{d}s} = -\frac{\sigma^2 k^2}{2}\hat{p}. \tag{6.80}$$

Equation (6.78) is a trivial one. The first characteristic function is given by (6.79), solving which leads to the solution $k(t) = k_0 e^{at}$. The second characteristic function is given by (6.80). Its solution is:

$$\hat{p}(k,t) = \hat{p}(k_0,0)\exp\left(-\frac{\sigma^2 k_0^2}{4a}(e^{2at}-1)\right). \tag{6.81}$$

Substituting $k_0 = ke^{-at}$ and the initial condition into (6.81) leads to the solution:

$$\hat{p}(k,t) = \exp\left(ik\overline{y}_0 e^{-at} - \frac{k^2}{2}\left(\overline{(x_0')^2}e^{-2at} + \frac{\sigma^2}{2a}(1-e^{-2at})\right)\right). \tag{6.82}$$

Applying an inverse Fourier transform, the solution using the y coordinate is given by the following Gaussian:

$$p(y,t) = \frac{1}{\sqrt{2\pi\Sigma(t)}}\exp\left(-\frac{(y-\mu(t))^2}{2\Sigma(t)}\right), \tag{6.83}$$

where:

$$\mu(t) = \overline{y}_0 e^{-at}, \tag{6.84}$$

$$\Sigma(t) = \overline{(x_0')^2}e^{-2at} + \frac{\sigma^2}{2a}(1-e^{-2at}). \tag{6.85}$$

The last step is to revert (6.83) to the variables in the x coordinate, which results in

$$p(x,t) = \frac{1}{\sqrt{2\pi\Sigma(t)}}\exp\left(-\frac{\left(x - \frac{f}{a} - (\overline{x}_0 - \frac{f}{a})e^{-at}\right)^2}{2\Sigma(t)}\right). \tag{6.86}$$

Clearly, the Gaussian distribution (6.86) resulting from directly solving the Fokker-Planck equation has the same time evolution of the mean and variance derived from the moment equations (6.67) and (6.72). □

Recall the solutions of the mean (6.67) and the variance (6.72) associated with the real-valued scalar SDE (6.64) (assuming f is a constant). As time evolves, each ensemble member of (6.64) continues to fluctuate due to the white noise, while the mean (6.67) and variance (6.72) converge to constant values. The PDF also stabilizes to a steady state. This long-term behavior is known as statistical equilibrium, a feature common in both linear and nonlinear SDEs with constant forcings. When the forcing is time-periodic, the long-term PDF is typically periodic as well, yielding a periodic statistical equilibrium. Since $a > 0$, the term e^{-at} vanishes as $t \to \infty$. The equilibrium solutions for the mean (6.67) and the variance (6.72) are thus given by $\bar{x}_\infty = f/a$ and $\mathrm{Var}(x)_\infty = \sigma^2/(2a)$, respectively. However, the equilibrium distribution alone is insufficient to uniquely determine the scalar SDE (6.64). Algebraically, (6.64) contains three parameters, while the equilibrium mean and variance provide only two constraints. Dynamically, equilibrium statistics lack temporal information, meaning different dynamical models can share the same equilibrium distribution. The temporal structure of (6.64) is governed by the parameter a, representing the damping rate of the system. Its inverse, $1/a$, is the decorrelation time, which quantifies the memory of the system. A large a results in strong damping and short memory, as the signal is quickly dominated by noise. Conversely, a small a implies a long decorrelation time, extending the potential predictability of the system. The decorrelation time for a stationary process is defined as [211]:

$$\tau_{corr} = \int_0^\infty R(s)\,\mathrm{d}s, \tag{6.87}$$

where $R(s)$ is the autocorrelation function (ACF),

$$R(s) = \frac{\mathbb{E}[(x_t - \bar{x}_\infty)(x_{t+s} - \bar{x}_\infty)]}{\mathrm{Var}(x)_\infty}. \tag{6.88}$$

The ACF measures the cross-covariance between points separated by a time lag s, typically decaying as the lag increases, reflecting weakening correlation. For (6.64), the ACF follows an exponential decay, $R(s) = \exp(-as)$, yielding a decorrelation time of $\tau_{corr} = 1/a$.

Example 6.16 (Linear Gaussian scalar model with different parameters). Figure 6.4 shows the simulation and statistics of the linear Gaussian scalar model (6.64) with two sets of parameter regimes. The damping parameter a is different in the two regimes, where the value in Regime 1 ($a = 0.5$) is smaller than that in Regime 2 ($a = 2$). Correspondingly, the ACF decays to zero more slowly in Regime 1. Similarly, the model realization towards the statistical equilibrium state is slower in Regime 1. The analytic solutions derived above are validated by a Monte Carlo simulation, where the model is run 10,000 times. □

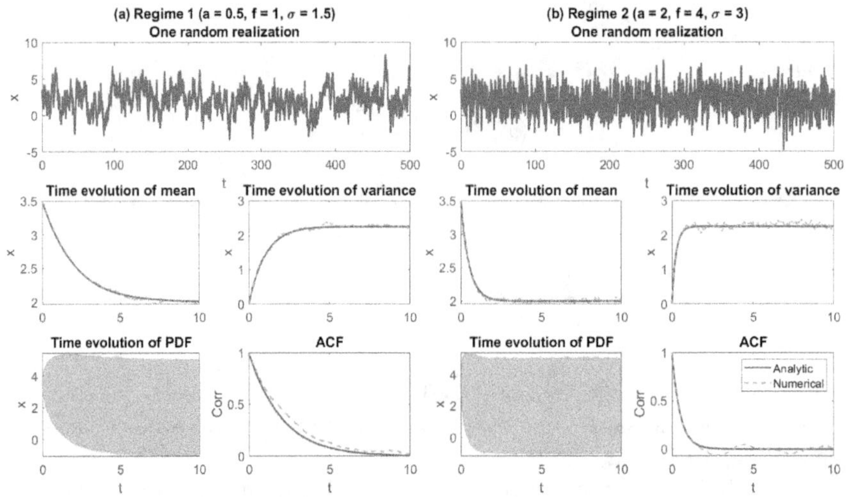

FIGURE 6.4: Linear Gaussian scalar model (6.64) with two sets of parameters (regimes). In each regime, one model realization and the statistics, including the time evolution of the mean, variance, and PDF, as well as the ACF are shown. The numerical results based on 10,000 samples are compared to the analytic solutions.

6.5.2 Linear non-Gaussian SDEs

In this book, we call an SDE linear non-Gaussian if its drift is linear in x but the diffusion amplitude depends on x (multiplicative noise). Such state-dependent noise typically destroys Gaussianity and can produce heavy-tailed stationary distributions. As was discussed previously, the multiplicative noise creates a state-dependent "mixture" of Gaussian inputs, leading to an overall non-Gaussian forcing term. This is the fundamental mechanism to lead to a non-Gaussian distribution of the state variable. The multiplicative noise will complicate the derivation of analytic solutions. Nevertheless, the linear structure of the system still provides a hope for us to obtain closed form solutions in some of these non-Gaussian SDEs.

Consider the following linear scalar SDE, in which the coefficient of the Wiener process W_b is a function of the state variable,

$$\mathrm{d}x(t) = [-ax(t) + f]\,\mathrm{d}t + bx(t)\,\mathrm{d}W_b(t) + c\,\mathrm{d}W_c(t). \tag{6.89}$$

In (6.89), W_b and W_c are independent scalar Wiener processes, and $a, b, c > 0$. The system degenerates to the geometric Brownian motion when $f = 0$ and $c = 0$.

The drift term is $\mu(x) = -ax + f$. Since the noises are independent, their contributions to the diffusion term add up. Thus, the total diffusion coefficient is $D(x) = b^2x^2 + c^2$. Therefore, the Fokker-Planck equation becomes:

$$\frac{\partial p(x,t)}{\partial t} = -\frac{\partial}{\partial x}[(-ax + f)p(x,t)] + \frac{1}{2}\frac{\partial^2}{\partial x^2}\left[(b^2x^2 + c^2)p(x,t)\right]. \tag{6.90}$$

Deriving the general time-dependent solution of the Fokker-Planck equation in (6.90) is challenging. Nevertheless, the equilibrium distribution $p_{\mathrm{eq}}(x)$ is easily available.

Example 6.17 (Stationary distribution of the linear non-Gaussian scalar SDE). The equilibrium distribution is obtained by setting $\frac{\partial p}{\partial t} = 0$, which gives:

$$-\frac{\mathrm{d}}{\mathrm{d}x}[(-ax + f)p_{\mathrm{eq}}(x)] + \frac{1}{2}\frac{\mathrm{d}^2}{\mathrm{d}x^2}\left[(b^2x^2 + c^2)p_{\mathrm{eq}}(x)\right] = 0. \tag{6.91}$$

Integrate once (assuming the solution vanishing at boundaries):

$$-(-ax + f)p_{\mathrm{eq}}(x) + \frac{1}{2}\frac{\mathrm{d}}{\mathrm{d}x}\left[(b^2x^2 + c^2)p_{\mathrm{eq}}(x)\right] = 0. \tag{6.92}$$

Let $q(x) = (b^2x^2 + c^2)p_{\mathrm{eq}}(x)$. Then $\frac{q'(x)}{q(x)} = \frac{2(-ax+f)}{b^2x^2+c^2}$. Taking the integration leads to:

$$\ln q(x) = -\frac{a}{b^2}\ln(b^2x^2 + c^2) + \frac{2f}{bc}\arctan\left(\frac{bx}{c}\right) + \text{const}. \tag{6.93}$$

Taking the exponential and solving for $p_{\mathrm{eq}}(x)$ gives the solution of the equilibrium distribution:

$$p_{\mathrm{eq}}(x) = C(b^2x^2 + c^2)^{-1-a/b^2}\exp\left(\frac{2f}{bc}\arctan\left(\frac{bx}{c}\right)\right), \tag{6.94}$$

where C is a normalization constant. □

The equilibrium distribution (6.94) of the linear non-Gaussian SDE can be non-Gaussian with fat algebraic tails. There are a few special cases of this model:

- No multiplicative noise ($b = 0$). The model has only additive noise and therefore the PDF is Gaussian:

$$p_{eq}(x) = \sqrt{\frac{a}{\pi c^2}}\exp\left(-\frac{a}{c^2}(x - f/a)^2\right). \tag{6.95}$$

- No additive noise ($c = 0$). The SDE reduces to $\mathrm{d}x = (-ax + f)\,\mathrm{d}t + bx\,\mathrm{d}W_b(t)$, whose diffusion degenerates at $x = 0$. Under a zero-flux condition, a stationary density (on an appropriate half-line) takes the form

$$p_{\mathrm{eq}}(x) = \frac{N_0}{|x|^{2+a/b^2}} \exp\left(-\frac{2f}{b^2 x}\right), \tag{6.96}$$

with the understanding that the support and normalizability depend on the sign of f and the boundary behavior at $x = 0$.

- No noise ($b = c = 0$). The model has no noise and therefore the system is a deterministic ODE with solution:

$$p_{eq}(x) = \delta(x - f/a), \tag{6.97}$$

that is, the solution is deterministically given by $x = f/a$.

Although the time evolution of the full PDF is hard to derive, the linear nature of the system allows us to derive the time evolution of the moments.

Example 6.18 (Time evolution of the moments). Assume b and c are constants, and the forcing f is a deterministic, bounded function of time. Let the damping coefficient $a(t) = a_0 + \widetilde{a}(t)$ be time-periodic with period T, where $\widetilde{a}(t)$ represents fluctuations around a positive mean a_0, defined by $a_0 \equiv \frac{1}{T}\int_{t_0}^{t_0+T} a(s)\mathrm{d}s > 0$. This ensures mean stability. A sufficient condition ensuring exponential decay of the homogeneous part (and hence boundedness of the n-th moment under bounded forcing) is:

$$-na_0 + \frac{1}{2}n(n-1)b^2 < 0. \tag{6.98}$$

To derive the evolution equation for the n-th moment, apply Itô's formula to x^n:

$$\mathrm{d}x^n = nx^{n-1}\mathrm{d}x + \frac{n(n-1)}{2}x^{n-2}(\mathrm{d}x)^2. \tag{6.99}$$

Substituting $\mathrm{d}x$ from (6.89) into (6.99) yields

$$\begin{aligned}
\mathrm{d}x^n &= nx^{n-1}\left[(-ax(t) + f)\,\mathrm{d}t + bx(t)\mathrm{d}W_b(t) + c\mathrm{d}W_c(t)\right] + \frac{n(n-1)}{2}x^{n-2}\left(b^2x^2 + c^2\right)\mathrm{d}t \\
&= \left(-na(t) + \frac{n(n-1)}{2}b^2\right)x^n\mathrm{d}t + \left(nf(t)x^{n-1} + \frac{n(n-1)}{2}c^2x^{n-2}\right)\mathrm{d}t \\
&\quad + bnx^n\mathrm{d}W_b(t) + nx^{n-1}c\mathrm{d}W_c(t).
\end{aligned} \tag{6.100}$$

Taking the ensemble average of (6.100) gives

$$\mathrm{d}\langle x^n(t)\rangle = \left(-na(t) + \frac{n(n-1)}{2}b^2\right)\langle x^n(t)\rangle\mathrm{d}t + nf(t)\langle x^{n-1}(t)\rangle\mathrm{d}t + \frac{n(n-1)}{2}c^2\langle x^{n-2}(t)\rangle\mathrm{d}t. \tag{6.101}$$

Integrating this yields the moment evolution equation:

$$\langle x^n(t)\rangle = \langle x_0^n\rangle e^{J_n(t_0,t)} + \int_{t_0}^{t} e^{J_n(s,t)}\left(nf(s)\langle x^{n-1}(s)\rangle + \frac{1}{2}n(n-1)c^2\langle x^{n-2}(s)\rangle\right)\mathrm{d}s, \tag{6.102}$$

where

$$J_n(s,t) = -n\int_s^t a(t')\mathrm{d}t' + \frac{1}{2}n(n-1)b^2(t-s). \tag{6.103}$$

Here, $\langle x^{n-k}(t)\rangle$ is defined as 1 for $n = k$ $(k = 1, 2)$ and 0 for $n - k < 0$. The stability condition (6.98) ensures that the exponential terms in (6.102) decay, requiring stronger damping a_0 for higher-order moments when b is fixed. This is given by the function $J_n(s,t)$ in (6.103), which represents the cumulative exponential growth or decay rate of the n-th moment. The damping term involving $a(t)$ suppresses growth, while the multiplicative noise term proportional to b^2 enhances it. For $\langle x^n(t)\rangle$ to remain bounded as $t \to \infty$, the average value of J_n must be negative, which leads to the condition (6.98).

□

We conclude this subsection with three remarks. First, the Fokker-Planck equation is a PDE, whereas moment equations are ODEs. Solving the latter is computationally simpler, and the resulting moments can be used to approximate the PDF via the maximum entropy principle (see Chapter 8). Note that theoretically an infinite number of moments would be required to fully reconstruct the PDF. Nevertheless, in practice, the leading moments often dominate the distribution. Therefore, the reconstructed PDF based on the leading few moments often provides a practical solution. Second, a moment-based dynamical system is a type of statistical reduced-order model, which is widely used in ensemble forecasting, data assimilation, and dynamical systems analysis. Third, for linear SDEs (with polynomial diffusion), each moment equation depends only on itself and lower-order moments. In other words, the moment equations form a closed system when truncated up to any order. However, the equations for lower-order moments associated with nonlinear SDEs inevitably introduce dependencies on higher-order moments. To close the system, closure methods become essential. Since complex or turbulent systems are intrinsically nonlinear, the closure problem has become a central topic in practice. Such a topic will be explored in the next subsection and Chapter 10.

6.5.3 Simple nonlinear SDEs

Consider a scalar non-Gaussian SDE with cubic nonlinearity:

$$dx(t) = [f + ax(t) + bx^2(t) - cx^3(t)]\,dt + \sigma\,dW_A(t), \tag{6.104}$$

where $c > 0$ is necessary to prevent finite-time blowup of the solution. The cubic nonlinearity interacting with additive noise produces rich dynamical features including non-Gaussian PDFs and potential bimodality.

The Fokker-Planck equation associated with (6.104) is:

$$\frac{\partial p}{\partial t} = -\frac{\partial}{\partial x}[(f + ax + bx^2 - cx^3)p] + \frac{1}{2}\sigma^2\frac{\partial^2 p}{\partial x^2}. \tag{6.105}$$

The equilibrium PDF, obtained by setting $\partial p/\partial t = 0$, takes the form:

$$p_{eq}(x) = N_0 \exp\left(\frac{2}{\sigma^2}\left(fx + \frac{a}{2}x^2 + \frac{b}{3}x^3 - \frac{c}{4}x^4\right)\right), \tag{6.106}$$

where N_0 is the normalization constant.

Example 6.19 (PDF of the cubic nonlinear SDE). Due to the cubic nonlinearity, the equilibrium distribution of the system (6.104) can exhibit various non-Gaussian features. Figure 6.5 shows the statistics for the following three regimes:

$$\begin{aligned}
\text{Highly skewed regime:}\quad & a=-2,\quad b=2,\quad c=1,\quad f=0.1,\quad \sigma=1\\
\text{Symmetric sub-Gaussian regime:}\quad & a=-1,\quad b=0,\quad c=0.2,\quad f=0,\quad \sigma=0.5\\
\text{Bimodal regime:}\quad & a=1,\quad b=0,\quad c=1,\quad f=0,\quad \sigma=0.6.
\end{aligned} \tag{6.107}$$

The parameter b can break the symmetry of the distribution and cause a skewed distribution. The model can also exhibit a sub-Gaussian distribution, the tails of which decay faster than Gaussian (so the kurtosis is less than the standard Gaussian value 3). Note that the highest-order $-cx^4$ term ($c > 0$) in (6.106) causes such a faster-than-Gaussian tail decay, suppressing the occurrence of fat tails in such a model. Different from the linear Gaussian model, this cubic nonlinear model can also generate bimodal distributions. The fundamental mechanism of the appearance of the bimodality can be traced back to the phase portraits discussed in Chapter 4, which are shown in the third row of Figure 6.5.

In the first two regimes, the deterministic drift $F(x) = f + ax + bx^2 - cx^3$ has either one real fixed point or three fixed points arranged so that only one is strongly attracting over the range where the stationary density concentrates, leading to an effectively unimodal PDF. In the bimodal regime, $F(x) = 0$ has three real fixed points with two stable equilibria separated by an unstable one, which supports a genuinely bimodal stationary distribution when the noise is not too large. By checking the first-order derivative, $x = \pm 1$ are stable fixed points while $x = 0$ is unstable. Therefore, depending on the initial condition, the trajectory of the deterministic system will converge to one of the stable fixed points. In the presence of stochastic noise, the solution fluctuates around the stable points, and with sufficiently large noise, transitions between both modes become possible. □

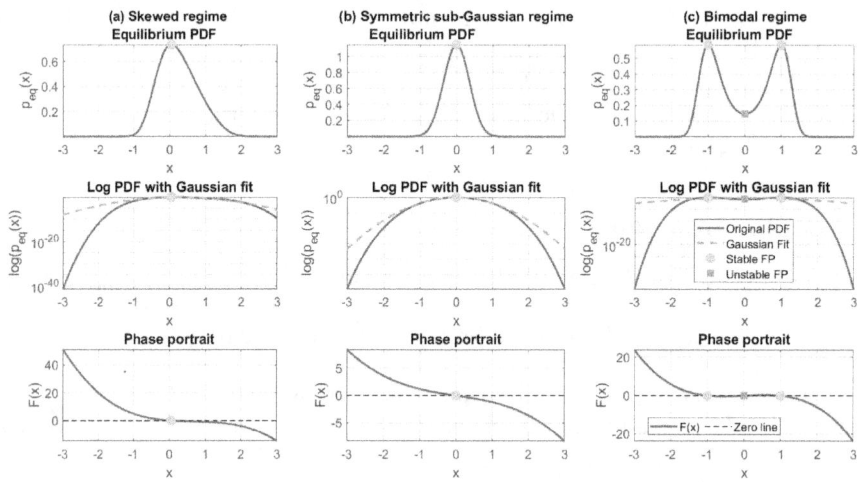

FIGURE 6.5: Statistics of the cubic model (6.104) in different dynamical regimes shown in (6.107). The first row shows the equilibrium PDF. The second row shows the equilibrium PDF in the logarithm scale together with the Gaussian fit. The third row shows the function $F(x) = ax + bx^2 - cx^3 + f$ and its zeros, which are the fixed points.

The moment equations can be derived in a similar way as those applied for the linear models. For the mean and second moment:

$$\begin{aligned} \mathrm{d}\overline{x} &= \left(f + a\overline{x} + b\overline{x^2} - c\overline{x^3}\right)\,\mathrm{d}t, \\ \mathrm{d}\overline{x^2} &= \left(\sigma^2 + 2f\overline{x} + 2a\overline{x^2} + 2b\overline{x^3} - 2c\overline{x^4}\right)\,\mathrm{d}t. \end{aligned} \tag{6.108}$$

However, since the right-hand side of the system contains a cubic term, the information of $(n+1)$- and $(n+2)$-th moments is required to solve the equation of the n-th moment, which means the system is never closed. This is a fundamental challenge for general nonlinear systems. Typically, closure methods are used to approximate higher-order moments by lower-order moments, which creates a solvable reduced-order system.

Example 6.20 (Quasi-Gaussian closure of the moments equations for the cubic nonlinear SDE). The quasi-Gaussian closure is one of the simplest and widely used closure methods. It considers only mean and variance, setting odd central moments to zero and approximating even moments via Gaussian relationships ($\overline{x'^4} = 3\overline{x'^2}^2$) [47].

To derive the moment equations, we first apply Reynolds decomposition by separating the state variable into its mean and fluctuation:

$$x(t) = \overline{x}(t) + x'(t), \quad \overline{x'} = 0. \tag{6.109}$$

Substituting this into (6.104) yields:

$$\mathrm{d}\overline{x} + \mathrm{d}x' = [f + a(\overline{x} + x') + b(\overline{x} + x')^2 - c(\overline{x} + x')^3]\,\mathrm{d}t + \sigma\,\mathrm{d}W_A(t). \tag{6.110}$$

Expanding the nonlinear terms up to third order:

$$\begin{aligned} (\overline{x} + x')^2 &= \overline{x}^2 + 2\overline{x}x' + x'^2, \\ (\overline{x} + x')^3 &= \overline{x}^3 + 3\overline{x}^2 x' + 3\overline{x}x'^2 + x'^3. \end{aligned} \tag{6.111}$$

The mean equation is obtained by taking the ensemble average of (6.110), which eliminates the fluctuation terms with zero mean and the Wiener process term:

$$\mathrm{d}\overline{x} = [f + a\overline{x} + b(\overline{x}^2 + \overline{x'^2}) - c(\overline{x}^3 + 3\overline{x}\overline{x'^2})]\,\mathrm{d}t, \tag{6.112}$$

where we have used the properties $\overline{x'} = 0$ and $\overline{x'^3} = 0$ (from quasi-Gaussian closure).

To derive the equation for the variance $\overline{x'^2}$, we first obtain the equation for the fluctuation x' by subtracting the mean equation from (6.110):

$$\mathrm{d}x' = [ax' + b(2\overline{x}x' + x'^2 - \overline{x'^2}) - c(3\overline{x}^2 x' + 3\overline{x}x'^2 + x'^3 - 3\overline{x}\overline{x'^2})]\,\mathrm{d}t + \sigma\,\mathrm{d}W_A(t). \tag{6.113}$$

Applying Itô's formula for the variance evolution:

$$\mathrm{d}\overline{x'^2} = \overline{2x'\,\mathrm{d}x'} + \overline{(\,\mathrm{d}x')^2}. \tag{6.114}$$

The first term comes from multiplying (6.113) by $2x'$ and taking the average:

$$\overline{2x'\,\mathrm{d}x'} = 2[a\overline{x'^2} + b(2\overline{x}\overline{x'^2} + \overline{x'^3} - \overline{x'^2}\overline{x'}) - c(3\overline{x}^2\overline{x'^2} + 3\overline{x}\overline{x'^3} + \overline{x'^4} - 3\overline{x}\overline{x'^2}\overline{x'})]\,\mathrm{d}t. \tag{6.115}$$

Applying quasi-Gaussian closure $\overline{x'} = \overline{x'^3} = 0$ and $\overline{x'^4} = 3\overline{x'^2}^2$, (6.115) is simplified to:

$$\overline{2x'\,\mathrm{d}x'} = 2[a + 2b\overline{x} - 3c\overline{x}^2]\overline{x'^2}\,\mathrm{d}t - 6c\overline{x}\overline{x'^2}^2\,\mathrm{d}t. \tag{6.116}$$

On the other hand, the second term in (6.114) leads to the quadratic variation $\overline{(\,\mathrm{d}x')^2} = \sigma^2\,\mathrm{d}t$. Combining both terms gives the variance equation:

$$\mathrm{d}\overline{x'^2} = \left(2[a + 2b\overline{x} - 3c\overline{x}^2 - 3c\overline{x'^2}]\overline{x'^2} + \sigma^2\right)\,\mathrm{d}t. \tag{6.117}$$

Figure 6.6 compares different methods for solving the time evolution of the mean and variance of the (6.104) in the three dynamical regimes with parameters in (6.107). In addition to the quasi-Gaussian closure and the truth (via the Monte Carlo simulation with 10,000 ensemble members), we have also included a system, called the bare truncated model, which sets $\overline{x'^4} = 0$ instead of connecting the fourth-order term with the second-order term as in the quasi-Gaussian closure. For the skewed and sub-Gaussian regimes, the quasi-Gaussian

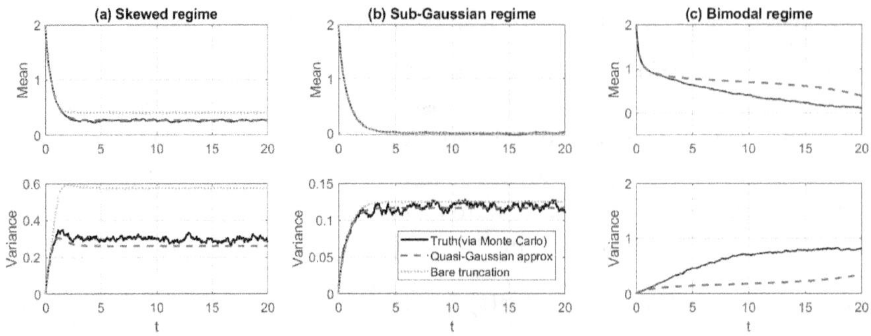

FIGURE 6.6: Time evolution of the mean and variance of the (6.104) in the three dynamical regimes with parameters in (6.107). In each panel, the results of the quasi-Gaussian approximation is compared to the true evolution (via Monte Carlo simulation with 10,000 ensemble members) and the bare truncated system. In the bimodal regime, the bare truncated system blows up quickly and therefore the curves are omitted.

closure almost captures the exact mean and variance, while the bare truncation has a significant error. This highlights the crucial role of approximating the fourth-order term, rather than completely ignoring it. The quasi-Gaussian approximation does not work well in the bimodal regime, which is as expected since the quasi-Gaussian approximation assumes the PDF has a single peak. Nevertheless, it still outperforms the bare truncation, which leads to a quick blowup in solution.

□

As a final remark, the nonlinear model in (6.104) can be expanded to have one further noise terms, called the correlated additive and multiplicative (CAM) noise,

$$\mathrm{d}x(t) = [f + ax(t) + bx^2(t) - cx^3(t)]\,\mathrm{d}t + \sigma\,\mathrm{d}W_A(t) + (A - Bx(t))\,\mathrm{d}W_C(t). \tag{6.118}$$

The model with CAM noise in (6.118) serves as a canonical model for low-frequency atmospheric variability, derived through stochastic mode reduction strategies. It has been applied in regression strategies for climate model data to build one-dimensional stochastic models for patterns like the North Atlantic Oscillation (NAO) [182].

6.5.4 Markov jump process

The Markov jump process is a powerful technique for modeling many natural phenomena. Unlike SDEs, where the state variables are continuous functions, the Markov jump processes have discrete (and often finite) states. Therefore, the Markov jump process is a concise mathematical modeling tool that is particularly useful in describing phenomena with diverse features that can be categorized into distinct clusters.

A finite-state Markov jump process is given by a continuous time stochastic process, which at any time t takes one of the discrete values in the finite set of states $S = \{s_i\}_{i\in\mathcal{N}}$, with $\mathcal{N} = \{1, 2, \ldots, N\}$. The set S is nontrivial if at least one of the s_i is different from the rest. In contrast to the SDEs, nontrivial Markov jump processes have discontinuous paths and do not satisfy the Fokker-Planck equation. Instead, the state probabilities satisfy the master equation (the Kolmogorov forward equation), which is derived from the differential version of the Kolmogorov equation upon the requirement that the process transitions between isolated states.

Example 6.21 (Two-state Markov jump process). Consider now the simplest case of a Markov jump process $X(t)$ which, at any time t, takes one of the only two values from the set of states $S = \{s_1, s_0\}$. Let ν be the rate of change from state s_1 to state s_0. Similarly, denote by μ the rate of change from s_0 to s_1. In addition to the Markov property of $X(t)$ in (6.11), the time homogeneity of the process is assumed:

$$P(X(t) = s_i | X(\tau) = s_j) = P(X(t-\tau) = s_i | X(0) = s_j), \qquad \forall\, t > \tau > 0. \tag{6.119}$$

Thus, the process X is fully determined by the transition probabilities $P(X(t) = s_i | X_0 = s_j)$, $i, j \in \{1, 0\}$.

The rates of change ν and μ define the following local transition probabilities with Δt being a small time increment,

$$\begin{aligned} p_{\Delta t}(s_1, s_0) &:= P(X_{t+\Delta t} = s_0 | X_t = s_1) = \nu \Delta t + o(\Delta t), \\ p_{\Delta t}(s_0, s_1) &:= P(X_{t+\Delta t} = s_1 | X_t = s_0) = \mu \Delta t + o(\Delta t), \\ p_{\Delta t}(s_1, s_1) &:= P(X_{t+\Delta t} = s_1 | X_t = s_1) = 1 - \nu \Delta t + o(\Delta t), \\ p_{\Delta t}(s_0, s_0) &:= P(X_{t+\Delta t} = s_0 | X_t = s_0) = 1 - \mu \Delta t + o(\Delta t), \end{aligned} \tag{6.120}$$

where the first and second entries in $p_{\Delta t}(\cdot, \cdot)$ are the starting and ending states, respectively.
□

The focus below is on deriving the master equation and the transition time for the two-state Markov jump process, which is particularly interesting because of its exact solvability.

Example 6.22 (Master equation for the two-state Markov jump process). Using (6.120) and the Markov property (6.11), the probability of finding the process in state s_0 at $t + \Delta t$ starting from s_1 at the initial time 0 is

$$\begin{aligned} p_{t+\Delta t}(s_1, s_0) &= p_t(s_1, s_0)\, p_{\Delta t}(s_0, s_0) + p_t(s_1, s_1)\, p_{\Delta t}(s_1, s_0) \\ &= p_t(s_1, s_0)(1 - \mu \Delta t) + p_t(s_1, s_1) \nu \Delta t + o(\Delta t), \end{aligned} \tag{6.121}$$

where the two possible paths passing through the two states at time t have been taken into consideration.

After regrouping terms, a finite difference equation is obtained

$$\frac{p_{t+\Delta t}(s_1, s_0) - p_t(s_1, s_0)}{\Delta t} = -\mu p_t(s_1, s_0) + \nu p_t(s_1, s_1) + o(1). \tag{6.122}$$

In the limit $\Delta t \to 0$, assuming that p is smooth with respect to t, one obtains

$$\frac{\partial}{\partial t} p_t(s_1, s_0) = -\mu p_t(s_1, s_0) + \nu p_t(s_1, s_1), \tag{6.123}$$

with the initial condition $p_t(s_1, s_0)\big|_{t=0} = 0$. Applying a similar procedure to the remaining transition probabilities yields

$$\frac{\partial}{\partial t} \hat{P}_t = \hat{P}_t A, \qquad \hat{P}_t\big|_{t=0} = I, \tag{6.124}$$

where I is a 2×2 identity matrix

$$\hat{P}_t = \begin{pmatrix} p_t(s_1, s_1) & p_t(s_1, s_0) \\ p_t(s_0, s_1) & p_t(s_0, s_0) \end{pmatrix}, \qquad A = \begin{pmatrix} -\nu & \nu \\ \mu & -\mu \end{pmatrix}. \tag{6.125}$$

Given that (6.124) is a linear system with constant coefficients, the transition probabilities are

$$\begin{aligned} p_t(s_1, s_1) &= \frac{\mu}{\nu + \mu} + \frac{\nu}{\nu + \mu} e^{-(\nu+\mu)t}, \qquad & p_t(s_1, s_0) &= \frac{\nu}{\nu + \mu}\left(1 - e^{-(\nu+\mu)t}\right), \\ p_t(s_0, s_0) &= \frac{\nu}{\nu + \mu} + \frac{\mu}{\nu + \mu} e^{-(\nu+\mu)t}, \qquad & p_t(s_0, s_1) &= \frac{\mu}{\nu + \mu}\left(1 - e^{-(\nu+\mu)t}\right). \end{aligned} \tag{6.126}$$

As $t \to \infty$, the process approaches the stationary distribution

$$\pi(s_1) = \frac{\mu}{\nu + \mu}, \qquad \pi(s_0) = \frac{\nu}{\nu + \mu}. \tag{6.127}$$

Finally, the stationary expectation of $X(t)$ is

$$\mathbb{E}_{eq}[X] = \frac{\mu s_1 + \nu s_0}{\nu + \mu}. \tag{6.128}$$

□

Example 6.23 (Transition time for the two-state Markov jump process). By definition, the two-state Markov jump process resides in one of the two states s_1 or s_0 and the switching rates between these states are ν (for the $s_1 \to s_0$ transition) and μ (for the $s_0 \to s_1$ transition). It is then important to understand the time instants when the process switches states. To this end, suppose that the system transitions to state s_1 at some time $t = t_0$. Define the time during which the system stays in state s_1 as

$$T_1 = t^* - t_0, \qquad t^* = \inf_{t \in [t_0, \infty)} \{t : X(t) = s_0\}. \tag{6.129}$$

Here, t^* is the switching time from s_1 to s_0, and T_1 is the residence time in s_1. Clearly, both t^* and T_1 are random variables.

The probability that the process remains in s_1 at time $t = t_0 + \Delta t$ is

$$P(T_1 > \Delta t) = P(t^* > t_0 + \Delta t) = P(X(t_0 + \Delta t) = s_1 \mid X(t_0) = s_1). \tag{6.130}$$

Then, one can write

$$\begin{aligned}
&P(X(t_0 + t + \Delta t) = s_1 | X(t_0) = s_1) \\
&= P(X(t_0 + t) = s_1 | X(t_0) = s_1) P(X(t_0 + t + \Delta t) = s_1 | X(t_0 + t) = s_1) \\
&= P(X(t_0 + t) = s_1 | X(t_0) = s_1) P_{\Delta t}(s_1, s_1) \\
&= P(X(t_0 + t) = s_1 | X(t_0) = s_1)(1 - \nu \Delta t + o(\Delta t)),
\end{aligned} \tag{6.131}$$

where the first equality is due to the Markov property of the process (6.11), the second equality is due to the homogeneity (6.119) and the last equality follows from (6.120).

In the limit $\Delta t \to 0$, the following master equation is reached

$$\frac{\mathrm{d}}{\mathrm{dt}} P(X(t_0 + t) = s_1 | X(t_0) = s_1) = -\nu P(X(t_0 + t) = s_1 | X(t_0) = s_1), \tag{6.132}$$

the solution of which is

$$P(X(t_0 + t) = s_1 | X(t_0) = s_1) = e^{-\nu t}. \tag{6.133}$$

It implies that T_1 satisfies an exponential distribution with mean $1/\nu$, and thus

$$P(T_1 < t) = 1 - e^{-\nu t}. \tag{6.134}$$

Similarly, the time T_0 the system spends in regime s_0 before switching to regime s_1 is a random variable with the exponential distribution function

$$P(T_0 < t) = 1 - e^{-\mu t}. \tag{6.135}$$

□

Example 6.24 (Numerical validation of the properties of the two-state Markov jump process). Figure 6.7 shows a simulated trajectory and the numerical validation of the theoretical results derived in (6.127)–(6.128) (equilibrium distribution and expectation) and (6.134)–(6.135) (transition distributions). The parameters used are as follows:

$$\mu = 0.5, \qquad \nu = 0.1, \qquad s_1 = 2.5, \qquad s_0 = \quad 0.05. \tag{6.136}$$

□

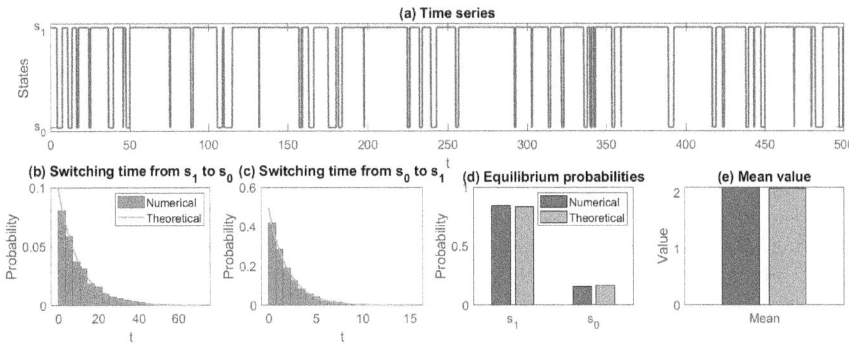

FIGURE 6.7: Two-state Markov jump process. Panel (a): one numerically simulated sample path. Panel (b)–(c): the distribution of the transition times. Panel (d)–(e): the equilibrium distribution and the expected value. Numerical solution based on a trajectory with 25,000 time units is compared to the theoretical values.

One natural question to ask is whether the Markov jump process can be combined with the SDEs to facilitate modeling and prediction. The following simple example demonstrates their advantages in developing a combined model.

Example 6.25 (Hybrid SDE-Markov jump process model and dispersive waves). Consider a one-dimensional flow field represented by a finite Fourier series, where each Fourier coefficient $\hat{u}_k(t)$ evolves according to a stochastic process. This process is a modified complex OU process, with a time-dependent damping coefficient $\gamma_k(t)$ governed by a two-state Markov jump process. The full solution is $u(x,t) = \sum_{k=-K}^{K} \hat{u}_k(t)e^{ikx}$, where

$$\begin{aligned} \frac{\mathrm{d}\hat{u}_k(t)}{\mathrm{d}t} &= \big(-\gamma_k(t) + i\omega_k\big)\hat{u}_k(t) + \sigma_k \dot{W}_k, \\ &\gamma_k(t) \text{ follows a two-state Markov jump process.} \end{aligned} \tag{6.137}$$

This hybrid model captures regime-switching behavior commonly observed in nature, where certain variables, such as the damping term $\gamma(t)$, can be approximated by discrete states. Here, $\gamma(t)$ switches between two values: $\gamma_1 = 2.5$ (strong damping) and $\gamma_0 = -0.05$ (anti-damping). When $\gamma(t) = \gamma_0 < 0$, the deterministic part induces exponential growth in amplitude, which is eventually limited by switching back to the damping state (γ_1). The alternation between these states produces intermittency, a notable feature of turbulent and wave-driven systems.

For the simulation, we adopt the parameters:

$$\mu = \nu = 0.2, \quad K = 3, \quad \omega_1 = 3, \quad \omega_2 = 1.5, \quad \omega_3 = 1 \quad (\text{in general } \omega_k = 3/k), \quad \sigma_k = 1. \tag{6.138}$$

The results are illustrated in Figure 6.8. During anti-damping phases ($\gamma = \gamma_0$), the amplitudes of $\hat{u}_k$ and $u(x,t)$ grow, exciting wave patterns. Notably, these waves are dispersive [233], meaning that their phase velocity $v_p = \omega_k/k$ differs from their group velocity

$v_g = d\omega_k / dk$. For the choice $\omega_k = 3/k$, we have $v_p = 3/k^2 > 0$ while $v_g = -3/k^2 < 0$, so the phase propagates toward positive x while energy propagates toward negative x. The dispersive waves are ubiquitous in fluid dynamics, optics, and plasma physics.

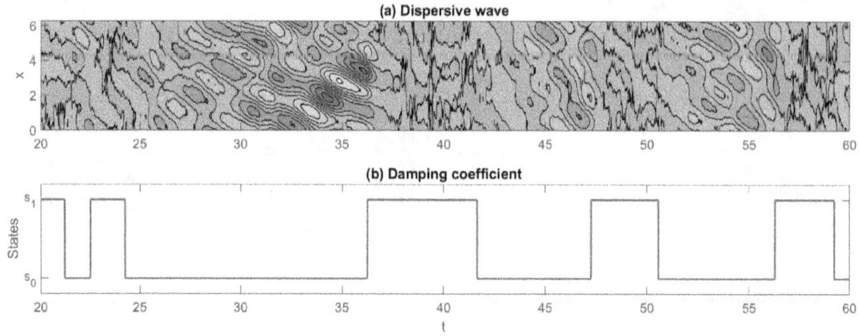

FIGURE 6.8: Dispersive waves generated by the hybrid SDE-Markov model (6.137). Panel (a): Spatial wave patterns. Panel (b): Time evolution of the two-state Markov process governing $\gamma(t)$.

□

6.6 Code listings for figures and examples

- `Chapter_6_Wiener_Processes.m`: generates Figure 6.1 in Example 6.1, which illustrates sample paths and statistical properties of Wiener processes.
- `Chapter_6_First_Passage_Time.m`: generates Figure 6.2 in Example 6.2, which illustrates the first-passage time for a Wiener process to reach a prescribed level.
- `Chapter_6_Linear_Additive_Multiplicative_Models.m`: generates Figure 6.3 in Example 6.11, which compares linear models with additive and multiplicative noise.
- `Chapter_6_Linear_Gaussian_SDEs.m`: generates Figure 6.4 in Example 6.16, which illustrates simulations and statistical properties of a linear Gaussian scalar model under different parameter settings.
- `Chapter_6_Cubic_Model.m`: generates Figure 6.5 in Example 6.19, which illustrates the probability density functions and phase portraits of a cubic model in different dynamical regimes.
- `Chapter_6_Quasi_Gaussian_Closure.m`: generates Figure 6.6 in Example 6.20, which compares the quasi-Gaussian closure with bare truncation in approximating the moment equations.
- `Chapter_6_Two_State_Markov_Jump_Process.m`: generates Figure 6.7 in Example 6.24, which illustrates the time series and statistical properties of a two-state Markov jump process.
- `Chapter_6_Two_State_Rossby.m`: generates Figure 6.8 in Example 6.25, which uses a two-state Markov jump process to generate Rossby waves.

7

Data Meets Models: An Introduction to Data Assimilation

7.1 Bayes' theorem review

Data assimilation seeks to optimally combine a numerical model with observations to improve state estimation [146, 97, 185, 165, 236]. The model is typically chaotic and has uncertainties, while the observations contain noise and are often available only for a subset of the state variables. Data assimilation was initially developed in the field of numerical weather prediction (NWP). These systems are usually chaotic or turbulent, where future states are sensitive to initial conditions. An accurate initialization is extremely important for the skillful forecast.

Data assimilation has become an essential tool in many applications, including state estimation, parameter estimation, dynamical interpolation of missing data, post-processing, control, and machine learning.

The mathematical principle of data assimilation lies in the Bayes' theorem [95, 32],

$$p(\mathbf{x} \mid \mathbf{y}) = \frac{p(\mathbf{x})p(\mathbf{y} \mid \mathbf{x})}{p(\mathbf{y})}, \tag{7.1}$$

which aims to compute the distribution of $\mathbf{x}$ conditioned on the known (or observed) event $\mathbf{y}$. Bayes' theorem is the fundamental building block of data assimilation, which optimally integrates prior knowledge from the model with observations represented as likelihoods. In such a context, (7.1) is often written in a slightly different way

$$p(\mathbf{x} \mid \mathbf{y}) \propto p(\mathbf{x})p(\mathbf{y} \mid \mathbf{x}), \tag{7.2}$$

where $\propto$ means proportional to. The denominator $p(\mathbf{y})$ in (7.1) is dropped because it acts as a normalization constant that makes the left-hand side a valid probability distribution (integrating to 1). Calculating the exact value of $p(\mathbf{y})$ can be computationally intractable or extremely expensive, especially in high-dimensional problems that often appear in data assimilation. However, for many algorithms, we don't need the full normalized posterior immediately. We just need to know its shape or how to draw samples from it. For general nonlinear data assimilation problems, ensemble and particle methods are the mainstream methods, which allow us to compute statistics, such as the mean and covariance, from posterior samples. By doing so, there is no need to know the full normalized distribution, which allows us to avoid explicitly computing the normalization constant $p(\mathbf{y})$.

The form in (7.2) highlights how the prior is updated by the likelihood to form the posterior.

DOI: 10.1201/9781003738459-7

Example 7.1 (Prior, likelihood, and posterior distributions in one-dimensional Gaussian case). Consider a Gaussian prior and Gaussian likelihood:

$$p(x) = \mathcal{N}(x \mid \mu_0, \sigma_0^2) = \frac{1}{\sqrt{2\pi\sigma_0^2}} \exp\left(-\frac{1}{2\sigma_0^2}(x-\mu_0)^2\right),$$

$$p(y \mid x) = \mathcal{N}(y \mid x, \sigma_y^2) = \frac{1}{\sqrt{2\pi\sigma_y^2}} \exp\left(-\frac{1}{2\sigma_y^2}(y-x)^2\right).$$

The goal is to find the posterior distribution $p(x \mid y)$. Using the proportional form:

$$\begin{aligned} p(x \mid y) &\propto p(x)\, p(y \mid x) \\ &\propto \exp\left(-\frac{(x-\mu_0)^2}{2\sigma_0^2}\right) \cdot \exp\left(-\frac{(y-x)^2}{2\sigma_y^2}\right) \\ &\propto \exp\left(-\frac{1}{2}\left[x^2\left(\frac{1}{\sigma_0^2}+\frac{1}{\sigma_y^2}\right) - 2x\left(\frac{\mu_0}{\sigma_0^2}+\frac{y}{\sigma_y^2}\right) + \left(\frac{\mu_0^2}{\sigma_0^2}+\frac{y^2}{\sigma_y^2}\right)\right]\right) \\ &\propto \exp\left(-\frac{1}{2}\frac{(x-\mu_n)^2}{\sigma_n^2}\right), \end{aligned}$$

where the posterior mean μ_n and variance σ_n^2 are given by

$$\mu_n = \sigma_n^2\left(\frac{\mu_0}{\sigma_0^2}+\frac{y}{\sigma_y^2}\right) = \frac{\frac{\mu_0}{\sigma_0^2}+\frac{y}{\sigma_y^2}}{\frac{1}{\sigma_0^2}+\frac{1}{\sigma_y^2}} = \frac{\sigma_y^2\mu_0+\sigma_0^2 y}{\sigma_0^2+\sigma_y^2}, \tag{7.3}$$

$$\sigma_n^2 = \frac{1}{\frac{1}{\sigma_0^2}+\frac{1}{\sigma_y^2}} = \frac{\sigma_0^2\sigma_y^2}{\sigma_0^2+\sigma_y^2}. \tag{7.4}$$

The result in (7.3)–(7.4) demonstrates that Gaussian prior combined with a Gaussian likelihood (with a linear observation operator) results in a Gaussian posterior. This mathematical convenience is a primary reason for the widespread use of Gaussian distributions in Bayesian methods and data assimilation. □

Example 7.2 (Understanding the Bayesian formula from the viewpoint of precision). Let us recall the one-dimensional Gaussian case in Example 7.1. The so-called precision is defined as the inverse of variance ($\tau = 1/\sigma^2$). The posterior variance simplifies to:

$$\frac{1}{\sigma_n^2} = \frac{1}{\sigma_0^2} + \frac{1}{\sigma_y^2} \quad \Rightarrow \quad \tau_n = \tau_0 + \tau_y. \tag{7.5}$$

This implies that the posterior precision is the sum of the prior precision and the data precision. Incorporating new data always increases the total precision (reduces the uncertainty) in the estimate.

The posterior mean can also be rewritten using precisions:

$$\mu_n = \left(\frac{\tau_0}{\tau_n}\right)\mu_0 + \left(\frac{\tau_y}{\tau_n}\right)y \tag{7.6}$$

This shows the posterior mean is a convex combination (a weighted average) of the prior mean μ_0 and the data y. The weights are the relative precisions. If the prior is very certain (τ_0 is high, σ_0^2 is low), the estimate leans toward μ_0. If the data is very precise (τ_y is high, σ_y^2 is low), the estimate leans toward y.

Figure 7.1 includes three cases, where the precisions in the prior and likelihood are different. The posterior distribution is closer to the one with a higher precision (or smaller uncertainty), which validates the facts presented above. □

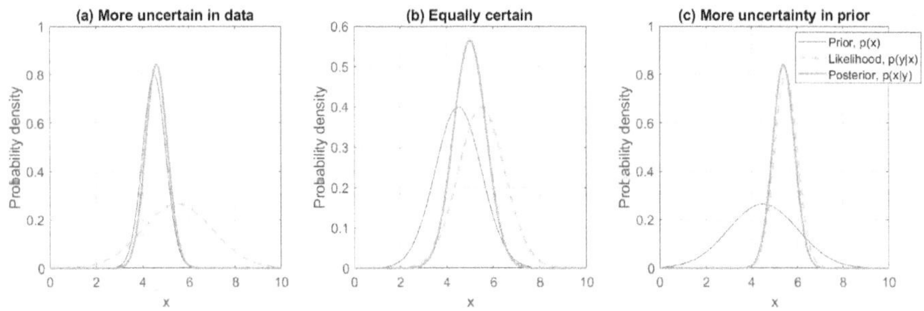

FIGURE 7.1: Bayes' formula for one-dimensional Gaussian distributions. The three panels show posterior distributions under different combinations of prior and likelihood precisions.

This one-dimensional Gaussian Bayesian update is the simplest prototype of data assimilation. Here, the random variable x represents the model state, $p(x)$ is the prior distribution obtained from the model, representing the prior knowledge, while y is an observation with observational error variance σ_y^2. The posterior distribution $p(x \mid y)$ optimally combines prior information and observational data. The posterior mean μ_n in (7.3) is the estimate obtained by balancing the model and the observation information, while the posterior variance σ_n^2 in (7.4) represents the reduced analysis uncertainty. As will be seen in the following, this update generalizes to the Kalman filter and other data assimilation methods, where the same precision-weighted structure governs the integration of model forecasts and observations.

7.2 Kalman filter

The derivation of the Gaussian posterior in Example 7.1 provides the foundational update step for a much more powerful algorithm: the Kalman filter. The Kalman filter can be viewed as a recursive Bayesian process that applies this same update to dynamic systems that evolve over time.

7.2.1 The method

The Kalman filter extends static Bayesian inference to a dynamic setting by introducing a time-evolving system state. The goal is to sequentially estimate this state by combining statistical predictions from a dynamical model with noisy observations of part of the state variables. The Kalman filter provides an exact recursive Bayesian solution only for linear models with Gaussian noise, under which the posterior distribution remains Gaussian and is fully characterized by its mean and covariance. For nonlinear or non-Gaussian settings, various extensions of the Kalman filter can be employed, which will be discussed in subsequent sections.

We assume a discrete-time linear state-space model consisting of a process (forecast) equation and an observation equation:

$$\text{Process (model) equation:} \quad \mathbf{x}_k = \mathbf{F}_k \mathbf{x}_{k-1} + \mathbf{f}_k + \mathbf{w}_k, \tag{7.7}$$

$$\text{Observation equation:} \quad \mathbf{y}_k = \mathbf{H}_k \mathbf{x}_k + \mathbf{v}_k, \tag{7.8}$$

where $\mathbf{x}_k \in \mathbb{R}^n$ denotes the model state at time k. The matrix $\mathbf{F}_k \in \mathbb{R}^{n \times n}$ is the state transition (or forecast) operator, which propagates the state from time $k-1$ to time k

according to the dynamical model, while $\mathbf{f}_k$ represents a known deterministic forcing. The term $\mathbf{w}_k$ is the process noise, which usually accounts for the contribution from unresolved scales or stochastic forcing. It is assumed to be Gaussian,

$$\mathbf{w}_k \sim \mathcal{N}(\mathbf{0}, \mathbf{Q}_k), \tag{7.9}$$

where the covariance matrix $\mathbf{Q}_k \in \mathbb{R}^{n \times n}$ quantifies uncertainty in the model dynamics. The observation vector $\mathbf{y}_k \in \mathbb{R}^p$ represents noisy measurements of the system state, where generally $p \neq n$. The observation operator $\mathbf{H}_k \in \mathbb{R}^{p \times n}$ maps the model state space to the observation space. The observational noise $\mathbf{v}_k$ is assumed Gaussian,

$$\mathbf{v}_k \sim \mathcal{N}(\mathbf{0}, \mathbf{R}_k), \tag{7.10}$$

with covariance matrix $\mathbf{R}_k \in \mathbb{R}^{p \times p}$ describing measurement errors and instrument uncertainty. The coupled model (7.7)–(7.8) defines a probabilistic dynamical system in which uncertainty is propagated forward in time by the model and reduced whenever new observations are assimilated. The Kalman filter provides a recursive Bayesian solution that alternates between forecast (model propagation) and analysis (data assimilation) steps, updating both the state estimate and its associated uncertainty.

It is worth emphasizing that the unknown true signal is one particular realization of (7.7). The Kalman filter does not access this realization directly; it only uses the statistical model and the noisy observations. In many synthetic data experiments, the true signal is often used to evaluate the accuracy of the state estimation.

The Kalman filter begins with the initialization for the state. As the state is characterized by Gaussian distributions, the initial state can be fully determined by its mean and covariance: $\hat{\mathbf{x}}_0^a = \mathbb{E}[\mathbf{x}_0]$, $\mathbf{P}_0^a = \mathbb{E}[(\mathbf{x}_0 - \hat{\mathbf{x}}_0^a)(\mathbf{x}_0 - \hat{\mathbf{x}}_0^a)^\mathtt{T}]$. Then, the two-step recursive procedure execute the following, for $k = 1, 2, \ldots$:

1. **Forecast step (prior prediction)**
   Predict the state and covariance at time k using a dynamical model. In the so-called perfect model scenario, the forecast model has the same structure and statistical properties as the model that generates the true signal in (7.7). Nevertheless, because the system contains stochastic process noise, the forecast cannot reproduce the exact realization of the true state. Instead, it provides a statistical prediction characterized by a mean and covariance. The forecast model is used to provide a statistical estimate of the state, known as the prior distribution:

$$\text{State prediction (prior mean):} \quad \hat{\mathbf{x}}_k^f = \mathbf{F}_k \hat{\mathbf{x}}_{k-1}^a + \mathbf{f}_k, \tag{7.11}$$

$$\text{Uncertainty prediction (prior covariance):} \quad \mathbf{P}_k^f = \mathbf{F}_k \mathbf{P}_{k-1}^a \mathbf{F}_k^\mathtt{T} + \mathbf{Q}_k, \tag{7.12}$$

   where the superscript f denotes forecast (the prior). The forecast step (7.11)–(7.12) uses the model dynamics to project the state and its uncertainty forward in time. The process noise $\mathbf{Q}_k$ accounts for the model uncertainty.

2. **Analysis step (posterior update)**
   When a new observation $\mathbf{y}_k$ arrives, the forecast is updated:

$$\text{State update (posterior mean):} \quad \hat{\mathbf{x}}_k^a = \hat{\mathbf{x}}_k^f + \mathbf{K}_k(\mathbf{y}_k - \mathbf{H}_k \hat{\mathbf{x}}_k^f), \tag{7.13}$$

$$\text{Uncertainty update (posterior covariance):} \quad \mathbf{P}_k^a = (\mathbf{I} - \mathbf{K}_k \mathbf{H}_k)\mathbf{P}_k^f, \tag{7.14}$$

   where $\mathbf{K}_k$ is the so-called Kalman gain,

$$\mathbf{K}_k = \mathbf{P}_k^f \mathbf{H}_k^\mathtt{T} (\mathbf{H}_k \mathbf{P}_k^f \mathbf{H}_k^\mathtt{T} + \mathbf{R}_k)^{-1}, \tag{7.15}$$

and the superscript a denotes analysis (the posterior). The term $(\mathbf{y}_k - \mathbf{H}_k\hat{\mathbf{x}}_k^f)$ is called the innovation. The analysis step (7.13)–(7.15) is the Bayesian core. The Kalman gain $\mathbf{K}_k$ optimally weights the confidence in the model forecast versus the new observations. It can be shown that

$$(\mathbf{P}_k^a)^{-1} = (\mathbf{P}_k^f)^{-1} + \mathbf{H}_k^\mathrm{T}\mathbf{R}_k^{-1}\mathbf{H}_k, \tag{7.16}$$

which is precisely as the precision-weighted average did in the one-dimensional case in Example 7.2.

The power of the Kalman filter lies in this recursive loop, which formalizes the process of continuous learning and prediction. The posterior (a) from one time step becomes the starting point for obtaining the prior (f) of the next:

$$\cdots \to \{\hat{\mathbf{x}}_{k-1}^a, \mathbf{P}_{k-1}^a\} \xrightarrow{\text{Forecast step}} \{\hat{\mathbf{x}}_k^f, \mathbf{P}_k^f\} \xrightarrow{\text{Analysis step}} \{\hat{\mathbf{x}}_k^a, \mathbf{P}_k^a\} \to \cdots \tag{7.17}$$

In essence, the Kalman filter is the operational realization of sequential Bayesian inference for linear Gaussian dynamical systems. It provides not only an optimal estimate of the system state in the minimum-variance sense, but also a rigorous quantification of uncertainty through the evolving covariance matrix. This dual capability makes the Kalman filter fundamental in applications such as tracking, navigation, signal processing, and, more broadly, as the conceptual foundation of modern data assimilation methods.

Example 7.3 (Limiting cases of the posterior state in the Kalman filter). The posterior mean in (7.13) is written in a form where it equals the prior mean plus a correction term from the observations. By a simple manipulation, (7.13) can be rewritten as

$$\hat{\mathbf{x}}_k^a = \hat{\mathbf{x}}_k^f + \mathbf{K}_k(\mathbf{y}_k - \mathbf{H}_k\hat{\mathbf{x}}_k^f) = (\mathbf{I} - \mathbf{K}_k\mathbf{H}_k)\hat{\mathbf{x}}_k^f + \mathbf{K}_k\mathbf{y}_k, \tag{7.18}$$

which shows that the analysis is a linear combination of the prior and the observation, weighted through the Kalman gain. Note that $\mathbf{H}_k$ maps the state of the prior $\hat{\mathbf{x}}_k^f$ to the space of observations $\mathbf{y}_k$. In the special case where $\mathbf{H}_k$ is an identity matrix, the prior state and the observations have the same unit. In such a situation, the coefficients in front of the prior and observation become $\mathbf{I} - \mathbf{K}_k$ and $\mathbf{K}_k$, respectively, which correspond to the standard weights.

The Kalman gain $\mathbf{K}_k$ controls the weighting between the prior model and the new data. Its behavior in two extreme cases reveals the filter's intuition.

- Case 1: Small observational noise ($\mathbf{R}_k \to \mathbf{0}$).
  When observations are very precise, namely, $\mathbf{R}_k$ is small, its inverse $\mathbf{R}_k^{-1}$ is large. In such a case, (7.15) becomes:

$$\mathbf{K}_k \to \mathbf{H}_k^{-1} \quad \text{(or the pseudoinverse } (\mathbf{H}_k^\mathrm{T}\mathbf{H}_k)^{-1}\mathbf{H}_k^\mathrm{T} \text{ if not square)} \tag{7.19}$$

  Substituting this into the update equation:

$$\hat{\mathbf{x}}_k^a = \hat{\mathbf{x}}_k^f + \mathbf{H}_k^{-1}(\mathbf{y}_k - \mathbf{H}_k\hat{\mathbf{x}}_k^f) = \mathbf{H}_k^{-1}\mathbf{y}_k. \tag{7.20}$$

  The filter completely trusts the precise observation. The posterior mean is simply the state implied by the observation itself, ignoring the prior forecast. The update equation effectively corrects the prior model value to exactly fit the data. In the fully observed case ($\mathbf{H}_k$ full rank), the posterior covariance collapses to zero in the limit $\mathbf{R}_k \to 0$.

- Case 2: Small process noise, namely, highly certain forecast ($\mathbf{P}_k^f \to \mathbf{0}$).
When the model forecast is very certain (low error, often due to small process noise $\mathbf{Q}_k$ over a short forecast), $\mathbf{P}_k^f$ is small.

$$\mathbf{K}_k \to \mathbf{P}_k^f\mathbf{H}_k^{\mathrm{T}}(\mathbf{H}_k\mathbf{P}_k^f\mathbf{H}_k^{\mathrm{T}} + \mathbf{R}_k)^{-1} \approx \mathbf{0} \tag{7.21}$$

Substituting this into the update equation:

$$\hat{\mathbf{x}}_k^a = \hat{\mathbf{x}}_k^f + \mathbf{0} \cdot (\mathbf{y}_k - \mathbf{H}_k\hat{\mathbf{x}}_k^f) = \hat{\mathbf{x}}_k^f. \tag{7.22}$$

The filter completely trusts the highly certain prior forecast. The observation is effectively ignored because the Kalman gain tends to zero. The posterior mean (and posterior variance) is unchanged from the prior forecast.

□

Example 7.4 (Long-term behavior of the posterior covariance). Recall the linear model with Gaussian noise (7.7)–(7.10) used in the Kalman filter. Let us assume now the forecast and observational operators and the noise covariance matrices are all constants over time, namely, $\mathbf{F}_k = \mathbf{F}$, $\mathbf{H}_k = \mathbf{H}$, $\mathbf{Q}_k = \mathbf{Q}$, and $\mathbf{R}_k = \mathbf{R}$. The covariance update is fully determined by the following three equations:

$$\mathbf{P}_k^f = \mathbf{F}\mathbf{P}_{k-1}^a\mathbf{F}^{\mathrm{T}} + \mathbf{Q}, \tag{7.23}$$

$$\mathbf{K}_k = \mathbf{P}_k^f\mathbf{H}^{\mathrm{T}}(\mathbf{H}\mathbf{P}_k^f\mathbf{H}^{\mathrm{T}} + \mathbf{R})^{-1}, \tag{7.24}$$

$$\mathbf{P}_k^a = (\mathbf{I} - \mathbf{K}_k\mathbf{H})\mathbf{P}_k^f. \tag{7.25}$$

Plugging (7.23) and (7.24) into (7.25), we obtain a time evolution equation of the posterior covariance going directly from $\mathbf{P}_{k-1}^a$ to $\mathbf{P}_k^a$ without the explicit dependence on the prior covariance and the Kalman gain:

$$\mathbf{P}_k^a = \mathbf{F}\mathbf{P}_{k-1}^a\mathbf{F}^{\mathrm{T}} + \mathbf{Q} - \mathbf{F}\mathbf{P}_{k-1}^a\mathbf{F}^{\mathrm{T}}\mathbf{H}^{\mathrm{T}}\Big(\mathbf{H}\mathbf{F}\mathbf{P}_{k-1}^a\mathbf{F}^{\mathrm{T}}\mathbf{H}^{\mathrm{T}} + \mathbf{H}\mathbf{Q}\mathbf{H}^{\mathrm{T}} + \mathbf{R}\Big)^{-1}\mathbf{H}\mathbf{F}\mathbf{P}_{k-1}^a\mathbf{F}^{\mathrm{T}}. \tag{7.26}$$

This equation is known as the (discrete-time) Riccati equation. Importantly, this evolution does not depend on the observed data $\mathbf{y}_k$, but only on the system parameters and the previous covariance. Since all coefficients in this matrix equation are constant, any steady-state solution $\mathbf{P}_\infty^a$, if it exists, must also be time-independent. In such cases, the posterior covariance converges to a constant matrix. Intuitively, at each step, the forecast model inflates uncertainty through $\mathbf{F}\mathbf{P}^a\mathbf{F}^{\mathrm{T}}$ and the process noise $\mathbf{Q}$, while the analysis step removes uncertainty according to the information content of the observations encoded in $\mathbf{H}$ and $\mathbf{R}$. The steady-state covariance $\mathbf{P}_\infty^a$ represents a balance between these two competing effects.
□

The Kalman filter is typically formulated for discrete-time observations and model forecasts. However, most dynamical models, such as those based on SDEs or chaotic ODEs, are inherently continuous in time. Nevertheless, these continuous models can naturally serve as the forecast component within the Kalman filter and other data assimilation frameworks.

Example 7.5 (One-dimensional Kalman filter with continuous SDE as the forecast model). Consider the following real-valued linear Gaussian SDE:

$$\mathrm{d}x = \big(-\gamma x + f_0\big)\,\mathrm{d}t + \sigma\,\mathrm{d}W, \tag{7.27}$$

where $\gamma > 0$ is the damping coefficient. The model has a constant deterministic forcing f_0 and a stochastic forcing $\sigma\,\mathrm{d}W$. Although the model in (7.27) is formulated in continuous time, whereas the process equation in the Kalman filter (7.7) is given in discrete time, the continuous-time model can be used naturally to generate forecast statistics over discrete time intervals. This connection enables a broad range of applications of the Kalman filter, since continuous-time dynamical systems, such as ODEs, PDEs, and SDEs, widely appear in science and engineering. To make the link between (7.27) and the discrete-time forecast step of the Kalman filter, let us focus on the exact path-wise solution from t_k to t_{k+1}, which can be expressed explicitly,

$$x(t_{k+1}) = x(t_k)e^{-\gamma(t_{k+1}-t_k)} + \frac{f_0}{\gamma}\left(1 - e^{-\gamma(t_{k+1}-t_k)}\right) + \sigma\int_{t_k}^{t_{k+1}} e^{-\gamma(t_{k+1}-s)}\,\mathrm{d}W(s). \quad (7.28)$$

Written in the discrete-step form (7.28) from t_k to t_{k+1}, the SDE can be directly mapped to the process equation (7.7). The three terms on the right-hand side of (7.28) correspond to those in (7.7), respectively. The formula in (7.28) leads to the forecast mean $\bar{x}(t_{k+1})$ and the forecast variance $r(t_{k+1})$ at t_{k+1} starting from t_k,

$$\begin{aligned}\bar{x}(t_{k+1}) &= \bar{x}(t_k)e^{-\gamma(t_{k+1}-t_k)} + \frac{f_0}{\gamma}\left(1 - e^{-\gamma(t_{k+1}-t_k)}\right),\\ r(t_{k+1}) &= r(t_k)e^{-2\gamma(t_{k+1}-t_k)} + \frac{\sigma^2}{2\gamma}\left(1 - e^{-2\gamma(t_{k+1}-t_k)}\right).\end{aligned}$$

As a simple illustration, consider the following parameters in (7.27):

$$\gamma = 0.2, \quad f_0 = 1, \quad \text{and} \quad \sigma = 0.5. \quad (7.29)$$

The observational time step is $\Delta t^{obs} = 0.1$.We vary R relative to Q (which drives the forecast uncertainty) to illustrate how the Kalman gain shifts trust between model and observations. See Figure 7.2. These three observational noise levels are chosen to represent fundamental scenarios in data assimilation, based on the relative uncertainty between model forecasts and measurements. The case where the observational noise variance R is much larger than the model's process noise variance Q ($R = 10Q$; Panel (a)) represents a situation with strong measurement noise where the filter result will trust the model forecast more than the observational data. The case where R is much smaller than Q ($R = 0.1Q$; Panel (c)) represents the opposite, where highly precise observations correct a relatively unreliable and uncertain model forecast. Finally, the balanced case where R is appropriately equal to Q (Panel (b)) demonstrates the filter's optimal integration of two information sources with comparable uncertainty, giving each roughly equal weight. The variation in the noise ratio R/Q illustrates how the Kalman gain adapts to balance trust between the model and the observations.

□

7.2.2 Observability

The performance of the Kalman filter relies critically on the observability of the underlying system.

A linear system is said to be observable if the initial state $\mathbf{x}_0 \in \mathbb{R}^n$ can be uniquely determined from a finite sequence of noise-free observations $\mathbf{y}_0, \mathbf{y}_1, \ldots, \mathbf{y}_n$ over a finite time interval, where all $\mathbf{y}_i \in \mathbb{R}^p$.

For a linear, time-invariant system without noise, defined by

$$\begin{aligned}\mathbf{x}_{k+1} &= \mathbf{A}\mathbf{x}_k\\ \mathbf{y}_k &= \mathbf{H}\mathbf{x}_k,\end{aligned}$$

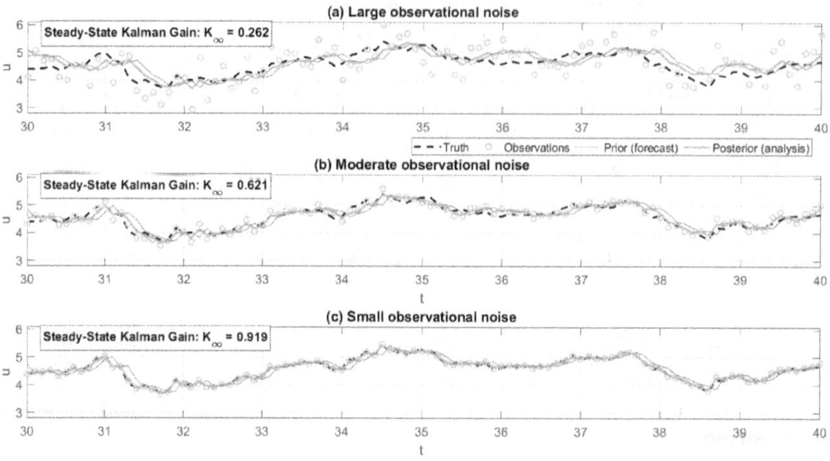

FIGURE 7.2: Kalman filter using (7.27) with different observational noise levels. Panel (a): $R = 10Q$, where R is the observational noise variance while Q is the model's process noise variance. Panel (b): $R \approx Q$. Panel (c): $R = 0.1Q$.

where $\mathbf{H} \in \mathbb{R}^{p \times n}$, the criterion for observability is based on the observability matrix $\mathcal{O}$:

$$\mathcal{O} = \begin{bmatrix} \mathbf{H} \\ \mathbf{HA} \\ \mathbf{HA}^2 \\ \vdots \\ \mathbf{HA}^{n-1} \end{bmatrix}.$$

The system is observable if and only if $\mathcal{O}$ has full column rank, i.e., $\text{rank}(\mathcal{O}) = n$. This means the sequence of observations must contain enough independent information to reconstruct the entire initial state.

Observability is crucial for the Kalman filter because it determines whether the sequence of observations contains enough independent information to estimate the entire system state. If a system is unobservable, it means one or more state variables cannot be inferred from the measurements. In other words, the observations do not help reduce the uncertainty in the state estimation of these variables. The components of the state lying in the unobservable subspace cannot be corrected by the observations; their posterior distribution coincides with the prior distribution.

Example 7.6 (Observability in the one-dimensional Kalman filter). Consider the one-dimensional process and one-dimensional observation (i.e., $n = p = 1$):

$$\text{Process equation:} \quad x_{k+1} = Fx_k + f + w_{k+1}, \tag{7.30}$$

$$\text{Observation equation:} \quad y_k = Hx_k + v_k, \tag{7.31}$$

In this special case, the observability matrix $\mathcal{O} = H$ degenerates to a scalar. The system is unobservable when $H = 0$. In such a situation, the Kalman gain becomes zero, and the posterior distribution coincides precisely with the prior. Intuitively, since $H = 0$, the observation y_k is independent of the state x_k, meaning the measurements provide no information to improve upon the prior estimate.

However, a more common and critical scenario arises when $H \approx 0$. The system is still technically observable, but the signal Hx_k contained in the observation is overwhelmed by the measurement noise v_k. Consequently, the information provided by y_k is negligible, and the Kalman gain remains very small. The filter effectively ignores the uninformative measurements, and the state estimate fails to improve significantly. This condition is known as a loss of practical observability. □

Example 7.7 (Observability for the non-directly measured variable). Consider a two-dimensional system where the state is $\mathbf{x}_k = (x_k^1, x_k^2)^{\mathrm{T}}$ and the forecast model is:

$$\begin{aligned} x_{k+1}^1 &= F_{11}x_k^1 + F_{12}x_k^2 + w_{k+1}^1, \\ x_{k+1}^2 &= F_{21}x_k^1 + F_{22}x_k^2 + w_{k+1}^2. \end{aligned}$$

Assume we can only observe the first variable: $y_k = x_k^1 = \mathbf{H}\mathbf{x}_k$, where $\mathbf{H} = \begin{bmatrix} 1 & 0 \end{bmatrix}$.

The observability matrix for this system is:

$$\mathcal{O} = \begin{bmatrix} \mathbf{H} \\ \mathbf{HA} \end{bmatrix} = \begin{bmatrix} 1 & 0 \\ F_{11} & F_{12} \end{bmatrix}. \tag{7.32}$$

This matrix has full rank (rank $= 2$) if and only if $F_{12} \neq 0$. The observation of x^1 alone is sufficient to estimate both states because the dynamics of x^2 (through F_{12}) influence the future evolution of the observed variable x^1. Therefore, if $F_{12} = 0$, the evolution of x^1 is independent of x^2, and the observation of x^1 contains no information about x^2. In this case, x^2 becomes unobservable from the measurements. Consequently, while the prior distribution of x^2 may still evolve according to the model dynamics, the observations of x^1 do not reduce the forecast uncertainty in x^2. The posterior uncertainty of x^2 therefore remains equal to its prior uncertainty. □

7.3 Nonlinear methods

The standard Kalman filter provides an optimal solution for state estimation in linear Gaussian systems. However, most dynamical systems in science and engineering are nonlinear with non-Gaussian statistics. Although the underlying principles for assimilating these systems, based on Bayes' theorem, remain the same, the nonlinearity introduces significant challenges in computing the posterior distribution. In this section, we briefly present three widely used practical approximate methods for nonlinear data assimilation: the extended Kalman filter (EKF), the ensemble Kalman filter (EnKF), and the particle filter (PF).

7.3.1 Extended Kalman filter (EKF)

The EKF is a direct extension of the Kalman filter to nonlinear systems [143]. The core idea is to linearize the nonlinear dynamics and observation operators around the current state estimate at each assimilation cycle.

Consider a nonlinear state-space model:

$$\begin{aligned} \mathbf{x}_k &= \mathcal{M}_{k-1}(\mathbf{x}_{k-1}) + \mathbf{w}_k, \\ \mathbf{y}_k &= \mathcal{H}_k(\mathbf{x}_k) + \mathbf{v}_k, \end{aligned} \tag{7.33}$$

where $\mathcal{M}$ and $\mathcal{H}$ are nonlinear functions. The EKF applies the standard Kalman filter equation for the mean

$$\hat{\mathbf{x}}_k^f = \mathcal{M}_{k-1}(\hat{\mathbf{x}}_{k-1}^a) \tag{7.34}$$

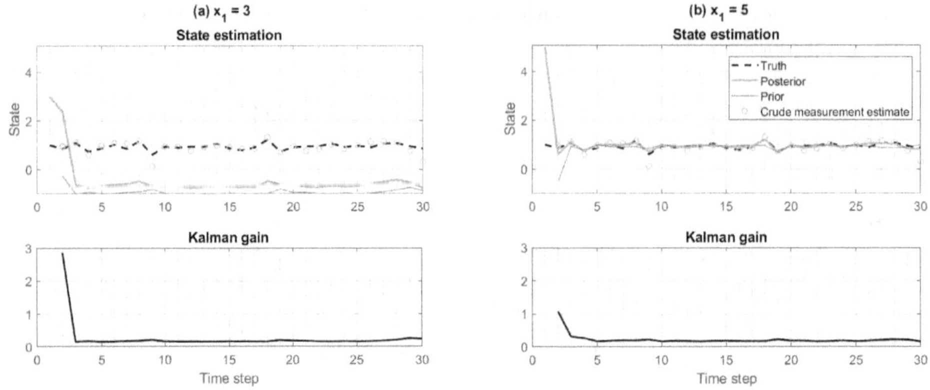

FIGURE 7.3: State estimation of (7.36) based on the extended Kalman filter (EKF) with different initial conditions. The crude measurement estimation is given by $x_k = y_k^{1/3}$, which projects the noisy observations into the state space.

but uses the Jacobian matrices of these nonlinear functions to propagate the covariance:

$$\mathbf{F}_k = \left.\frac{\partial \mathcal{M}_{k-1}}{\partial \mathbf{x}}\right|_{\hat{\mathbf{x}}_{k-1}^a} \qquad \text{and} \qquad \mathbf{H}_k = \left.\frac{\partial \mathcal{H}_k}{\partial \mathbf{x}}\right|_{\hat{\mathbf{x}}_k^f}. \tag{7.35}$$

These Jacobians are then used in the forecast and update equations as in the standard Kalman filter (7.12) and (7.15). While simple in concept, the EKF has limitations. The linearization can introduce significant errors for highly nonlinear systems, and the computational cost of calculating Jacobians can be prohibitive for high-dimensional models.

Example 7.8 (Limitation of the EKF). Let us consider the following scalar model,

$$\begin{aligned} x_k &= \sin(2x_{k-1}) + w_k, \qquad w_k \sim \mathcal{N}(0, Q), \\ y_k &= x_k^3 + v_k, \qquad v_k \sim \mathcal{N}(0, R). \end{aligned} \tag{7.36}$$

where x_k is the state variable at time t_k while y_k is the observations. The nonlinear observational process contains a cubic function of the state variable. Figure 7.3 shows the state estimation using the EKF starting from two different initial conditions, where $Q = 0.01$ and $R = 0.1$ are used. When the initial value is $x_1 = 3$ (Panel (a)), the EKF exhibits a pathological behavior. The first few forecasts and updates are as follows

$$\hat{x}_1^a = 3 \xrightarrow{\text{Forecast}} \hat{x}_2^f = -0.30 \xrightarrow{\text{Analysis}} \hat{x}_2^a = 2.36 \xrightarrow{\text{Forecast}} \hat{x}_3^f = -1.00 \xrightarrow{\text{Analysis}} \hat{x}_3^a = -0.62... \tag{7.37}$$

Afterwards, the prior mean $\hat{x}_k^f$ stays near the value -1 and the posterior mean $\hat{x}_k^a$ near the value -0.6. However, the truth is around 1. The difference between the posterior mean and the truth is significantly higher than the posterior standard deviation, which is less than 0.1. Therefore, the filter is not skillful. The Kalman gain is approximately $K = 0.18$, indicating that the filtering state places more trust in the forecast, with very limited correction from observations. In fact, the posterior mean state is closer to the truth than the prior, but the corrections are insignificant. When this slightly corrected posterior is fed back into the dynamics, it produces a prior that is again far from the truth, creating a stable but inaccurate loop. The root of the problem lies in the covariance propagation based on the linearized operator. Although the forecast mean is computed using the full nonlinear model $\hat{x}_k^f = \sin(2\hat{x}_{k-1}^a)$, the forecast uncertainty is propagated using the Jacobian

$$F_k = 2\cos(2\hat{x}_{k-1}^a). \tag{7.38}$$

When $\hat{x}_{k-1}^a \approx -0.7$, the local linearization poorly represents the curvature of the nonlinear function $\sin(2x)$ in that region. As a result, the forecast variance is severely underestimated. The small forecast variance leads to a small Kalman gain, preventing the observations from effectively correcting the biased forecast. As a result, the prior state is only slightly different from the previous posterior step. What is worse, the forecast $\sin(2\hat{x}_k^a) = -0.9854$ shifts a little bit further towards negative values compared to the previous posterior. Since the Kalman gain is small, the contribution from the observation is weak. The correction towards the positive side due to the observation nearly cancels the negative shift introduced in the prior state. Therefore, the posterior state remains almost unchanged at each assimilated cycle.

Note that the EKF does not always fail. For example, starting from $x_1 = 5$ (Panel (b)), which is even farther from the truth than in the previous case, the posterior state estimate synchronizes quickly to values near the true state. This is because after a few assimilative cycles, the incorrect forecast happens to be close to the truth, which is about $\hat{x}_k^a \approx 1$. In this case, the true F_k and its approximation are given by

$$\sin(2\hat{x}_k^a) = 0.9093 \quad \text{and} \quad 2\cos(2\hat{x}_k^a)\hat{x}_k^a = -0.8323, \tag{7.39}$$

respectively, which have about the same amplitude. As a result, the forecast uncertainty, which depends on F_k^2, is pretty accurate. In addition, the forecast error (the forecast value of 0.9093 compared to the truth of 1) is insignificant. The resulting small forecast error can mostly be corrected by the observations.

This example illustrates that the EKF may succeed in certain dynamical regimes but fail catastrophically in others. The results here demonstrate its strong dependence on local linearization, and the method may lack robustness for highly nonlinear systems. □

7.3.2 Ensemble Kalman filter (EnKF)

The EnKF is designed for nonlinear systems and is effective in high-dimensional cases [97]. Instead of propagating a single covariance matrix, the EnKF uses an ensemble of state vectors (typically tens to hundreds of members in large-scale applications) to represent the prior and posterior distributions.

The algorithm proceeds as follows:

1. **Forecast step:** Each ensemble member $\mathbf{x}_{(i),k-1}^a$ is propagated forward by the full nonlinear model:

$$\mathbf{x}_{(i),k}^f = \mathcal{M}_{k-1}(\mathbf{x}_{(i),k-1}^a) + \mathbf{w}_{(i),k}. \tag{7.40}$$

   The prior mean $\hat{\mathbf{x}}_k^f$ and covariance $\mathbf{P}_k^f$ are estimated empirically from the ensemble.

2. **Analysis step:** Each member is updated individually using a perturbed observation $\mathbf{y}_{(i),k} = \mathbf{y}_k + \mathbf{v}_{(i),k}$ such that

$$\mathbf{x}_{(i),k}^a = \mathbf{x}_{(i),k}^f + \mathbf{K}_k(\mathbf{y}_{(i),k} - \mathcal{H}_k(\mathbf{x}_{(i),k}^f)). \tag{7.41}$$

   The Kalman gain $\mathbf{K}_k$ is computed using the ensemble-estimated prior covariance $\mathbf{P}_k^f$.

The EnKF avoids the need for tangent linear models and can naturally handle nonlinear dynamics. However, when the ensemble size is small, stochastic observation perturbations introduce additional sampling errors, which often lead to various numerical instability issues. In practice, several empirical strategies are often used to mitigate these issues. Covariance inflation [14, 280] is commonly used to compensate for the underestimation of ensemble spread and to prevent filter divergence. On the other hand, deterministic square-root formulations, such as the ensemble transform Kalman filter (ETKF) [36, 141] and the ensemble adjustment Kalman filter (EAKF) [11], eliminate the need for perturbed observations. They update the ensemble deterministically, deterministically matching the Kalman analysis covariance implied by the ensemble statistics. These ensemble square-root filter (EnSRF) variants [280, 256] typically reduce sampling noise and often achieve good performance with smaller ensemble sizes. For high-dimensional systems where only limited ensemble sizes are computationally feasible, localization techniques are also essential to mitigate spurious long-range correlations arising from sampling errors. Localization restricts the influence of an observation to a neighborhood around its physical location, based on the intuition that distant state variables are often only weakly related. By eliminating unrealistic long-distance correlations, localization helps retain physically meaningful updates and significantly improves filter stability and accuracy. Practical localization approaches [141, 13, 12, 103, 54, 139] have been widely incorporated into ensemble-based methods and are widely used to solve operational-scale problems.

Example 7.9 (Comparison between the EKF and the EnKF). Figure 7.4 compares the performance of the EKF and the ETKF (as an EnKF method) for the case shown in Panel (a) in Figure 7.3, where the underlying system and the nonlinear observations are given by (7.36). The same noise amplitudes $Q = 0.01$ and $R = 0.1$ as in Example 7.8 are used here. Panel (a) in Figure 7.4 demonstrates that the ETKF captures the truth quite well, which contrasts with the EKF. Panel (b) shows the error evolution as a function of the ensemble size in the ETKF. With only two ensemble members, the ETKF diverges and produces errors comparable to those of the EKF. Nevertheless, even with an ensemble size of 5, the ETKF works effectively. This surprisingly small ensemble size demonstrates a key advantage of the ETKF over both the EKF and the stochastic EnKF. Panel (c) shows the time evolution of the posterior variance. Notably, the posterior variances in both EKF and ETKF vary over time even after the solutions reach statistical equilibrium. For EKF, the covariance depends on the Jacobian $\mathbf{F}_k$ in (7.35), which differs at different time instants. Similarly, for the EnKF, the variance depends on the state and changes over time. This is fundamentally different from the linear Gaussian cases with constant coefficients using the Kalman filter, as was shown in Example 7.4.

The skillful performance of ETKF with very few ensemble members is due to several key factors. First, as a deterministic square-root filter, the ETKF eliminates the additional sampling noise associated with perturbed observations in the stochastic EnKF, although sampling errors due to finite ensemble size still remain. Second, the ensemble transform mechanism optimally preserves the ensemble spread while updating the mean, providing better uncertainty quantification than naive sampling approaches. Third, for low-dimensional systems like the one considered here, even a small ensemble can adequately capture the principal directions of uncertainty in the state space. The transformation matrix in the ETKF effectively rotates and scales the ensemble to match the analysis covariance, based solely on the limited number of ensemble members. This computational advantage makes the ETKF particularly useful in many practical applications where model evaluations are expensive. □

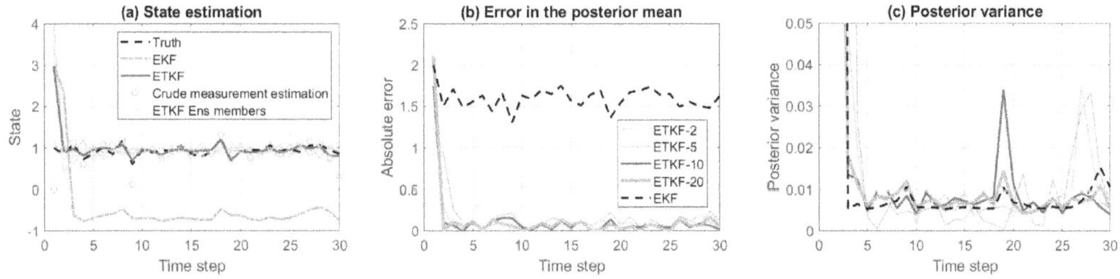

FIGURE 7.4: Comparison of ETF and ETKF (as an EnKF method) with different ensemble sizes based on the model in (7.36). Panel (a): state estimation as a function of time. Panel (b): absolute error in the posterior mean. Panel (c): posterior variance.

7.3.3 Particle filter (PF)

The PF, or sequential Monte Carlo (SMC) method, is a fully nonlinear and non-Gaussian data assimilation scheme [88, 172]. It represents the PDF $p(\mathbf{x}_k|\mathbf{y}_{1:k})$ by a set of random samples (particles) with associated weights $\{\mathbf{x}_k^{(i)}, w_k^{(i)}\}_{i=1}^N$.

The standard algorithm involves:

1. **Forecast step:** Particles are propagated by the nonlinear model: $\mathbf{x}^f_{(i),k} \sim p(\mathbf{x}_k|\mathbf{x}^a_{(i),k-1})$.
2. **Analysis step:** Weights are updated based on the likelihood of the new observation: $w_k^{(i)} \propto w_{k-1}^{(i)} \cdot p(\mathbf{y}_k|\mathbf{x}^f_{(i),k})$.
3. **Resampling:** To mitigate the degeneracy problem (where most weights become negligible), particles are resampled with replacement according to their weights. This focuses the computational effort on the most likely regions of state space.

The PF is asymptotically exact as the number of particles $N \to \infty$. However, it suffers from the "curse of dimensionality", which requires an exponentially large number of particles to accurately represent the posterior in high-dimensional state spaces.

The choice of method depends on several factors, including computational cost and the degree of nonlinearity/non-Gaussianity in the system. The EnKF has yielded numerous successes in high-dimensional geophysical applications, such as numerical weather prediction, due to its balance of efficiency and effectiveness for moderately nonlinear problems. The PF remains a powerful tool for highly nonlinear but low-dimensional problems where capturing non-Gaussian features is critical.

7.4 Continuous-time assimilation

The data assimilation methods discussed so far involve discrete-time observations. The forecast model is advanced forward in time over a specified interval, and the state is updated whenever a new observation becomes available. With ongoing improvements in measurement technology, observations can arrive at high frequency in many applications, and they can be viewed as continuous-time signals. Therefore, continuous-time data assimilation becomes an essential technique for better incorporating information from these continuous measurements. On the theoretical side, continuous-time data assimilation also facilitates mathematical analysis by allowing the filtering problem to be formulated as a stochastic differential equation. This advances the use of tools from stochastic calculus, dynamical systems, and control theory to analyze the properties of the time evolution of the posterior statistics, and to systematically study stability, consistency, and long-time behavior.

7.4.1 Kalman-Bucy filter

The continuous-time analogue of the Kalman filter is known as the Kalman-Bucy filter [145]. Consider a continuous-time linear system with continuous observations:

$$\text{Process equation:} \quad d\mathbf{x}(t) = \mathbf{A}(t)\mathbf{x}(t)\,dt + \mathbf{f}(t)\,dt + \mathbf{Q}^{1/2}(t)\,d\mathbf{W}(t), \tag{7.42}$$

$$\text{Observation equation:} \quad d\mathbf{y}(t) = \mathbf{H}(t)\mathbf{x}(t)\,dt + \mathbf{R}^{1/2}(t)\,d\mathbf{V}(t), \tag{7.43}$$

where $\mathbf{W}(t)$ and $\mathbf{V}(t)$ are independent Wiener processes, $\mathbf{Q}(t)$ is the process noise covariance matrix, and $\mathbf{R}(t)$ is the covariance matrix describing the observational uncertainty. The Kalman-Bucy filter requires that both the process equation (7.42) and the observation equation (7.43) be linear.

The Kalman-Bucy filter provides the optimal estimate $\hat{\mathbf{x}}(t)$ and its error covariance $\mathbf{P}(t)$ through the following differential equations:

$$\text{State estimate:} \quad d\hat{\mathbf{x}}(t) = \mathbf{A}(t)\hat{\mathbf{x}}(t)\,dt + \mathbf{f}(t)\,dt + \mathbf{K}(t)\left[d\mathbf{y}(t) - \mathbf{H}(t)\hat{\mathbf{x}}(t)\,dt\right], \tag{7.44}$$

$$\text{Covariance:} \quad d\mathbf{P}(t) = \left[\mathbf{A}(t)\mathbf{P}(t) + \mathbf{P}(t)\mathbf{A}^{\mathrm{T}}(t) + \mathbf{Q}(t) - \mathbf{P}(t)\mathbf{H}^{\mathrm{T}}(t)\mathbf{R}^{-1}(t)\mathbf{H}(t)\mathbf{P}(t)\right] dt, \tag{7.45}$$

where the Kalman gain $\mathbf{K}(t) = \mathbf{P}(t)\mathbf{H}^{\mathrm{T}}(t)\mathbf{R}^{-1}(t)$ now appears as a continuous feedback term. The state mean $\hat{\mathbf{x}}(t)$ and covariance $\mathbf{P}(t)$ therefore represent the posterior statistics at each instant in time. In contrast to the discrete-time Kalman filter, where state estimation proceeds through alternating forecast (prior) and analysis (posterior) steps at observation times, the Kalman-Bucy filter operates with observations available continuously in time. As a result, the update is no longer performed through discrete analysis steps. Instead, the posterior mean and covariance evolve smoothly according to (7.44)–(7.45), in which the effects of forecasting and assimilation are combined into a single set of continuous-time differential equations. This formulation eliminates the explicit distinction between prior and posterior updates and provides a natural continuous-time analogue of the Kalman filter. Note that the time evolution of the covariance in (7.45) is a Riccati equation, which is an analog to the discrete-time Riccati equation in (7.26) derived for the Kalman filter.

An important conceptual difference from the discrete-time Kalman filter lies in the interpretation of the observation process. In the discrete formulation (7.8), observations $\mathbf{y}_k$ are treated as deterministic measurements corrupted by additive measurement noise $\mathbf{v}_k$. In contrast, the continuous-time observation (7.43) represents $\mathbf{y}(t)$ as a genuine stochastic process where the noise term $\mathbf{R}^{1/2}(t)\,d\mathbf{V}(t)$ is intrinsic to the process itself. Therefore, the uncertainty in continuous-time observations arises from the inherent fluctuations of the stochastic process $\mathbf{y}(t)$, rather than from externally added measurement noise.

7.4.2 Ensemble Kalman-Bucy filter

For nonlinear systems, the ensemble method can be extended to continuous time, leading to the ensemble Kalman-Bucy filter (EnKBF) [30]. Consider the nonlinear system:

$$d\mathbf{x}(t) = \mathbf{f}(\mathbf{x}(t), t)\,dt + \mathbf{Q}^{1/2}(t)\,d\mathbf{W}(t), \tag{7.46}$$

$$d\mathbf{y}(t) = \mathbf{h}(\mathbf{x}(t), t)\,dt + \mathbf{R}^{1/2}(t)\,d\mathbf{V}(t), \tag{7.47}$$

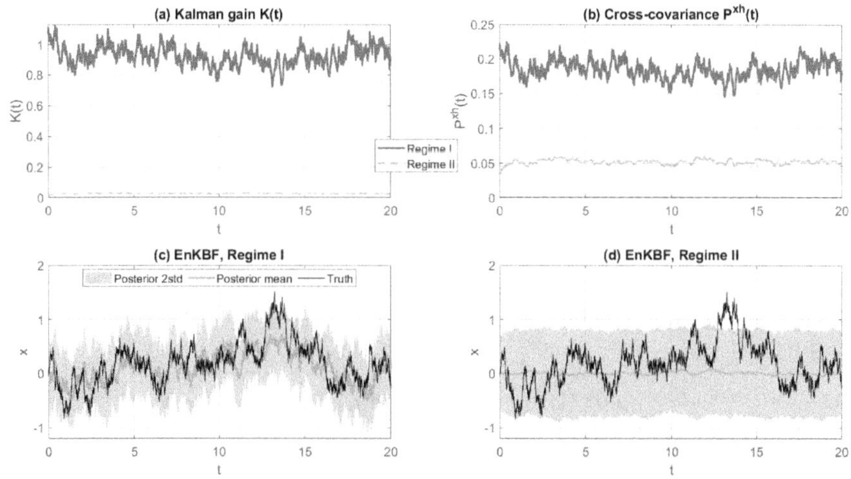

FIGURE 7.5: Applying the EnKBF to (7.51)–(7.52). Panel (a): time evolution of the Kalman gain $K(t)$. Panel (b): time evolution of the cross-covariance $P^{xh}(t)$. Panels (c)–(d): the true trajectory of x, the posterior ensemble mean, and the posterior two standard deviations, shown for the two regimes, respectively. The two regimes are defined in (7.54).

The EnKBF updates each ensemble member $\mathbf{x}^{(i)}(t)$ according to:

$$\mathrm{d}\mathbf{x}^{(i)}(t) = \mathbf{f}(\mathbf{x}^{(i)}(t), t)\,\mathrm{d}t + \mathbf{Q}^{1/2}(t)\,\mathrm{d}\mathbf{W}^{(i)}(t) + \mathbf{K}(t)\left[\mathrm{d}\mathbf{y}(t) + \mathbf{R}^{1/2}(t)\,\mathrm{d}\mathbf{V}^{(i)}(t) - \mathbf{h}(\mathbf{x}^{(i)}(t), t)\,\mathrm{d}t\right], \tag{7.48}$$

where $\mathbf{V}^{(i)}(t)$ are independent Wiener processes for each particle, and the Kalman gain is computed from the ensemble statistics:

$$\mathbf{K}(t) = \mathbf{P}^{xh}(t)\mathbf{R}^{-1}(t), \tag{7.49}$$

with $\mathbf{P}^{xh}(t)$ being the sample cross-covariance between the state and observation. The matrix $\mathbf{P}^{xh}(t)$ can be computed as follows. For each ensemble member $\mathbf{x}^{(i)}(t)$, we define the predicted observation $\mathbf{h}^{(i)}(t) \equiv \mathbf{h}\big(\mathbf{x}^{(i)}(t), t\big)$, for $i = 1, \ldots, M$. The ensemble means in the state space and observation space are $\bar{\mathbf{x}}(t) = \frac{1}{M}\sum_{i=1}^{M}\mathbf{x}^{(i)}(t)$ and $\bar{\mathbf{h}}(t) = \frac{1}{M}\sum_{i=1}^{M}\mathbf{h}^{(i)}(t)$. The sample cross-covariance is then given by

$$\mathbf{P}^{xh}(t) = \frac{1}{M-1}\sum_{i=1}^{M}\left(\mathbf{x}^{(i)}(t) - \bar{\mathbf{x}}(t)\right)\left(\mathbf{h}^{(i)}(t) - \bar{\mathbf{h}}(t)\right)^{\mathtt{T}}. \tag{7.50}$$

The matrix $\mathbf{P}^{xh}(t)$ characterizes how deviations in the state variables are correlated with deviations in the predicted observations.

Example 7.10 (EnKFB with strong and weak correlations between state variables and observations).

Figure 7.5 demonstrates a numerical example of the above EnKBF applied to a one-dimensional nonlinear system with continuous-time observations. The true state evolves according to the following nonlinear scalar SDE:

$$\mathrm{d}x(t) = \big(-a\,x(t) - b\,x^3(t)\big)\;\mathrm{d}t + \sqrt{q}\;\mathrm{d}W(t), \tag{7.51}$$

□

which is mean-reverting due to the linear and cubic damping terms. Observations are generated continuously in time via

$$dy(t) = h(x(t))\,dt + \sqrt{r}\,dV(t), \qquad h(x) = cx + dx^3, \tag{7.52}$$

where $W(t)$ and $V(t)$ are independent Wiener processes. The EnKBF propagates an ensemble of particles using the stochastic, perturbed-observation update

$$dx^{(i)}(t) = f(x^{(i)}(t))\,dt + \sqrt{q}\,dW^{(i)}(t) + K(t)\big[dy(t) + \sqrt{r}\,dV^{(i)}(t) - h(x^{(i)}(t))\,dt\big], \tag{7.53}$$

with the Kalman gain given by $K(t) = P^{xh}(t)/r$, where $P^{xh}(t) = \text{cov}\big(x(t), h(x(t))\big)$ is the ensemble cross-covariance between the state and the predicted observation. Two regimes are studied here

$$\begin{aligned} &\text{Regime I (strongly observed case):} \qquad c = 2.0,\\ &\text{Regime II (weakly observed case):} \qquad c = 0.3. \end{aligned} \tag{7.54}$$

The other parameters are the same in the two regimes, which are $a = 1$, $b = 0.5$, $q = 0.4$, $d = 0.05$, and $r = 0.2$. The ensemble size is 500, which is sufficient for this scalar case to eliminate most of the sampling errors.

Panels (a) and (b) show the time evolution of the Kalman gain $K(t)$ and the cross-covariance $P^{xh}(t)$, respectively, for these two regimes. In Regime I (strongly observed case), although the observation operator has the same structure in both cases, the magnitude of the cross-covariance $P^{xh}(t)$ is significantly smaller in Regime II, resulting in a reduced Kalman gain.

Panels (c) and (d) compare the true state trajectory with the posterior ensemble mean for the two regimes, respectively. The associated uncertainty, characterized by the posterior two standard deviations around the posterior mean, is also included. In the strongly observed regime (Regime I), the posterior mean closely tracks the true state. In contrast, a much larger error between the posterior mean and the truth is found in the weakly observed regime (Regime II). Specifically, the posterior mean is close to the prior mean, which is zero. This means the observations provide minimal correction to the prior error, consistent with the small Kalman gain in this regime. These results demonstrate that, even when the state-observation relationship is structurally similar, the magnitude of the state-observation cross-covariance, and hence the Kalman gain, plays a crucial role in determining the effectiveness of the EnKBF update. This is also related to the effective observability discussed in Section 7.2.2.

7.4.3 Continuous-time conditional Gaussian nonlinear filter

For a special class of nonlinear systems, known as conditional Gaussian nonlinear systems (CGNS), exact filtering equations can be derived [66]. Consider the following system:

$$d\mathbf{x} = \big(\mathbf{a_0}(\mathbf{y}, t) + \mathbf{a_1}(\mathbf{y}, t)\mathbf{x}\big)\,dt + \mathbf{\Sigma_x}(\mathbf{y}, t)\,d\mathbf{W_x}, \tag{7.55}$$

$$d\mathbf{y} = \big(\mathbf{A_0}(\mathbf{y}, t) + \mathbf{A_1}(\mathbf{y}, t)\mathbf{x}\big)\,dt + \mathbf{\Sigma_y}(\mathbf{y}, t)\,d\mathbf{W_y}, \tag{7.56}$$

where a time series of $\mathbf{y}$ is available as observations. In (7.55)–(7.56), the matrices $\mathbf{A_1}$, $\mathbf{a_1}$, $\mathbf{\Sigma_y}$ and $\mathbf{\Sigma_x}$, as well as the vectors $\mathbf{A_0}$ and $\mathbf{a_0}$ can all be nonlinear functions of $\mathbf{y}$. The key insight is that although the full system is nonlinear, the dynamics of $\mathbf{x}$ are conditionally linear and Gaussian once the trajectory of $\mathbf{y}$ is given. As a result, conditioned on the history of observations $\mathcal{U}_t = \{\mathbf{y}(s) : 0 \le s \le t\}$, the posterior distribution of $\mathbf{x}(t)$ remains Gaussian:

$$p(\mathbf{x}(t) \mid \mathcal{U}_t) = \mathcal{N}(\hat{\mathbf{x}}(t), \mathbf{P}(t)), \tag{7.57}$$

where $\boldsymbol{\mu}(t)$ and $\mathbf{R}(t)$ satisfy the following differential equations:

$$\mathrm{d}\hat{\mathbf{x}} = \mathbf{a}_0 + \mathbf{a}_1\hat{\mathbf{x}} + \mathbf{K}\left(\mathrm{d}\mathbf{y} - (\mathbf{A}_0 + \mathbf{A}_1\hat{\mathbf{x}})\,\mathrm{d}t\right), \tag{7.58a}$$

$$\mathrm{d}\mathbf{P} = \left(\mathbf{a}_1\mathbf{P} + \mathbf{P}\mathbf{a}_1^* + \boldsymbol{\Sigma}_\mathbf{x}\boldsymbol{\Sigma}_\mathbf{x}^* - \mathbf{K}\mathbf{A}_1\mathbf{P}\right)\mathrm{d}t, \tag{7.58b}$$

with $\mathbf{K}$ being the Kalman gain,

$$\mathbf{K} = \mathbf{P}\mathbf{A}_1^*(\boldsymbol{\Sigma}_\mathbf{y}\boldsymbol{\Sigma}_\mathbf{y}^*)^{-1}. \tag{7.59}$$

Thus, the conditional Gaussian framework provides an exact nonlinear filter for the unobserved state variables $\mathbf{x}$. Notably, the conditional Gaussian nonlinear filter is fundamentally different from the Kalman-Bucy filter, which applies only to fully linear systems. In fact, the Kalman-Bucy filter can be viewed as a special case of the conditional Gaussian nonlinear filter in which all coefficients are independent of the state and observations. It is also worth noting that (7.58b) is a random Riccati equation since the coefficients of the matrices and vectors depend on the random time series $\mathbf{y}(t)$, which is different from the constant-coefficient Riccati equation in the Kalman-Bucy filter (7.45) (and the Kalman filter). Because the coefficients depend on the stochastic trajectory $\mathbf{y}(t)$, the Riccati equation is random. As a result, the covariance $\mathbf{P}(t)$ generally does not converge to a deterministic steady-state matrix, even if the underlying system is statistically stationary. This is a feature of nonlinear data assimilation and has been shown in Example 7.9.

From an intuitive perspective, CGNS describes a situation in which the full system is nonlinear, but a subset of variables, $\mathbf{y}$, is directly observed and serves as a time-dependent driver for the remaining variables, $\mathbf{x}$. The filtering problem can thus be interpreted as recovering the hidden components of the system (e.g., unresolved variables) by continuously observing a subset of state variables (e.g., variables at large-scale resolved grids). Importantly, this structure of the CGNS allows exact filtering without linearization or ensemble approximations.

Example 7.11 (Recovering the observed variable in the noisy Lorenz 63 model via the conditional Gaussian nonlinear filter). The noisy version of the Lorenz 63 model reads:

$$\begin{aligned}\frac{\mathrm{d}x}{\mathrm{d}t} &= \sigma(y-x) + \sigma_x\dot{W}_x,\\ \frac{\mathrm{d}y}{\mathrm{d}t} &= x(\rho - z) - y + \sigma_y\dot{W}_y,\\ \frac{\mathrm{d}z}{\mathrm{d}t} &= xy - \beta z + \sigma_z\dot{W}_z.\end{aligned} \tag{7.60}$$

The noisy Lorenz 63 model in (7.60) fits into the CGNS framework by setting $\mathbf{y} = x$ and $\mathbf{x} = (y, z)^\mathrm{T}$ or $\mathbf{y} = (y, z)^\mathrm{T}$ and $\mathbf{x} = x$ in (7.55)–(7.56). For the first case,

$$\begin{aligned}&\mathbf{a}_0(\mathbf{y}) = \begin{pmatrix}0\\0\end{pmatrix}, \qquad \mathbf{a}_1(\mathbf{y}) = \begin{pmatrix}-1 & -x\\ x & -\beta\end{pmatrix}, \qquad \boldsymbol{\Sigma}_\mathbf{x} = \begin{pmatrix}\sigma_y & 0\\ 0 & \sigma_z\end{pmatrix},\\ &\mathbf{A}_0(\mathbf{y}) = -\sigma x, \qquad \mathbf{A}_1(\mathbf{y}) = \begin{pmatrix}\sigma & 0\end{pmatrix}, \qquad \boldsymbol{\Sigma}_\mathbf{y} = \sigma_x.\end{aligned} \tag{7.61}$$

For the second case,

$$\begin{aligned}&a_0(\mathbf{y}) = \sigma y, \qquad a_1(\mathbf{y}) = -\sigma, \qquad \Sigma_\mathbf{x} = \sigma_x,\\ &\mathbf{A}_0(\mathbf{y}) = \begin{pmatrix}-y\\ -\beta z\end{pmatrix}, \qquad \mathbf{A}_1(\mathbf{y}) = \begin{pmatrix}\rho - z\\ y\end{pmatrix}, \qquad \boldsymbol{\Sigma}_\mathbf{y} = \begin{pmatrix}\sigma_y & 0\\ 0 & \sigma_z\end{pmatrix}.\end{aligned} \tag{7.62}$$

Figure 7.6 includes a numerical illustration. Here, the standard parameters in the deterministic part $\sigma = 10$, $\rho = 28$, $\beta = 8/3$ are used while the noise coefficients are $\sigma_x = \sigma_y = \sigma_z = 10$. Panel (a) shows the data assimilation result of the first case, in which x is the only observed variable. The recovered trajectory of y closely follows the truth, whereas the estimation error for z is more significant. The difference in recovering y and z can be explained by the structure of the dynamical coupling. The variable y enters directly into the evolution equation of the observed variable x. Therefore, the variations in y are immediately reflected in the observational time series and can be effectively incorporated into the filtering update. In contrast, the variable z influences the observations only indirectly, through its effect on y and subsequently on x. As observational information propagates through this multi-step dynamical pathway, uncertainty accumulates, and the effective information content available for estimating z decreases. Consequently, the posterior uncertainty and the estimation error for z are larger than those for y. Panel (b) shows the second case, in which both y and z are observed. Since the hidden variable x appears linearly and directly in the observation equations for (y, z), the observational information clearly plays a significant role in the state estimation of x. As a result, the recovery of x is comparatively accurate. As a final remark, it is seen that the posterior uncertainty varies over time in both cases, which again illustrates this feature of nonlinear data assimilation.

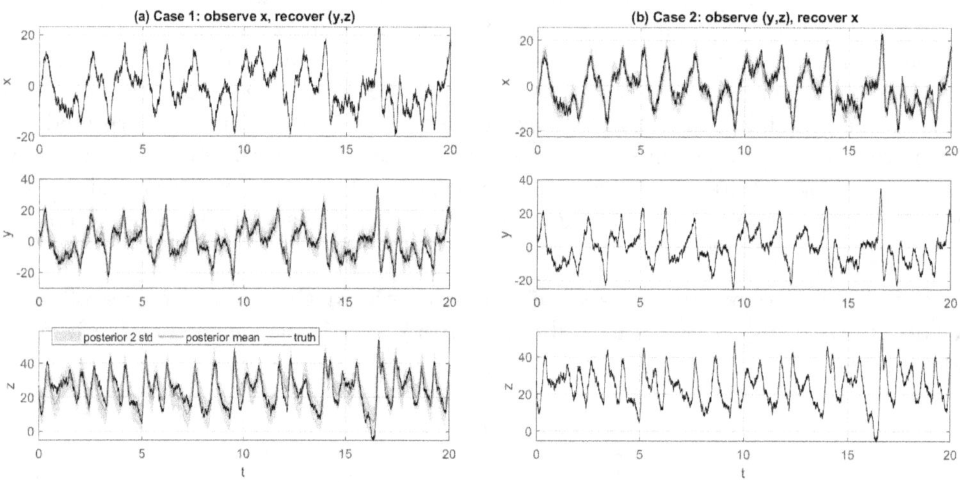

FIGURE 7.6: Recovering the unobserved state using the conditional Gaussian nonlinear filter. Panel (a): observing x and recovering y and z. Panel (b): observing y and z and recovering x. The true signals are presented by thin curves, while the posterior mean time series are shown in thicker curves. The shading areas indicate two posterior standard deviations.

□

Example 7.12 (Continuous-time Lagrangian data assimilation). Another example of continuous-time nonlinear data assimilation is Lagrangian data assimilation. Instead of relying on measurements taken at fixed locations, Lagrangian data assimilation uses the trajectories of moving tracers, such as drifters or floats, as observational data to infer the underlying flow field. It plays an important role in state estimation and prediction in geophysics, ocean and atmospheric sciences, and hydrology, especially when traditional Eulerian observations (e.g., from fixed stations) are sparse or unavailable. In many applications, tracer positions are recorded at time intervals that are much shorter than the characteristic time scale of the flow. As a result, the observational data can be naturally viewed as being available almost continuously in time, making Lagrangian data assimilation well-suited for a continuous-time formulation.

Interestingly, we will show that by appropriately designing an approximate forecast model for the underlying flow field, the continuous-time nonlinear Lagrangian data assimilation will fit into the conditional Gaussian nonlinear filtering framework. This provides a useful analytical tool for understanding the behavior of the Lagrangian data assimilation.

Let us consider a two-dimensional random flow field $\mathbf{u} = (u, v)^{\mathrm{T}}$ modeled by a finite number of Fourier modes with random amplitudes in a periodic domain $(0, 2\pi]^2$,

$$\mathbf{u}(\mathbf{x}, t) = \sum_{\mathbf{k}\in\mathbf{K}} \hat{v}_{\mathbf{k}}(t) \cdot e^{i\mathbf{k}\cdot\mathbf{x}} \cdot \mathbf{r}_{\mathbf{k}}, \tag{7.63}$$

where the Fourier wavenumber $\mathbf{k} = (k_1, k_2)^{\mathrm{T}}$ belongs to a finite set $\mathbf{K}$, $\hat{v}_{\mathbf{k}}(t)$ is the Fourier coefficient, $e^{i\mathbf{k}\cdot\mathbf{x}}$ the Fourier basis and $\mathbf{r}_{\mathbf{k}}$ the direction vector that connects the two velocity components. The velocity field modeled in (7.63) fits precisely into the general Fourier representation framework described in Section 5.1.2.

In general, the governing equation of $\hat{v}_{\mathbf{k}}(t)$ can be rigorously derived by applying the Fourier decomposition to the underlying flow model that is usually described by a complex PDE. The resulting equation of $\hat{v}_{\mathbf{k}}(t)$ is nonlinear if the underlying PDE is nonlinear. Nevertheless, since the process model used in data assimilation simply serves the purpose of forecasting, there is no need to require such a model to be exactly the same as the underlying dynamical system, which is even unknown in most cases. Following this logic, we can use a simplified forecast model in the Lagrangian data assimilation. One natural way is to adopt the following OU process to characterize each $\hat{v}_{\mathbf{k}}(t)$,

$$\frac{\mathrm{d}\hat{v}_{\mathbf{k}}(t)}{\mathrm{d}t} = -d_{\mathbf{k}}\hat{v}_{\mathbf{k}}(t) + f_{\mathbf{k}}(t) + \sigma_{\mathbf{k}}\dot{W}^{v}_{\mathbf{k}}(t). \tag{7.64}$$

Since the approximate forecast model mainly aims to capture the forecast statistics, using stochastic noise to replace complicated nonlinear terms is an effective way to significantly simplify the equation and help accelerate data assimilation. The details of using systematic stochastic tools to build reduced-order or simplified models will be presented in Chapter 10. The stochastic equation in (7.64), albeit linear, can mimic the statistical behavior of many turbulent flow fields. Therefore, it is an appropriate approximate forecast model in many scenarios.

On the other hand, the observations are given by the trajectories of L noisy Lagrangian tracers,

$$\frac{\mathrm{d}\mathbf{x}_l(t)}{\mathrm{d}t} = \mathbf{u}(\mathbf{x}_l(t), t) + \sigma_x\dot{W}^{x}_l(t) = \sum_{\mathbf{k}\in\mathbf{K}} \hat{v}_{\mathbf{k}}(t) \cdot e^{i\mathbf{k}\cdot\mathbf{x}_l(t)} \cdot \mathbf{r}_{\mathbf{k}} + \sigma_x\dot{W}^{x}_l(t), \tag{7.65}$$

where $l = 1, \ldots, L$. The equation in (7.65) is essentially the kinematic definition of velocity, which states that the time derivative of the displacement equals the velocity. The additional stochastic noise often represents the contribution to the velocity field from the unresolved small-scale modes that are truncated and excluded from the set $\mathbf{K}$. The contribution from these modes introduces uncertainty in characterizing the motion of the tracers. It is worth noting that $\mathbf{x}$ appears on the exponential function of the right-hand side of the observational process (7.65). Therefore, Lagrangian data assimilation is intrinsically nonlinear.

Collecting all the velocity components $\mathbf{U} = (\hat{v}_1, \ldots \hat{v}_{|\mathbf{K}|})^{\mathrm{T}}$ and all the tracer trajectories $\mathbf{X} = (x_{1,x}, x_{1,y}, \ldots, x_{L,x}, x_{L,y})^{\mathrm{T}}$, the abstract form of the Lagrangian data assimilation is given by [69, 64]

$$\begin{aligned}\text{Observations:} \quad & \frac{d\mathbf{X}}{dt} = \mathbf{N}(\mathbf{X})\mathbf{U} + \boldsymbol{\Sigma}_{\mathbf{X}}\dot{\mathbf{W}}_{\mathbf{X}}, \\ \text{Underlying flow:} \quad & \frac{d\mathbf{U}}{dt} = -\boldsymbol{\Gamma}\mathbf{U} + \mathbf{F}(t) + \boldsymbol{\Sigma}_{\mathbf{U}}\dot{\mathbf{W}}_{\mathbf{U}},\end{aligned} \tag{7.66}$$

where $\mathbf{N}(\mathbf{X})$ contains all the exponential terms on the right-hand side of (7.65). The goal is to compute the posterior distribution $p(\mathbf{U}(t)|\mathbf{X}(s \le t))$. Once the Fourier coefficients in $\mathbf{U}(t)$ are estimated, they can be uniquely mapped to recover the velocity field in physical space via (7.63). Note that without any observations, the best estimation of the flow field at any time instant is provided by the equilibrium distribution of the underlying flow field, which can be computed by running the underlying flow field model using an ensemble method or by analytic formulae if the system is analytically solvable. However, because most flow field models are strongly turbulent, the forecast state usually contains high uncertainties. With observed tracer trajectories, uncertainty can be reduced, and the estimate is guided by the information they provide.

The Lagrangian data assimilation problem in (7.66) is nonlinear due to the term $\mathbf{N}(\mathbf{X})$. Nevertheless, it belongs to the CGNS (7.55)–(7.56) when the forecast model for the flow field is modeled by a linear stochastic system, which allows the use of closed analytic formulae to find the time evolution of the posterior distribution. □

As a final remark, although the conditional Gaussian nonlinear data asssimilation presented here is for the continuous-time case, there is a discrete-time analogue for such a nonlinear data assimilation framework [171].

7.5 Approximate computational methods

While the nonlinear extensions of the Kalman filter discussed above work well, and some of them provide optimal estimates under certain assumptions, their computational cost can become prohibitive for high-dimensional systems. This challenge has motivated the development of approximate data assimilation methods that trade some degree of statistical optimality for improved computational efficiency and scalability. In this section, we will discuss two widely used approximate approaches: three-dimensional variational data assimilation (3DVAR) and nudging (also known as Newtonian relaxation). In addition, we briefly highlight the role of simplified or approximate forecast models in making data assimilation feasible for large-scale systems.

7.5.1 3DVAR assimilation

The 3DVAR method formulates data assimilation as an optimization problem that seeks the state $\mathbf{x}$ that minimizes a cost function balancing the model background and observations:

$$J(\mathbf{x}) = \frac{1}{2}(\mathbf{x} - \mathbf{x}^b)^{\mathtt{T}}\mathbf{B}^{-1}(\mathbf{x} - \mathbf{x}^b) + \frac{1}{2}(\mathbf{y} - \mathcal{H}(\mathbf{x}))^{\mathtt{T}}\mathbf{R}^{-1}(\mathbf{y} - \mathcal{H}(\mathbf{x})), \tag{7.67}$$

where $\mathbf{x}^b$ is the background state (typically a short-term forecast). The matrix $\mathbf{B}$ is the background covariance matrix, which characterizes the statistical structure of uncertainty in the background state. In practical 3DVAR implementations, $\mathbf{B}$ is often prescribed in advance. It is frequently assumed to be time-invariant and constructed based on the equilibrium statistics of the system, simplified models, or empirical tuning. The vector $\mathbf{y}$ is the observation vector, $\mathcal{H}$ is the observation operator, and $\mathbf{R}$ is the observation error covariance matrix.

The minimizing solution $\mathbf{x}^a$ satisfies:

$$\mathbf{x}^a = \mathbf{x}^b + \mathbf{K}(\mathbf{y} - \mathcal{H}(\mathbf{x}^b)), \tag{7.68}$$

where the gain matrix $\mathbf{K}$ is given by:

$$\mathbf{K} = \mathbf{B}\mathbf{H}^{\mathrm{T}}(\mathbf{H}\mathbf{B}\mathbf{H}^{\mathrm{T}} + \mathbf{R})^{-1}, \tag{7.69}$$

with $\mathbf{H}$ being the linearized observation operator.

Example 7.13 (3DVAR as a steady-state Kalman filter). For a linear time-invariant system, if the background covariance $\mathbf{B}$ is chosen as the steady-state (i.e., equilibrium) solution of the Kalman filter covariance equation, then 3DVAR becomes equivalent to the steady-state Kalman filter. Consider a linear system with constant parameters. The steady-state Kalman filter gain $\mathbf{K}_\infty$ satisfies:

$$\mathbf{K}_\infty = \mathbf{P}_\infty\mathbf{H}^{\mathrm{T}}(\mathbf{H}\mathbf{P}_\infty\mathbf{H}^{\mathrm{T}} + \mathbf{R})^{-1}, \tag{7.70}$$

where $\mathbf{P}_\infty$ is the steady-state forecast covariance. If we set $\mathbf{B} = \mathbf{P}_\infty$ in 3DVAR, the methods become identical. This connection explains why 3DVAR often performs well in practice when the system statistics are approximately stationary. □

In essence, 3DVar can be viewed as a special case of the Kalman filter in which the forecast covariance matrix $\mathbf{P}_k^f$ is replaced by a predefined constant background covariance matrix $\mathbf{B}$. Note that in most of the nonlinear data assimilation methods, the error covariance evolves dynamically as the system propagates forward in time. This time-dependent covariance reflects the dynamical changes of the forecast uncertainty and therefore plays a central role in determining how observational information is transferred into state corrections. Thus, using a pre-determined constant background covariance inevitably introduces error. The main computational advantage of 3DVAR over the nonlinear extensions of the Kalman filter is that it avoids the expensive covariance propagation step, where only the posterior mean needs to be updated. It, therefore, provides significant savings, especially for high-dimensional systems. Yet, this comes at the cost of using a static background covariance $\mathbf{B}$ that does not reflect the time-dependent variation of the uncertainties.

7.5.2 Nudging (Newtonian relaxation)

Nudging is one of the simplest and most intuitive data assimilation methods, in which observations are incorporated by adding a relaxation term to the model equations that drives the state toward the observed values:

$$\frac{\mathrm{d}\mathbf{x}}{\mathrm{d}t} = \mathbf{f}(\mathbf{x}, t) + \mathbf{G}\big(\mathbf{y} - \mathcal{H}(\mathbf{x})\big), \tag{7.71}$$

where $\mathbf{G}$ is the nudging gain matrix that controls the strength of the relaxation toward the observations. In many cases, $\mathbf{G}$ is chosen as a scalar multiple of the identity matrix. More generally, $\mathbf{G}$ may be a full matrix, allowing for different nudging strengths across state variables and cross-variable coupling, in which case the method can be viewed as closely related to continuous-time variational data assimilation, such as 3DVAR. Due to its mathematical form, nudging is also known as Newtonian relaxation. The method also admits a discrete-time formulation.

Example 7.14 (Optimal nudging for linear systems). For the linear system (7.42)–(7.43), the nudging formulation can be directly related to the Kalman-Bucy filter (7.44). In particular, when the nudging gain is chosen as $\mathbf{G} = \mathbf{P}\mathbf{H}^{\mathrm{T}}\mathbf{R}^{-1}$, where $\mathbf{P}$ is the state error covariance, the nudging equation becomes identical to the Kalman-Bucy state update. In this sense, nudging with this choice of gain is optimal and recovers the continuous-time Kalman filter. However, in practice the covariance $\mathbf{P}$ is unknown and must be evolved through a Riccati equation. As a result, nudging is typically implemented with a constant or empirically tuned gain matrix $\mathbf{G}$, based on physical insight or numerical experimentation, rather than the exact Kalman gain. For conditional Gaussian systems, a similar correspondence holds. When the covariance is held fixed, the nudging term in (7.58) coincides exactly with the Kalman gain term in the corresponding filtering equations. □

Nudging has several computational advantages. First, it is straightforward to implement, requiring only minor modifications to the existing model code. Second, due to the predetermined nudging coefficient, the data assimilation process does not require matrix inversions or covariance calculations, making the method extremely efficient. In addition, the relaxation term can have a stabilizing effect on the model dynamics, thereby enhancing numerical stability. However, the method has clear drawbacks. Without flow-dependent covariance information, the gain matrix $\mathbf{G}$ cannot optimally weight observations according to their uncertainties. The continuous formulation of nudging also does not naturally handle asynchronous observations. Finally, if applied too strongly, nudging can suppress physically realistic variability and introduce dynamical imbalances.

7.5.3 Model approximation for efficient data assimilation

Beyond algorithmic simplifications such as 3DVAR and nudging, another powerful and widely used method for improving computational efficiency in data assimilation is the development of suitable approximate models that are significantly simpler and faster to run than the original system. These approximate models are designed to reproduce the essential forecast statistics of the full model. Therefore, they can be used to generate accurate prior distributions in place of the original forecast model. When the forecast statistics of the original system are approximately Gaussian, a linear stochastic model such as an OU process can be a natural and effective surrogate. In other situations, approximate models can be employed for the entire data assimilation procedure (the so-called end-to-end approach) rather than only for the forecast step. One natural strategy is to simplify a general nonlinear system into a CGNS. In such a case, closed-form analytical expressions are available for efficient nonlinear data assimilation, eliminating the need for ensemble-based methods. The systematic development of approximate models will be discussed in Chapter 10.

7.6 Multi-model assimilation

In many applications, several imperfect models are available for the same physical system. Each model may have different resolutions, physical parameterizations, or missing physics, so that no single model can be trusted on its own. At the same time, observations are typically sparse, noisy, and only partially constrain the state. Multi-model data assimilation (MM-DA) provides a systematic Bayesian framework for integrating all available dynamical information with observations for optimal state estimation [24, 289].

7.6.1 Basic idea

Denote by $\mathbf{x} \in \mathbb{R}^n$ the true state, and let $\mathbf{x}_f^{(m)}$ be the forecast from model m mapped into a common reference space. Assuming unbiased Gaussian errors, the MM-DA posterior covariance $\mathbf{P}_a$ and mean $\mathbf{x}_a$ are

$$\mathbf{P}_a^{-1} = \sum_{m=1}^{M} \mathbf{Q}_m^{-1} + \mathbf{H}^\top \mathbf{R}^{-1}\mathbf{H}, \tag{7.72}$$

$$\mathbf{x}_a = \mathbf{P}_a \left(\sum_{m=1}^{M} \mathbf{Q}_m^{-1} \mathbf{x}_f^{(m)} + \mathbf{H}^\top \mathbf{R}^{-1}\mathbf{y} \right), \tag{7.73}$$

where $\mathbf{Q}_m$ is the forecast-error covariance of model m, and $\mathbf{R}$ is the observation-error covariance. These expressions assume that the forecast errors of the different models are mutually independent and independent of the observational errors. If cross-model error correlations are present, additional off-diagonal covariance terms must be included in the precision matrix. These expressions directly generalize the classical Kalman filter update: when $M = 1$, the formulae reduce exactly to the standard analysis covariance $\mathbf{P}_a^{-1} = \mathbf{Q}^{-1} + \mathbf{H}^\top \mathbf{R}^{-1}\mathbf{H}$ and the usual posterior mean $\mathbf{x}_a = \mathbf{P}_a(\mathbf{Q}^{-1}\mathbf{x}_f + \mathbf{H}^\top \mathbf{R}^{-1}\mathbf{y})$. Thus, MM-DA can be viewed as performing a Kalman update in which the "forecast" is replaced by a collection of independent forecasts, and the total forecast precision is simply the sum of the model precisions.

This interpretation also clarifies the weighting: models with large forecast uncertainty (large $\mathbf{Q}_m$) receive a small weight, while models with small forecast uncertainty carry a larger influence, and the observations act as an additional information source of weight $\mathbf{R}^{-1}$. When the models differ substantially in quality, MM-DA automatically downweights the less skillful models while still extracting useful information from all of them. Applications span numerical weather prediction, climate projections, multi-fidelity engineering models, and emerging combinations of physical models with machine-learning or data-driven surrogates.

Example 7.15 (Multi-model assimilation with two biased linear models). To illustrate the behavior of MM-DA under structural model biases, we consider a one-dimensional setting in which the true system and the forecast models all evolve according to OU dynamics. The true signal satisfies

$$\frac{\mathrm{d}x^{\text{true}}}{\mathrm{d}t} = -a_{\text{true}}\big(x^{\text{true}} - \mu_{\text{true}}\big) + \sigma_{\text{true}} \dot{W}_t, \qquad a_{\text{true}} = 0.4, \quad \mu_{\text{true}} = 1.0, \quad \sigma_{\text{true}} = 0.3. \tag{7.74}$$

Two imperfect models are available. Model 1 is biased low and uses the OU dynamics

$$\frac{\mathrm{d}x^{(1)}}{\mathrm{d}t} = -a_1\big(x^{(1)} - \mu_1\big) + \sigma_1 \dot{W}_t^{(1)}, \qquad a_1 = 0.2, \quad \mu_1 = -1.5, \quad \sigma_1 = 0.6, \tag{7.75}$$

whereas Model 2 uses the dynamics

$$\frac{\mathrm{d}x^{(2)}}{\mathrm{d}t} = -a_2\big(x^{(2)} - \mu_2\big) + \sigma_2 \dot{W}_t^{(2)}, \qquad a_2 = 0.6, \quad \sigma_2 = 0.7. \tag{7.76}$$

Figure 7.7 includes two experiments with different equilibrium mean of the second model, with $\mu_2 = 3.0$ and $\mu_2 = -3.0$, respectively. Noisy observations of the truth are collected every $\Delta t_{\text{obs}} = 0.5$ time units:

$$y_k = x^{\text{true}}(t_k) + \varepsilon_k, \qquad \varepsilon_k \sim \mathcal{N}(0, R), \qquad R = (0.2)^2. \tag{7.77}$$

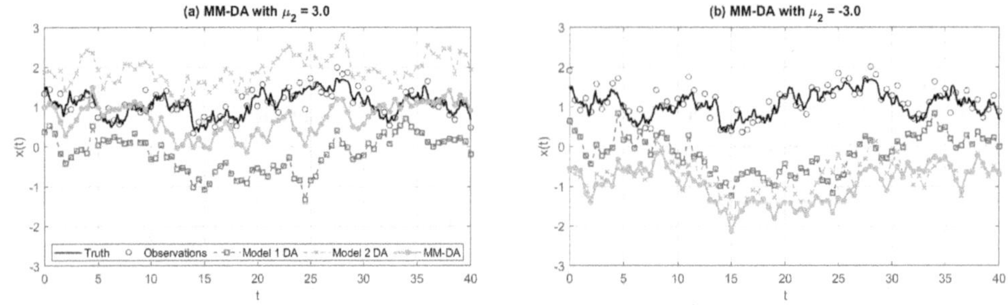

FIGURE 7.7: MM-DA using biased OU models. The truth is given by (7.74) and the observations (black circles) are taken every 0.5 time units with noise standard deviation 0.2. Two biased models are used for the MM-DA. Model 1 (squares) (7.75) is consistently biased low. Model 2 (7.76) uses two different equilibrium means. Panel (a): Model 2 with $\mu_2 = 3.0$, producing a strong positive bias. Panel (b): Model 2 with $\mu_2 = -3.0$, producing a strong negative bias.

Since all three systems (truth, Model 1, and Model 2) are linear OU processes with Gaussian noise, the posterior updates at observational instants can be computed exactly using Kalman filter formulae. In the single-model case, assimilating only model m together with the observation yields the posterior mean

$$x_{a,k}^{(m)} = \frac{x_k^{(m)}/Q_m + y_k/R}{1/Q_m + 1/R}, \qquad m = 1, 2, \tag{7.78}$$

where Q_m denotes the forecast-error variance of Model m at the observational time t_k and R is the observation-error variance. In contrast, the multi-model DA (MM-DA) update treats both model forecasts and the observation as independent information sources and performs a precision-weighted fusion:

$$x_k^{\text{MMDA}} = \frac{x_k^{(1)}/Q_1 + x_k^{(2)}/Q_2 + y_k/R}{1/Q_1 + 1/Q_2 + 1/R}. \tag{7.79}$$

These expressions indicate how MM-DA generalizes the usual Kalman update, in which each model and the observation contribute according to their estimated precisions.

Panel (a) of Figure 7.7 examines the case $\mu_2 = 3.0$, where the two models contain structurally different biases: Model 1 consistently underestimates the true state, whereas Model 2 strongly overestimates it. In this setting, each single-model DA produces an estimate lying between its model forecast and the observation, but still inherits a substantial portion of the underlying model bias. By contrast, the MM-DA estimate leverages the complementary, oppositely signed biases of the two models. The positive and negative deviations partially cancel in (7.79), allowing MM-DA to recover an analysis that is noticeably closer to the truth than either single-model DA.

Panel (b) considers the opposite situation with $\mu_2 = -3.0$, in which both models are biased low. Because the model errors now have the same sign, each single-model DA remains biased downward despite the observational correction. Moreover, the MM-DA estimate also shifts further downward and, at many observation times, performs worse than either single-model DA. This is consistent with the update formulae (7.78)–(7.79): MM-DA incorporates an additional biased model forecast, which increases the total forecast precision and therefore decreases the relative weight placed on the observation. Consequently, the observational information becomes less effective at correcting the shared model bias.

This example highlights a crucial practical topic. While MM-DA can improve state estimation when model errors differ structurally, it may fail to enhance and can even degrade performance when all participating models share similar biases. Choosing an appropriate model ensemble is thus crucial for successful multi-model assimilation.

□

7.6.2 Integrating models with different complexity and resolution

A central challenge in many practical scenarios is how to effectively integrate models of widely varying complexity and resolution to improve state estimation, prediction, and uncertainty quantification. High-resolution operational models provide detailed physical fields and have achieved remarkable success in many applications. However, they often exhibit persistent structural biases, particularly in representing extreme events, long-term statistics, and regime transitions. In contrast, coarse-grained idealized models isolate essential dynamical mechanisms, can be rigorously analyzed, and are often carefully calibrated to reproduce specific statistical or dynamical features with high fidelity. These complementary strengths naturally motivate the development of an MM-DA framework [29].

In many practical settings, especially in geophysics, observational data are often very limited. Consequently, one powerful strategy is to use the idealized model to generate high-quality surrogate time series of key variables. These surrogate data serve as pseudo-observations that provide constraints during assimilation into the high-resolution model. The technical challenge lies in finding a suitable state space in which the high-resolution model can be advanced forward while remaining meaningfully constrained by the information in the low-resolution surrogate. Machine learning–based latent representations provide a promising approach to constructing such a shared state space. A lower-dimensional latent space can be learned so that the operational model can propagate states efficiently and the latent coordinates retain the dominant physical information necessary for accurate multimodel coupling.

The resulting assimilated state inherits the rich spatial structure and full physical complexity of the operational model, while being globally informed by the calibrated statistical and dynamical accuracy of the idealized model. This hybrid product is uniquely valuable in both scientific and operational contexts. Large ensembles can be generated efficiently from the MM-DA, allowing robust uncertainty quantification and improved estimation of rare extreme events that are poorly sampled in observational records. Furthermore, the physically consistent, high-resolution output allows detailed process-level investigations, such as energy transfer, regime transition mechanisms, or teleconnection pathways, that would otherwise be obscured by biases in the original operational model. As a result, MM-DA serves as a powerful bridge between theoretical insight and high-resolution realism.

7.7 Smoother

7.7.1 Kalman smoother (Rauch-Tung-Striebel smoother)

To present the idea of smoothing, we will focus on the linear Gaussian case. A similar procedure can be generated for other nonlinear data assimilation cases [223]. For nonlinear cases, ensemble Kalman smoother can be naturally applied [98, 237].

The Kalman filter produces an online estimate of the state. At each time k, it uses only the observations available up to that time, $\mathbf{y}_{1:k} = \{\mathbf{y}_1, \ldots, \mathbf{y}_k\}$, and returns the filtered posterior distribution,

$$p(\mathbf{x}_k \mid \mathbf{y}_{1:k}) = \mathcal{N}(\hat{\mathbf{x}}_k^a, \mathbf{P}_k^a). \tag{7.80}$$

However, in many applications, we are interested in reconstructing the state over an entire time window after all observations in that window have been collected. This motivates the Kalman smoother (also known as the Rauch-Tung-Striebel smoother) [223], which computes the so-called smoothed posterior distribution,

$$p(\mathbf{x}_k \mid \mathbf{y}_{1:N}) = \mathcal{N}(\hat{\mathbf{x}}_k^s, \mathbf{P}_k^s), \qquad k = 0, 1, \ldots, N. \tag{7.81}$$

Intuitively, smoothing is "filtering plus hindsight". Future observations $\mathbf{y}_{k+1:N}$ help refine past states $\mathbf{x}_k$ by ruling out the filter-estimated possible states that are inconsistent with what happens later. Although smoother utilizes future information and is not for the initialization of the real-time forecast, it has many important applications. It is widely used for retrospective state reconstruction, which is called reanalysis in atmospheric and ocean science. It also provides improved estimates of hidden variables and parameters.

Recall the linear Gaussian system in (7.7)–(7.8). After running the Kalman filter forward in time, we have at each k:

$$\begin{aligned} &(\hat{\mathbf{x}}_k^f, \mathbf{P}_k^f) \quad \text{(forecast/prior)}, \\ &(\hat{\mathbf{x}}_k^a, \mathbf{P}_k^a) \quad \text{(analysis/filtered posterior)}. \end{aligned} \tag{7.82}$$

The smoother adds

$$(\hat{\mathbf{x}}_k^s, \mathbf{P}_k^s) \quad \text{(smoothed posterior)}. \tag{7.83}$$

Because the model describes how the state moves from time k to time $k+1$, knowing the state more accurately at a later time also tells us something about what the state must have been earlier. The Kalman smoother uses this idea by working backward in time, starting from the final estimate and refining earlier states.

The Kalman smoother starts with running a forward pass, which is exactly the Kalman filter. It is used to compute the filter posterior states in the entire interval $\{(\hat{\mathbf{x}}_k^f, \mathbf{P}_k^f), (\hat{\mathbf{x}}_k^a, \mathbf{P}_k^a)\}_{k=0}^N$. This is followed by running a backward pass to improve the state estimation. The initial value of this backward pass is the filter solution at the final time, where the filter and smoother estimates are the same:

$$\hat{\mathbf{x}}_N^s = \hat{\mathbf{x}}_N^a, \qquad \mathbf{P}_N^s = \mathbf{P}_N^a. \tag{7.84}$$

Then for $k = N-1, N-2, \ldots, 0$, define the smoother gain

$$\mathbf{G}_k = \mathbf{P}_k^a \mathbf{F}_{k+1}^{\mathtt{T}} \left(\mathbf{P}_{k+1}^f\right)^{-1}. \tag{7.85}$$

The Kalman smoother updates the mean and covariance as

$$\hat{\mathbf{x}}_k^s = \hat{\mathbf{x}}_k^a + \mathbf{G}_k\left(\hat{\mathbf{x}}_{k+1}^s - \hat{\mathbf{x}}_{k+1}^f\right), \tag{7.86}$$

$$\mathbf{P}_k^s = \mathbf{P}_k^a + \mathbf{G}_k\left(\mathbf{P}_{k+1}^s - \mathbf{P}_{k+1}^f\right)\mathbf{G}_k^{\mathtt{T}}. \tag{7.87}$$

The term $\hat{\mathbf{x}}_{k+1}^f = \mathbf{F}_{k+1}\hat{\mathbf{x}}_k^a + \mathbf{f}_{k+1}$ is what the model expected at time $k+1$ based on information up to time k. After seeing all future data, the smoother produces a refined estimate $\hat{\mathbf{x}}_{k+1}^s$. Their difference $\hat{\mathbf{x}}_{k+1}^s - \hat{\mathbf{x}}_{k+1}^f$ acts like a backward innovation, which is a mismatch between predicted and retrospectively inferred future state. The smoother gain $\mathbf{G}_k$ maps this mismatch back to time k through the dynamical coupling $\mathbf{F}_{k+1}$ and the uncertainty levels encoded in $\mathbf{P}_k^a$ and $\mathbf{P}_{k+1}^f$.

Similarly, (7.87) says that the smoothed uncertainty at time k equals the filtered uncertainty plus a correction that depends on how much the future uncertainty changed when going from $\mathbf{P}^f_{k+1}$ (one-step prediction) to $\mathbf{P}^s_{k+1}$ (hindsight-informed). Since conditioning on more data cannot increase uncertainty in the linear Gaussian setting, smoothing typically reduces uncertainty compared to filtering.

7.7.2 Fixed-lag and online smoother

The standard Kalman smoother uses all observations in a time window to improve past state estimates. In many practical situations, such as long simulations or streaming observations, waiting until the final time is either inconvenient or impossible. Moreover, because dynamical systems typically have finite memory, observations far in the future often provide little additional information for estimating the current state. Fixed-lag smoother addresses these issues by improving the estimate of the state at time k using only observations up to time $k + L$, where L is a prescribed lag [10]. Intuitively, the fixed-lag smoother allows the estimate to benefit from a limited amount of future information, capturing most of the smoothing gain while significantly reducing both computational cost and latency.

Online smoothers are closely related to fixed-lag smoothers and are designed for settings in which observations arrive sequentially and estimates must be updated continuously [292, 147]. Rather than waiting for all future data, an online smoother maintains a moving time window and repeatedly refines recent state estimates as new observations become available. States that fall outside the window are no longer updated and are treated as final. In this way, future information is still used to improve past estimates, but only over a limited horizon. It is worth noting that, although online smoothers operate sequentially, they should not be interpreted as real-time forecasting methods, since their estimates rely on observations from future times within the lag window. Compared with pure filtering, online smoothing provides more accurate state estimates, while remaining computationally efficient and suitable for long-running or near-real-time applications.

7.7.3 Continuous-time conditional Gaussian nonlinear smoother

Smoothing can also be developed for continuous-time data assimilation. Here, we will focus on the CGNS in Section 7.4.3. Recall that the Kalman-Bucy filter is a special case of the conditional Gaussian nonlinear filter when the model becomes linear and Gaussian, namely when all coefficients in (7.55)–(7.56) are independent of the observed variable $\mathbf{y}$. Similarly, the conditional Gaussian nonlinear smoother reduces to the classical Kalman-Bucy smoother for linear continuous-time systems. Thus, the Kalman-Bucy smoother is a special case of the conditional Gaussian nonlinear smoother within this more general nonlinear framework.

The conditional Gaussian nonlinear filter in Section 7.4.3 provides the optimal online estimate of the hidden state $\mathbf{x}(t)$ conditioned on the observation history $\mathcal{U}_t = \{\mathbf{y}(s) : 0 \leq s \leq t\}$. When the entire observation record on a finite interval $[0, T]$ is available, the state estimate can be further improved by incorporating future information. This leads to the conditional Gaussian nonlinear smoothing problem. It seeks the posterior distribution

$$p\big(\mathbf{x}(t) \mid \mathcal{U}_T\big) = \mathcal{N}\big(\hat{\mathbf{x}}^s(t), \mathbf{P}^s(t)\big), \qquad 0 \leq t \leq T. \tag{7.88}$$

which is Gaussian with $\hat{\mathbf{x}}^s(t)$ and $\mathbf{P}^s(t)$ denoting the smoothed mean and covariance, respectively.

The conditional Gaussian smoother follows a similar two-pass procedure as the Kalman smoother. First, the conditional Gaussian filter (7.58) is run forward in time on $[0, T]$ to obtain the filtered mean $\hat{\mathbf{x}}(t)$ and covariance $\mathbf{P}(t)$. Second, a backward system is solved to incorporate information from future observations. This backward correction removes state trajectories that are consistent with early observations but incompatible with later observations.

The smoothed mean $\hat{\mathbf{x}}^s(t)$ and covariance $\mathbf{P}^s(t)$ satisfy the following backward equations [171]:

$$\overleftarrow{\mathrm{d}\hat{\mathbf{x}}^s} = \left[-\mathbf{a}_0 - \mathbf{a}_1\hat{\mathbf{x}}^s + \mathbf{\Sigma}_{\mathbf{x}}\mathbf{\Sigma}_{\mathbf{x}}^* \mathbf{P}^{-1}(\hat{\mathbf{x}} - \hat{\mathbf{x}}^s)\right] \mathrm{d}t, \tag{7.89}$$

$$\overleftarrow{\mathrm{d}\mathbf{P}^s} = \left[- (\mathbf{a}_1 + \mathbf{\Sigma}_{\mathbf{x}}\mathbf{\Sigma}_{\mathbf{x}}^*\mathbf{P}^{-1})\mathbf{P}^s - \mathbf{P}^s(\mathbf{a}_1^* + \mathbf{\Sigma}_{\mathbf{x}}\mathbf{\Sigma}_{\mathbf{x}}^*\mathbf{P}^{-1}) + \mathbf{\Sigma}_{\mathbf{x}}\mathbf{\Sigma}_{\mathbf{x}}^*\right] \mathrm{d}t, \tag{7.90}$$

where $\hat{\mathbf{x}}(t)$ and $\mathbf{P}(t)$ are the filtered mean and covariance obtained from (7.58). The notation $\overleftarrow{\mathrm{d}(\cdot)}$ indicates that (7.89)–(7.90) are solved backward in time from $t = T$ to $t = 0$, with terminal conditions

$$\hat{\mathbf{x}}^s(T) = \hat{\mathbf{x}}(T), \qquad \mathbf{P}^s(T) = \mathbf{P}(T). \tag{7.91}$$

In (7.89)–(7.90), the forward filter collects information from past observations, while the backward equations propagate information from future observations through the conditionally linear dynamics of $\mathbf{x}$. As in filtering, the covariance equation is again a random Riccati equation because its coefficients depend on the observed trajectory $\mathbf{y}(t)$, and therefore there is typically no steady-state solution for the covariance.

Associated with the conditional Gaussian nonlinear smoother, a closed-form conditional sampling formula can be derived. While the smoother provides the optimal posterior mean $\hat{\mathbf{x}}^s(t)$ and covariance $\mathbf{P}^s(t)$ at each time instant, conditional sampling generates entire state trajectories that are consistent with both the temporal dynamics and the full observation history. These sampled trajectories preserve pathwise temporal correlations and can be regarded as continuous-time analogs of ensemble members in ensemble Kalman smoothers.

Conditioned on a given observation record $\mathcal{U}_T = \{\mathbf{y}(s) : 0 \leq s \leq T\}$, the optimal conditional sampling strategy for the hidden state $\mathbf{x}(t)$ satisfies the backward stochastic differential equation [15, 18],

$$\overleftarrow{\mathrm{d}\mathbf{x}} = \overleftarrow{\mathrm{d}\hat{\mathbf{x}}^s} - \left(\mathbf{a}_1 + \mathbf{\Sigma}_{\mathbf{x}}\mathbf{\Sigma}_{\mathbf{x}}^*\mathbf{P}^{-1}\right)(\mathbf{x} - \hat{\mathbf{x}}^s)\, \mathrm{d}t + \mathbf{\Sigma}_{\mathbf{x}}\, \mathrm{d}\mathbf{W}_{\mathbf{x}}(t), \tag{7.92}$$

where $\mathbf{W}_{\mathbf{x}}(t)$ is a Wiener process independent of the observation noise, and $\hat{\mathbf{x}}^s(t)$ and $\mathbf{P}(t)$ are given by the smoother and filter equations (7.89)–(7.90) and (7.58), respectively.

The sampling equation (7.92) ensures that, for each fixed time t, the marginal distribution of the sampled trajectories is exactly

$$\mathbf{x}(t) \mid \mathcal{U}_T \sim \mathcal{N}(\hat{\mathbf{x}}^s(t), \mathbf{P}^s(t)), \tag{7.93}$$

while simultaneously preserving the correct temporal dependence implied by the conditional dynamics. Compared with sampling based solely on the filter solution, smoother-based conditional sampling incorporates future observational information and therefore yields trajectories that are statistically closer to the true hidden states.

In practice, these conditionally sampled trajectories can be used for uncertainty quantification, nonlinear functionals of the state, rare-event diagnostics, and parameter estimation problems where pathwise consistency is essential.

Example 7.16 (Comparison of filtering, smoothing, and sampling). Consider the following dyad model:

$$\begin{aligned} \mathrm{d}u &= \left[(-d_u + cv)u + F_u\right] \mathrm{d}t + \sigma_u\, \mathrm{d}W_u, \\ \mathrm{d}v &= \left[- d_v v - cu^2 + F_v\right] \mathrm{d}t + \sigma_v\, \mathrm{d}W_v, \end{aligned} \tag{7.94}$$

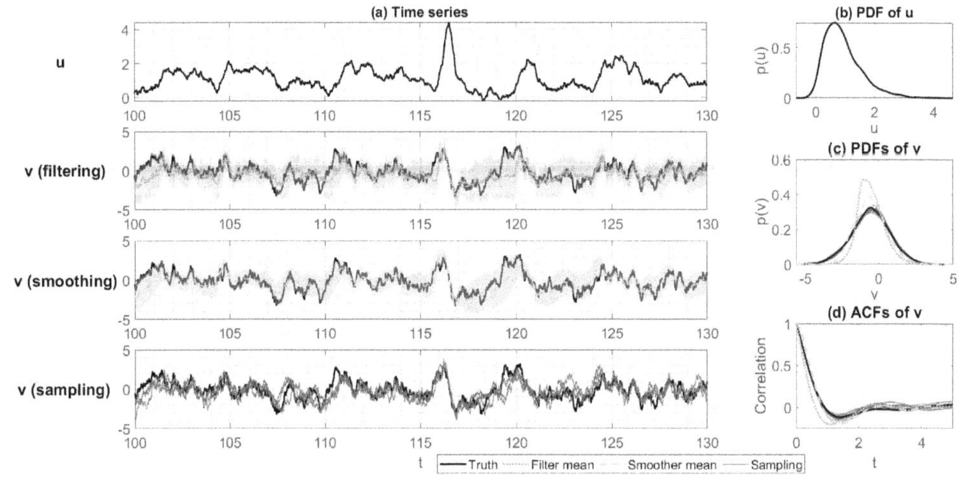

FIGURE 7.8: Comparison of filtering, smoothing, and sampling results of variable v in the dyad model (7.94) with parameters (7.95). Panel (a): time series. The truth is shown as a thick solid curve. The dotted and dashed curves show the filter and smoother mean states, while the shaded areas show the mean plus and minus two posterior standard deviations. The sampled trajectories of v are shown as thin curves. Panel (b): non-Gaussian PDF of the observed variable u. Panel (c): PDFs of v. Panel (d): ACFs of v.

where W_u and W_v are independent Wiener processes. We use the parameters

$$d_u = d_v = 0.8, \quad c = 1.2, \quad F_u = F_v = 1, \quad \sigma_u = 0.5, \quad \text{and} \quad \sigma_v = 2. \tag{7.95}$$

Given the observational time series of u, the conditional dynamics of v are linear with additive noise, and thus the conditional distribution of v is Gaussian. Therefore, the model fits into the conditional Gaussian data assimilation framework. With these parameters, the time series of u exhibits intermittent extreme events and a non-Gaussian PDF. Figure 7.8 compares the filtering, smoothing, and conditional sampling results for v.

Overall, the smoother mean captures the truth more accurately than the filter mean. In particular, the filter does not incorporate future information from the observed variable u, and it may miss the timing of the growth phase in v that triggers observed events (e.g., around $t = 116$). The posterior uncertainty in the smoother estimate is also smaller than that in the filter.

However, neither the filter mean nor the smoother mean is itself a typical realization of v, and therefore their time series generally do not reproduce the long-term statistics of the truth (e.g., PDFs and ACFs). In contrast, the smoother-based conditional sampling trajectories are genuine posterior trajectories conditioned on the full observation record, and an ensemble of such samples can recover the statistical features of v while remaining dynamically consistent with the observations.

□

7.7.4 Dynamical interpolation

Missing or sparse data appear in almost all scientific and engineering applications. Standard interpolation methods, such as linear, quadratic, or spline interpolation, rely solely on fitting deterministic functions directly to the data. These methods are effective for simple curves in low-complexity settings. However, they lack physical interpretability. In particular, they often fail to recover important dynamical characteristics, such as oscillatory behavior, intermittency, or multiscale variability.

One crucial application of data assimilation is to reconstruct such missing information through dynamical interpolation. Dynamical interpolation is typically implemented using the smoother data assimilation technique, in which a physical model provides a rough estimate of the states, while observations serve as constraints that correct the trajectory. This is similar to classical interpolation, such as linear interpolation, except that the rule of a simple deterministic linear function is replaced by a complex dynamical model that captures the underlying physics. As a result, the interpolated states from the smoother can often remain consistent with both the underlying dynamics and the available measurements.

Example 7.17 (Standard vs. dynamical interpolation in a noisy oscillatory system). Consider a stochastic harmonic oscillator,

$$\frac{\mathrm{d}}{\mathrm{d}t}\begin{pmatrix} x \\ y \end{pmatrix} = \begin{pmatrix} -\gamma & -\omega \\ \omega & -\gamma \end{pmatrix}\begin{pmatrix} x \\ y \end{pmatrix} + \sigma\begin{pmatrix} \dot{W}_x \\ \dot{W}_y \end{pmatrix}, \tag{7.96}$$

where $\omega = 1$ is the rotation frequency, $\gamma = 0.02$ is a small damping coefficient, and $\dot{W}_x$ and $\dot{W}_y$ are independent Gaussian white noises with noise coefficient $\sigma = 0.01$. The system evolves over $t \in [0, 6\pi]$ with a time step $\Delta t = 0.05$.

Noisy measurements of the full state are taken only once every 50 time steps ($\Delta t^{obs} = 2.5$) with the observational noise level being $r^o = 0.015$ for both the x and y components. Such sparse observations create long data gaps where standard interpolation methods may not work effectively.

Figure 7.9 compares the reconstruction from the three interpolation methods: (a) linear interpolation, (b) cubic spline interpolation, and (c) dynamical interpolation. In the linear interpolation, each state component is interpolated piecewise linearly using only the observation points. A standard cubic spline is used to obtain a twice continuously differentiable curve that fits the observed data points [241]. In the dynamical interpolation, a standard Kalman smoother is used to estimate the full trajectory by combining the stochastic model (7.96) and the noisy observations.

The top row of Figure 7.9 compares the true and interpolated $x(t)$ time series, while the bottom row shows these trajectories in phase space (x, y). Due to sparse observations, the resulting curves from linear and spline interpolations deviate significantly from the truth, especially in phase space, clearly indicating that these methods fail to capture the underlying dynamic structures. The spline is smoother than linear interpolation, but because it is purely data-fitted, it distorts the oscillatory structure, especially during long data gaps. In contrast, dynamical interpolation via the Kalman smoother enforces dynamical consistency with the damped oscillator dynamics, even in the presence of noise and sparse measurements.

□

Dynamical interpolation is closely connected to the concept of reanalysis in atmospheric and oceanic sciences. Reanalysis aims to reconstruct the historical evolution of the coupled Earth system by combining sparse and heterogeneous observations, such as satellite radiances, radiosonde profiles, buoy measurements, and aircraft reports, with numerical weather and climate models through advanced data assimilation schemes. Because observational coverage varies dramatically in space and time and many components of the Earth system are unobserved, reanalysis relies on the underlying dynamical equations (e.g., Navier-Stokes equations, thermodynamic constraints, and parameterized physical processes) to fill gaps in both space and time and to infer unmeasured variables. In this sense, reanalysis can be viewed as a large-scale form of dynamical interpolation: the governing physics and statistical

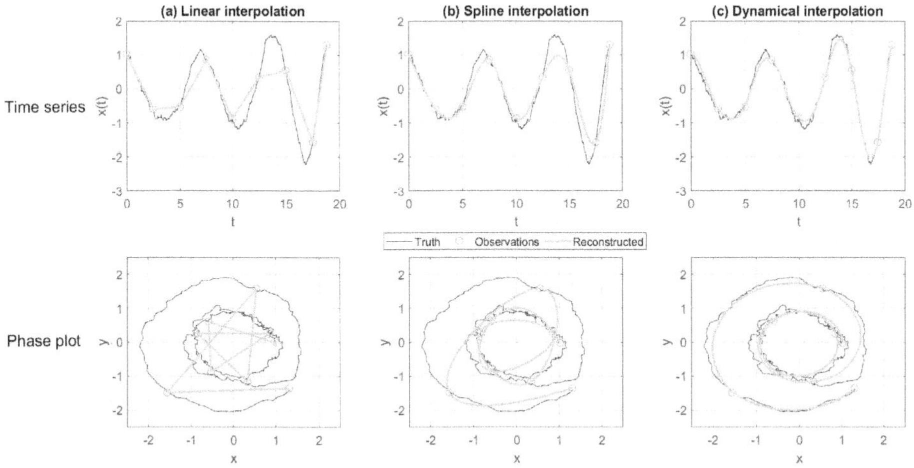

FIGURE 7.9: Comparison of three interpolation strategies ((a) linear interpolation, (b) cubic spline interpolation, and (c) dynamical interpolation) for a noisy damped oscillator. Top row: time series of $x(t)$. Bottom row: phase-space trajectories (x, y). Observations are sparse and corrupted by noise.

uncertainty models guide the interpolation between observations, leading to a gridded product that is physically consistent, multivariate, and globally complete. This interpretation highlights the broad utility of dynamical interpolation in geoscience, from small mechanistic examples to operational environmental reconstructions.

Example 7.18 (Advantages of dynamical interpolation in the marginal ice zone of the Arctic). The marginal ice zone (MIZ) is a broad transition region between compact sea ice and open ocean, where ice behaves in a brittle and highly discontinuous manner [276]. Individual floes drift, deform, and rotate in ways that cannot be adequately described by a continuum field. Lagrangian observations of these floes, extracted from satellite imagery, provide valuable information about coupled sea ice, ocean, and atmospheric processes. However, optical satellite images of the Arctic are frequently obscured by cloud cover, leading to substantial gaps in observed floe trajectories.

Traditional gap-filling methods face significant limitations. Simple linear interpolation cannot recover the curvature of floe trajectories, and even high-order spline methods fail to capture the underlying physics. Running a full operational ice–ocean model to fill missing data is computationally expensive. Nevertheless, appropriate reduced-order dynamical models that are both statistically accurate and computationally efficient can dramatically accelerate dynamical interpolation while maintaining high fidelity in the reconstructed floe motion.

A key advantage of dynamical interpolation is that it exploits the coupled ocean–ice dynamics. As a result, it not only reconstructs missing floe positions but also recovers the unobserved ocean field, a task that standard geometric interpolation methods cannot achieve. Moreover, by augmenting the state with key unknown parameters (e.g., floe thickness or drag coefficients), dynamical interpolation naturally provides parameter estimates alongside state reconstruction, which is a feature that again lacks in standard interpolation methods.

Dynamical interpolation has been applied to ice floe data in the Beaufort Sea [80]. The results demonstrate its success in reconstructing locations, curvatures, rotational motions, and strongly non-Gaussian statistics of missing floes. The results from dynamical interpolation can provide comprehensive datasets that advance our understanding of Arctic ocean and atmospheric dynamics. □

7.8 Model identification and parameter estimation

7.8.1 Parameter estimation via state augmentation

In many dynamical systems, certain physical parameters are not directly observed and are only partially known. Data assimilation provides a systematic framework for jointly estimating these unknown parameters with the state variables, while accounting for their uncertainties [99].

One widely used parameter estimation method via data assimilation is state augmentation, in which the parameters are treated as additional components of the state vector. Denote by $\mathbf{x}_k$ the original state vector and assume the dynamical system depends on an unknown parameter vector $\boldsymbol{\theta}$. The augmented state is defined as

$$\mathbf{z}_k = \begin{pmatrix} \mathbf{x}_k \\ \boldsymbol{\theta} \end{pmatrix}. \tag{7.97}$$

If the parameter is assumed constant in time, a common modeling choice is to prescribe a trivial dynamics [165],

$$\frac{\mathrm{d}\boldsymbol{\theta}}{\mathrm{d}t} = 0, \tag{7.98}$$

which reflects the assumption that the true parameter does not evolve dynamically. Importantly, this equation does not imply that the estimate of $\boldsymbol{\theta}$ remains fixed. Rather, parameter updates occur indirectly through the statistical coupling between the state variables and the parameters in the augmented system. To enable such updates, an initial uncertainty must be assigned to $\boldsymbol{\theta}$. If the initial parameter covariance is zero, then the parameters have no cross-covariance with the observed state variables, and observational information cannot be propagated to the parameter estimates. In contrast, a nonzero initial uncertainty allows the filter to exploit state-parameter correlations, so that observations of the state variables can progressively reduce the uncertainty in $\boldsymbol{\theta}$. Standard filtering methods applied to the augmented state can then jointly estimate both $\mathbf{x}_k$ and $\boldsymbol{\theta}$ in real time.

Below, we illustrate the parameter estimation method with the standard state augmentation (7.98) using a one-dimensional linear model.

Example 7.19 (Parameter estimation with the standard state augmentation). Consider the following scalar linear stochastic differential equation

$$\frac{\mathrm{d}x}{\mathrm{d}t} = -ax + f + \sigma \dot{W}_x, \tag{7.99}$$

where $a > 0$ is the damping parameter, f is an unknown constant forcing term, and $\sigma > 0$ is the noise coefficient. The state is observed at a discrete time instant through

$$y_k = x(t_k) + \eta_k, \qquad \eta_k \sim \mathcal{N}(0, r). \tag{7.100}$$

To estimate the unknown forcing f, we augment the state and treat f as a new state following (7.98). Some initial uncertainty is assigned to f such that the filter will help update

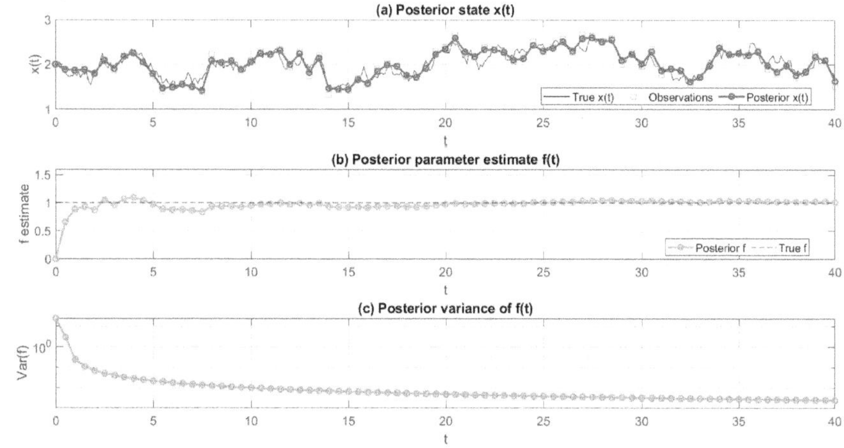

FIGURE 7.10: Parameter estimation using the state augmentation in (7.101). Panel (a): Posterior estimates of the state $x(t)$. Panel (b): Posterior estimates of the forcing f. Panel (c): Posterior variance of f, which decreases monotonically at observation times, reflecting increasing confidence in the parameter estimate.

f to eliminate the error due to its initial guessed value. The augmented dynamics become [63]

$$\frac{\mathrm{d}}{\mathrm{d}t}\begin{pmatrix} x \\ f \end{pmatrix} = \begin{pmatrix} -ax + f \\ 0 \end{pmatrix} + \begin{pmatrix} \sigma \\ 0 \end{pmatrix} \dot{W}_x. \tag{7.101}$$

Applying a standard Kalman filter to this system yields an evolving posterior distribution for (x, f) at each time instant. As more observations arrive, the estimator gradually identifies the correct forcing value f while simultaneously tracking the state x.

Figure 7.10 illustrates the parameter estimation results using the state augmentation in (7.101). The true parameter f is $f = 1$ while the other parameters are $a = 0.5$, $\sigma = 0.3$, and $r = 0.01$. The initial state and the uncertainty of the estimated f are 0 and 25, respectively, and the observational frequency is $\Delta t = 0.5$. Because the forcing f directly drives the dynamics of $x(t)$, the two variables are tightly coupled, and uncertainty about f directly propagates into uncertainty about x. Panel (a) shows that the posterior state estimate aligns well with the truth at the observation instants. Panel (b) demonstrates that, as more data become available, the filter identifies the true value of the forcing, and the parameter uncertainty decays steadily as seen in the posterior variance shown in Panel (c).

□

Example 7.20 (Convergence rate of a constant parameter under standard state augmentation). Let us fix a time step Δt and consider the linear SDE (7.99) in which the (unknown) forcing f is assumed to be constant in time over the assimilation window. In the standard state-augmentation approach, this modeling choice is encoded by the trivial parameter dynamics $f_{k+1} = f_k$ (equivalently $\mathrm{d}f/\mathrm{d}t = 0$), while f is treated as a random variable with an informative prior.

The exact discrete-time dynamics are

$$x_{k+1} = \phi x_k + \gamma f + w_k, \qquad w_k \sim \mathcal{N}(0, q), \tag{7.102}$$

where $\phi = e^{-a\Delta t}$, $\gamma = (1 - e^{-a\Delta t})/a$, and $q = \sigma^2(1 - e^{-2a\Delta t})/(2a)$. To isolate the parameter-learning mechanism, suppose for the moment that we observe the state exactly, namely,

$y_k = x_k$. Then the increments

$$\Delta x_k := x_{k+1} - \phi x_k = \gamma f + w_k, \qquad w_k \sim \mathcal{N}(0, q),$$

provide N independent linear measurements of f. Equivalently, $z_k := \Delta x_k/\gamma$ satisfies

$$z_k = f + \varepsilon_k, \qquad \varepsilon_k \sim \mathcal{N}\left(0, \frac{q}{\gamma^2}\right),$$

so that averaging N such measurements reduces the effective noise level by the usual $1/\sqrt{N}$ law.

With a Gaussian prior $f \sim \mathcal{N}(m_0, P_0)$, according to (7.16) and Example 8.16 in Chapter 8, the posterior is Gaussian with variance

$$P_N = \left(P_0^{-1} + N\gamma^2/q\right)^{-1}, \tag{7.103}$$

and the posterior mean is a weighted summation of the prior and the likelihood,

$$m_N = P_N\left(P_0^{-1}m_0 + \frac{\gamma}{q}\sum_{k=0}^{N-1}\Delta x_k\right) = (1-\lambda_N)m_0 + \lambda_N \bar{z}_N, \qquad \lambda_N = \frac{P_0}{P_0 + q/(\gamma^2 N)}, \tag{7.104}$$

where $\bar{z}_N = \frac{1}{N}\sum_{k=0}^{N-1} z_k$. Thus, m_N is a shrinkage estimator that interpolates between the prior mean and the sample average of the N noisy measurements of f. As $N \to \infty$, $\lambda_N \to 1$ and the data dominate the prior. Moreover,

$$P_N \sim \frac{q}{\gamma^2 N}, \qquad \text{so} \qquad \sqrt{\mathbb{E}(m_N - f)^2} = \sqrt{P_N} = \mathcal{O}(N^{-1/2}), \tag{7.105}$$

i.e., the posterior mean converges at the standard Monte Carlo (or central-limit) rate because information about f accumulates essentially linearly in the number of independent observation times.

With noisy observations $y_k = x_k + \eta_k$, the algebra is more complicated, but the same convergence rate in (7.105) holds. □

While (7.98) provides a natural and intuitive way to augment the state by treating the parameter vector $\boldsymbol{\theta}$ as constant, it is often useful in practice to incorporate additional prior knowledge about the statistics of $\boldsymbol{\theta}$ through a nontrivial evolution model [63]. A common choice is to assume that each component of $\boldsymbol{\theta}$, denoted by θ, follows an OU process,

$$\frac{\mathrm{d}\theta}{\mathrm{d}t} = -d_\theta\,(\theta - \hat{\theta}) + \sigma_\theta\,\dot{W}_\theta, \tag{7.106}$$

where $d_\theta > 0$ is a prescribed damping rate, $\hat{\theta}$ is a prescribed long-term mean, and σ_θ, which is also prescribed, controls the stochastic variability. This choice encodes the belief that θ should remain close to a reference value $\hat{\theta}$ while still allowing moderate fluctuations, which is used to determine the parameter value θ via the online filtering process.

The OU dynamics (7.106) have a well-known steady-state distribution,

$$\theta \sim \mathcal{N}\left(\hat{\theta}, \frac{\sigma_\theta^2}{2d_\theta}\right), \tag{7.107}$$

so prior knowledge of the mean or variance of θ can be built directly into the forecast model by choosing $(d_\theta, \hat{\theta}, \sigma_\theta)$ accordingly. Furthermore, the OU drift d_θ enforces an intrinsic exponential contraction toward $\hat{\theta}$, which stabilizes the augmented state dynamics and mitigates the risk of parameter drift in long assimilation windows. In contrast to the static prior $\mathrm{d}\theta/\mathrm{d}t = 0$, which yields only algebraic learning of θ from accumulated observational information, the OU prior introduces a physically meaningful memory time scale $1/d_\theta$ and leads to an exponential convergence of the posterior mean toward its equilibrium value once combined with the Kalman update. This makes the OU-based parameter evolution model particularly advantageous when partial prior information is available or when slow parameter variability is expected in the underlying physical system.

Example 7.21 (Parameter estimation with state augmentation using an informative prior). Still consider the linear model (7.99) and the linear observation (7.100). Now assume that f obeys an OU prior,

$$\frac{\mathrm{d}f}{\mathrm{d}t} = -d_f(f - \hat{f}) + \sigma_f \dot{W}_f. \tag{7.108}$$

The augmented dynamics become

$$\frac{\mathrm{d}}{\mathrm{d}t}\begin{pmatrix} x \\ f \end{pmatrix} = \begin{pmatrix} -ax + f \\ -d_f(f - \hat{f}) \end{pmatrix} + \begin{pmatrix} \sigma & 0 \\ 0 & \sigma_f \end{pmatrix}\begin{pmatrix} \dot{W}_x \\ \dot{W}_f \end{pmatrix}. \tag{7.109}$$

Figure 7.11 illustrates parameter estimation using the augmented system (7.109). In this experiment, the setup is the same as that in Example 7.19. The additional parameter values are $d_f = 0.1$, $\hat{f} = 0.5$, and $\sigma_f = 0.5$. The filter employs the prior model (7.108), which encodes prior knowledge in the equilibrium mean $\hat{f}$ and introduces stochastic variability through σ_f. Here, the prior mean $\hat{f}$ is chosen to be different from the truth, which allows us to introduce a prior bias in parameter estimation. The OU prior produces a competition between data-driven correction and the mean-reverting drift toward $\hat{f}$, so the posterior estimates of f in Panel (b) do not converge exactly to the truth as in Example 7.19. Instead, the time series of the estimated f fluctuates around the truth. Despite the prior bias, the observational data on x helps push the parameter f towards values near its true value. The posterior variance in Panel (c) decreases at each observation time but eventually stabilizes at a positive value reflecting the continual injection of uncertainty from the OU process. Despite this additional complexity, the posterior state estimates shown in Panel (a) remain accurate at the observation instants. In practice, the time-averaged value of f once it reaches statistical equilibrium can be used as an estimator of f. It is also worth noting that the convergence of the posterior variance of f is faster than the case in Example 7.19.

□

Example 7.22 (Convergence rate of the parameter with state augmentation using an informative prior). The OU process of f in (7.108) written in discrete time becomes

$$f_{k+1} = \phi_f f_k + \gamma_f + w_k^f, \qquad \phi_f = e^{-d_f \Delta t}, \;\; \gamma_f = (1 - \phi_f)\hat{f}, \;\; w_k^f \sim \mathcal{N}(0, q_f). \tag{7.110}$$

Along with the equation for x, these two equations lead to a linear Gaussian state-space model for the augmented state $(x_k, f_k)^\mathrm{T}$. The augmented Kalman filter covariance $\mathbf{P}_k$ then satisfies a Riccati recursion, as shown in (7.26),

$$\mathbf{P}_{k+1} = \mathbf{A}\mathbf{P}_k\mathbf{A}^\mathrm{T} + \mathbf{Q} - \mathbf{A}\mathbf{P}_k\mathbf{H}^\mathrm{T}\left(\mathbf{H}\mathbf{P}_k\mathbf{H}^\mathrm{T} + \mathbf{R}\right)^{-1}\mathbf{H}\mathbf{P}_k\mathbf{A}^\mathrm{T}, \tag{7.111}$$

where $\mathbf{A}$ and $\mathbf{Q}$ encode the coupled dynamics of x and f, and $\mathbf{H}, \mathbf{R}$ come from the observation model $y_k = x_k + \eta_k$. Under mild observability and stabilizability conditions, $\mathbf{P}_k$ converges to a unique steady-state covariance $\mathbf{P}_\infty$, and the mean error $\mathbf{e}_k = \mathbb{E}[\hat{\mathbf{z}}_k - \mathbf{z}_k]$ evolves approximately as

$$\mathbf{e}_{k+1} = (\mathbf{A} - \mathbf{K}_\infty\mathbf{H}\mathbf{A})\,\mathbf{e}_k, \tag{7.112}$$

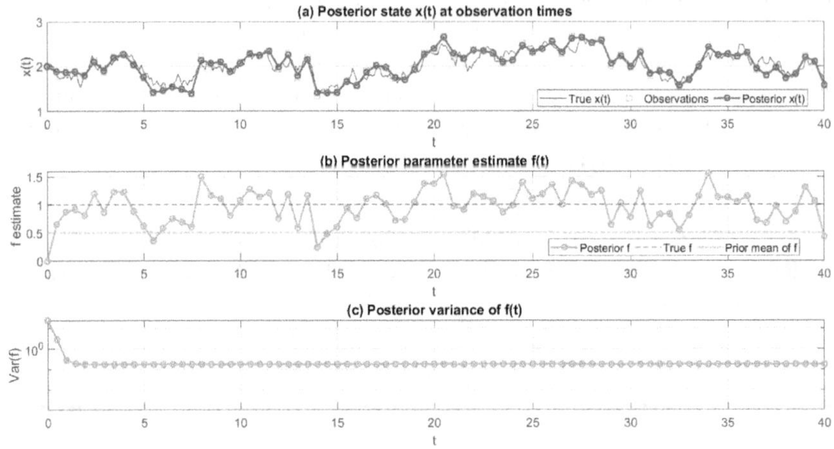

FIGURE 7.11: Parameter estimation using the augmented system (7.109). Panel (a): Posterior estimates of $x(t)$. Panel (b): Posterior estimates of the parameter f. Panel (c): Posterior variance of f decreases at observation times and approaches a nonzero steady level determined by the balance between observational information and the stochastic OU forcing.

with $\mathbf{K}_\infty$ the steady Kalman gain and $\mathbf{z}_k = (x_k, f_k)^\mathrm{T}$. The eigenvalues of $\mathbf{A} - \mathbf{K}_\infty \mathbf{HA}$ lie strictly inside the unit circle, so $|\mathbf{e}_k| \sim \exp(-\lambda_{\text{eff}} k \Delta t)$ for some $\lambda_{\text{eff}} > 0$. In particular, the f-component of the filter mean converges exponentially fast to its equilibrium value, with an effective rate λ_{eff} determined jointly by the OU drift d_f and the informativeness of the observations. This exponential convergence under the OU prior stands in sharp contrast to the algebraic $N^{-1/2}$ rate obtained in the static prior case and is validated by Panel (c) of Figure 7.11.

□

7.8.2 Parameter estimation via the EM algorithm

Different from state augmentation approaches, parameters in a dynamical model can also be estimated via regression-based methods such as least squares or by maximum likelihood estimation. When the system state is fully observed, these standard techniques can often be implemented directly. However, parameter estimation becomes substantially more challenging when the system is only partially observed, for example, the observations are noisy or only a subset of the state variables are available. In such cases, the likelihood of the observations involves an integral over all possible unobserved state trajectories, arising from the unobserved system states, which typically leads to a high-dimensional and intractable optimization problem. The expectation-maximization (EM) algorithm [197] provides an efficient framework for approximating the maximum likelihood estimate in this setting by iteratively alternating between state inference and parameter updates.

In a linear Gaussian model, the key advantage is that the conditional distribution of the state trajectory given the observations is Gaussian and can be computed exactly using a Kalman smoother. Each EM iteration consists of an E-step, in which a Kalman smoother is applied (using the current parameter values) to compute conditional means and covariances of the state trajectory, and an M-step, in which the parameters are updated by maximizing the expected complete-data log-likelihood evaluated under the smoothed distribution from the E-step. This process can also be generalized to the CGNS, which involves highly nonlinear systems, but closed analytic formulae remain available for the EM parameter estimation algorithm. For general nonlinear systems, ensemble methods are needed to estimate the unobserved states.

For the convenience of presentation, the general procedure of the EM algorithm is discussed using a discrete-time linear Gaussian system:

$$\mathbf{x}_{k+1} = \mathbf{F}(\boldsymbol{\theta})\,\mathbf{x}_k + \mathbf{u}(\boldsymbol{\theta}) + \boldsymbol{\xi}_k, \qquad \boldsymbol{\xi}_k \sim \mathcal{N}(\mathbf{0}, \mathbf{Q}), \tag{7.113}$$

$$\mathbf{y}_k = \mathbf{H}\,\mathbf{x}_k + \boldsymbol{\eta}_k, \qquad \boldsymbol{\eta}_k \sim \mathcal{N}(\mathbf{0}, \mathbf{R}), \tag{7.114}$$

where $\mathbf{F}(\boldsymbol{\theta})$ and $\mathbf{u}(\boldsymbol{\theta})$ are functions of the parameters $\boldsymbol{\theta}$. Only the noisy observations $\{\mathbf{y}_k\}_{k=0}^K$ are known. The state trajectory $\{\mathbf{x}_k\}_{k=0}^K$ and parameters $\boldsymbol{\theta}$ are unknown. The goal is to find the maximum likelihood estimate

$$\widehat{\boldsymbol{\theta}} = \arg\max_{\boldsymbol{\theta}} \log p(\mathbf{y}_{0:K} \mid \boldsymbol{\theta}). \tag{7.115}$$

The starting point of the EM algorithm is the incomplete-data log-likelihood, in which only the observations $\mathbf{y}_{0:K}$ are available,

$$\mathcal{L}(\boldsymbol{\theta}) := \log p(\mathbf{y}_{0:K} \mid \boldsymbol{\theta}). \tag{7.116}$$

This quantity is typically difficult to optimize directly because the unobserved states $\mathbf{x}_{0:K}$ have been marginalized out,

$$p(\mathbf{y}_{0:K} \mid \boldsymbol{\theta}) = \int p(\mathbf{x}_{0:K}, \mathbf{y}_{0:K} \mid \boldsymbol{\theta})\,\mathrm{d}\mathbf{x}_{0:K}. \tag{7.117}$$

The EM algorithm constructs a tractable lower bound via Jensen's inequality. Introduce an arbitrary density $q(\mathbf{x}_{0:K})$ over the latent variables and write

$$\log p(\mathbf{y}_{0:K} \mid \boldsymbol{\theta}) = \log \int q(\mathbf{x}_{0:K})\,\frac{p(\mathbf{x}_{0:K}, \mathbf{y}_{0:K} \mid \boldsymbol{\theta})}{q(\mathbf{x}_{0:K})}\,\mathrm{d}\mathbf{x}_{0:K}. \tag{7.118}$$

Applying Jensen's inequality yields the classical EM lower bound,

$$\log p(\mathbf{y}_{0:K} \mid \boldsymbol{\theta}) \;\geq\; \int q(\mathbf{x}_{0:K}) \log \frac{p(\mathbf{x}_{0:K}, \mathbf{y}_{0:K} \mid \boldsymbol{\theta})}{q(\mathbf{x}_{0:K})}\,\mathrm{d}\mathbf{x}_{0:K} \;=\; \mathcal{F}(q, \boldsymbol{\theta}), \tag{7.119}$$

where $\mathcal{F}(q, \boldsymbol{\theta})$ is sometimes called the free energy. The key observation is that this lower bound becomes tight when $q(\mathbf{x}_{0:K})$ is chosen as the posterior distribution under the current parameter estimate,

$$q^\star(\mathbf{x}_{0:K}) := p(\mathbf{x}_{0:K} \mid \mathbf{y}_{0:K}, \boldsymbol{\theta}^{(m)}). \tag{7.120}$$

Substituting this optimal $q^\star$ into the free-energy expression simplifies it dramatically, because the entropy term $-\int q^\star \log q^\star$ no longer depends on $\boldsymbol{\theta}$. The only remaining term that involves $\boldsymbol{\theta}$ is

$$\mathcal{Q}(\boldsymbol{\theta}; \boldsymbol{\theta}^{(m)}) := \int p(\mathbf{x}_{0:K} \mid \mathbf{y}_{0:K}, \boldsymbol{\theta}^{(m)})\, \log p(\mathbf{x}_{0:K}, \mathbf{y}_{0:K} \mid \boldsymbol{\theta})\,\mathrm{d}\mathbf{x}_{0:K}. \tag{7.121}$$

Equivalently, written in expectation form,

$$\mathcal{Q}(\boldsymbol{\theta}; \boldsymbol{\theta}^{(m)}) := \mathbb{E}\Big[\log p(\mathbf{x}_{0:K}, \mathbf{y}_{0:K} \mid \boldsymbol{\theta}) \,\Big|\, \mathbf{y}_{0:K}, \boldsymbol{\theta}^{(m)}\Big]. \tag{7.122}$$

where the expectation is taken with respect to the conditional distribution of $\mathbf{x}_{0:K}$ given the observations and the current parameter estimate $\boldsymbol{\theta}^{(m)}$. This function is the quantity maximized in the M-step. Hence, the EM algorithm can be interpreted as an alternating optimization of the tight lower bound:

$$\text{E-step: } q \leftarrow p(\mathbf{x}_{0:K} \mid \mathbf{y}_{0:K}, \boldsymbol{\theta}^{(m)}), \qquad \text{M-step: } \boldsymbol{\theta}^{(m+1)} \leftarrow \arg\max_{\boldsymbol{\theta}} \mathcal{Q}(\boldsymbol{\theta}; \boldsymbol{\theta}^{(m)}). \tag{7.123}$$

In other words, one EM iteration consists of:

- **E-step:** compute $\mathcal{Q}(\boldsymbol{\theta}; \boldsymbol{\theta}^{(m)})$ by evaluating the state of the unobserved variables using a Kalman smoother;
- **M-step:** update the parameters by

$$\boldsymbol{\theta}^{(m+1)} = \arg\max_{\boldsymbol{\theta}} \mathcal{Q}(\boldsymbol{\theta}; \boldsymbol{\theta}^{(m)}). \tag{7.124}$$

Because the model is linear and Gaussian, the conditional distribution $p(\mathbf{x}_{0:K} \mid \mathbf{y}_{0:K}, \boldsymbol{\theta}^{(m)})$ is Gaussian, and the Kalman smoother yields its means and covariances:

$$\widehat{\mathbf{x}}_k = \mathbb{E}[\mathbf{x}_k \mid \mathbf{y}_{0:K}, \boldsymbol{\theta}^{(m)}], \qquad \widehat{\mathbf{P}}_k = \operatorname{Cov}(\mathbf{x}_k \mid \mathbf{y}_{0:K}, \boldsymbol{\theta}^{(m)}). \tag{7.125}$$

These moments associated with the smoother estimate allow us to compute the expectations that appear in $\mathcal{Q}$.

Assume that the drift term can be written in the linear form regarding the parameters

$$\mathbf{F}(\boldsymbol{\theta})\,\mathbf{x}_k + \mathbf{u}(\boldsymbol{\theta}) = \mathbf{M}_k\,\boldsymbol{\theta} + \mathbf{S}_k, \tag{7.126}$$

where $\mathbf{M}_k$ and $\mathbf{S}_k$ are known matrices and vectors (possibly depending on $\mathbf{x}_k$ but not on $\boldsymbol{\theta}$). This is the case in many dynamical systems. Then the state equation becomes

$$\mathbf{x}_{k+1} = \mathbf{M}_k\,\boldsymbol{\theta} + \mathbf{S}_k + \boldsymbol{\xi}_k. \tag{7.127}$$

Under the Gaussian assumption, the contribution of the state equation to the complete-data log-likelihood is a quadratic form in the residual $\mathbf{x}_{k+1} - \mathbf{M}_k\boldsymbol{\theta} - \mathbf{S}_k$. Taking the conditional expectation with respect to the smoother distribution as the state is estimated by such a Gaussian distribution, one obtains the likelihood objective of the form

$$\mathcal{Q}(\boldsymbol{\theta}; \boldsymbol{\theta}^{(m)}) = -\frac{1}{2}\sum_{k=0}^{K-1} \Big\langle \big(\mathbf{x}_{k+1} - \mathbf{M}_k\boldsymbol{\theta} - \mathbf{S}_k\big)^{\mathrm{T}} \mathbf{Q}^{-1} \big(\mathbf{x}_{k+1} - \mathbf{M}_k\boldsymbol{\theta} - \mathbf{S}_k\big) \Big\rangle + \text{const}, \tag{7.128}$$

where $\langle\cdot\rangle$ denotes expectation with respect to $p(\mathbf{x}_{0:K} \mid \mathbf{y}_{0:K}, \boldsymbol{\theta}^{(m)})$. Differentiating with respect to $\boldsymbol{\theta}$ and setting the result to zero yields the linear system

$$\mathbf{D}\,\boldsymbol{\theta}^{(m+1)} = \mathbf{c}, \qquad \boldsymbol{\theta}^{(m+1)} = \mathbf{D}^{-1}\mathbf{c}, \tag{7.129}$$

with

$$\mathbf{D} = \sum_{k=0}^{K-1} \langle \mathbf{M}_k^{\mathrm{T}} \mathbf{Q}^{-1} \mathbf{M}_k \rangle, \qquad \text{and} \qquad \mathbf{c} = \sum_{k=0}^{K-1} \langle \mathbf{M}_k^{\mathrm{T}} \mathbf{Q}^{-1} (\mathbf{x}_{k+1} - \mathbf{S}_k) \rangle. \tag{7.130}$$

Here the expectations $\langle\cdot\rangle$ involve moments such as $\langle \mathbf{x}_{k+1} \rangle = \widehat{\mathbf{x}}_{k+1}$ and $\langle \mathbf{x}_{k+1}\mathbf{x}_{k+1}^{\mathrm{T}} \rangle = \widehat{\mathbf{x}}_{k+1}\widehat{\mathbf{x}}_{k+1}^{\mathrm{T}} + \widehat{\mathbf{P}}_{k+1}$.

Example 7.23 (Estimating the forcing parameter in a one-dimensional OU process using the EM algorithm).

Let us illustrate the EM algorithm using a simple scalar OU model. The exact time evolution from t_k to t_{k+1} can be written down explicitly:

$$x_{k+1} = \phi x_k + \gamma \theta + \sqrt{Q}\, \xi_k, \qquad \xi_k \sim \mathcal{N}(0, 1), \tag{7.131}$$

with known damping rate $a = 0.5$, time step $\Delta t = 0.05$, and process noise standard deviation $\sigma = 0.3$. The discrete-time coefficients are

$$\phi = e^{-a\Delta t}, \qquad \gamma = \frac{1 - e^{-a\Delta t}}{a}, \qquad Q = \frac{\sigma^2}{2a}\left(1 - e^{-2a\Delta t}\right). \tag{7.132}$$

The unknown parameter is the constant forcing θ, and the true value is $\theta_{\text{true}} = 1.0$. Noisy observations are available every 10 time steps,

$$y_k = x_k + \eta_k, \qquad \eta_k \sim \mathcal{N}(0, R), \tag{7.133}$$

where we consider two regimes:

$$\text{Regime 1: } R = 0.5^2, \qquad \text{Regime 2: } R = 1.5^2. \tag{7.134}$$

A larger observation noise variance reduces information content and slows EM convergence.

Figure 7.12 shows the EM parameter estimation algorithm with 30 EM iterations. The initial guess of the parameter $\theta^{(0)} = -3.0$ is set to be far from the truth. Panel (a) shows Regime I with $R = 0.5^2$. In such a case, the observations are sufficiently informative that the smoother estimate becomes accurate after only one or two iterations, causing the EM parameter estimate to jump rapidly to the vicinity of the truth. In contrast, Panel (b) shows Regime II with $R = 1.5^2$, which leads to noisier observations. The smoother estimate initially has a large uncertainty, and the EM updates absorb information only gradually, resulting in a slow but steady convergence of $\theta^{(m)}$ toward $\theta_{\text{true}} = 1$.

□

The above methodology is illustrated in the linear Gaussian case, where closed-form formulae are available for state estimation using the smoother. For general nonlinear dynamical systems, state estimation often requires ensemble methods and approximations to non-Gaussian conditional distributions using only the mean and covariance, which may introduce additional errors. Nevertheless, it is worth noting that closed-form analytic formulae exist for the nonlinear smoother estimates using the CGNS [60]. As was discussed in Section 7.4.3, the CGNS covers a wide range of nonlinear systems in realistic situations. Therefore, it is useful to generalize the above procedure to the CGNS. In the EM framework, the E-step uses this smoother to obtain the conditional means and covariances of the unobserved variables, while the M-step updates the parameters by maximizing the expected complete-data log-likelihood, which reduces to a quadratic optimization problem.

Example 7.24 (Exploiting EM algorithm to estimate parameters in a nonlinear system with the CGNS structure). Consider the dyad model in (7.94) with parameters $d_u, d_v, F_u, \sigma_u, \sigma_v$ fixed and (c, F_v) unknown. We observe $u(t)$ continuously (no additional measurement noise), while $v(t)$ is unobserved.

In the dyad model, we treat the drift parameters $\boldsymbol{\theta} = (c, F_v)^{\mathtt{T}}$ as unknown, while $d_u, d_v, F_u, \sigma_u, \sigma_v$ are fixed. After Euler-Maruyama discretization with step Δt, the (u, v)-equations are

$$\begin{aligned} u_{k+1} &= u_k + \left[(-d_u + c\, v_k)u_k + F_u\right]\Delta t + \sigma_u \sqrt{\Delta t}\, \xi_k, \\ v_{k+1} &= v_k + \left[-d_v v_k - c\, u_k^2 + F_v\right]\Delta t + \sigma_v \sqrt{\Delta t}\, \eta_k, \end{aligned} \tag{7.135}$$

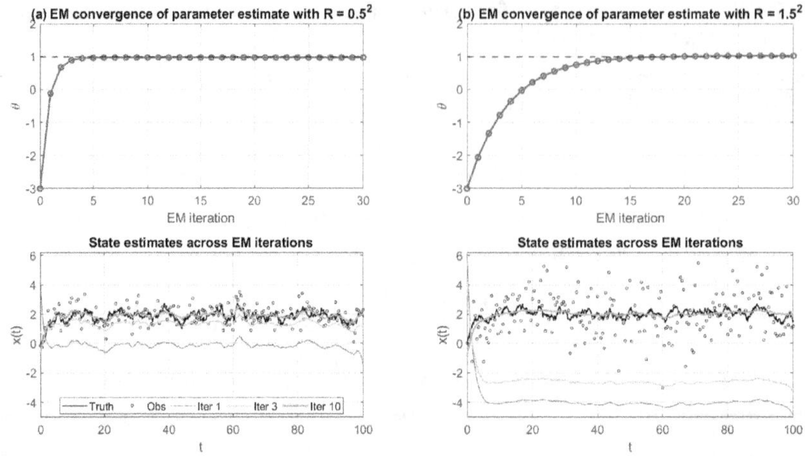

FIGURE 7.12: EM parameter estimation for a partially observed OU process. Top row: Convergence of the EM estimate of θ in the low-noise regime ($R = 0.5^2$) and high-noise regime ($R = 1.5^2$), both started from a poor initial guess $\theta^{(0)} = -3.0$. Bottom row: Smoothed state estimates at iterations 1, 2, and 30 plotted against the truth and noisy observations. Small observation noise yields near-instantaneous recovery of the latent trajectory and hence rapid convergence of θ, while large observation noise leads to slow but consistent improvement across iterations.

with independent standard Gaussians ξ_k, η_k. This can be written in the linear-in-parameter form

$$\mathbf{z}_{k+1} = \mathbf{M}_k\,\boldsymbol{\theta} + \mathbf{S}_k + \boldsymbol{\varepsilon}_k, \qquad \mathbf{z}_{k+1} = \begin{pmatrix} u_{k+1} \\ v_{k+1} \end{pmatrix}, \quad \boldsymbol{\varepsilon}_k \sim \mathcal{N}(\mathbf{0}, \mathbf{Q}), \tag{7.136}$$

where

$$\mathbf{M}_k = \begin{pmatrix} \Delta t\, u_k v_k & 0 \\ -\Delta t\, u_k^2 & \Delta t \end{pmatrix}, \qquad \mathbf{S}_k = \begin{pmatrix} u_k + (-d_u u_k + F_u)\Delta t \\ v_k + (-d_v v_k)\Delta t \end{pmatrix}, \qquad \mathbf{Q} = \begin{pmatrix} \sigma_u^2 \Delta t & 0 \\ 0 & \sigma_v^2 \Delta t \end{pmatrix} \tag{7.137}$$

In the EM algorithm, the angle brackets $\langle\cdot\rangle$ denote conditional expectations with respect to the smoothing distribution $p(v_k, v_{k+1} \mid u_{0:K}, \boldsymbol{\theta}^{(m)})$. Let

$$\hat{v}_k := \mathbb{E}(v_k \mid u_{0:K}, \boldsymbol{\theta}^{(m)}), \qquad P_k := \mathrm{Var}(v_k \mid u_{0:K}, \boldsymbol{\theta}^{(m)}), \tag{7.138}$$

so that, according to the Reynolds decomposition of a Gaussian random variable,

$$\langle v_k \rangle = \hat{v}_k, \qquad \langle v_k^2 \rangle = \hat{v}_k^2 + P_k. \tag{7.139}$$

With these definitions, the matrices in (7.129)–(7.130) specialize to

$$\mathbf{D} = \sum_{k=0}^{K-1} \langle \mathbf{M}_k^{\mathrm{T}} \mathbf{Q}^{-1} \mathbf{M}_k \rangle = \sum_{k=0}^{K-1} \Delta t \begin{pmatrix} \dfrac{u_k^2 \langle v_k^2 \rangle}{\sigma_u^2} + \dfrac{u_k^4}{\sigma_v^2} & -\dfrac{u_k^2}{\sigma_v^2} \\ -\dfrac{u_k^2}{\sigma_v^2} & \dfrac{1}{\sigma_v^2} \end{pmatrix}, \tag{7.140}$$

and

$$\mathbf{c} = \sum_{k=0}^{K-1} \left\langle \mathbf{M}_k^{\mathrm{T}} \mathbf{Q}^{-1} (\mathbf{z}_{k+1} - \mathbf{S}_k) \right\rangle. \tag{7.141}$$

Writing

$$r_{u,k} := u_{k+1} - u_k - (-d_u u_k + F_u)\Delta t, \qquad r_{v,k} := v_{k+1} - v_k + d_v v_k \Delta t, \tag{7.142}$$

we have

$$\mathbf{z}_{k+1} - \mathbf{S}_k = \begin{pmatrix} r_{u,k} \\ r_{v,k} \end{pmatrix}, \tag{7.143}$$

and a straightforward computation gives

$$c_1 := (\mathbf{c})_1 = \sum_{k=0}^{K-1} \left\langle \frac{\Delta t\, u_k v_k}{\sigma_u^2 \Delta t} r_{u,k} + \frac{-\Delta t\, u_k^2}{\sigma_v^2 \Delta t} r_{v,k} \right\rangle = \sum_{k=0}^{K-1} \left[\frac{u_k}{\sigma_u^2} r_{u,k}\, \hat{v}_k - \frac{u_k^2}{\sigma_v^2} \left(\hat{v}_{k+1} - \hat{v}_k + d_v \hat{v}_k \Delta t \right) \right],$$

$$c_2 := (\mathbf{c})_2 = \sum_{k=0}^{K-1} \left\langle \frac{\Delta t}{\sigma_v^2 \Delta t} r_{v,k} \right\rangle = \sum_{k=0}^{K-1} \frac{1}{\sigma_v^2} \left(\hat{v}_{k+1} - \hat{v}_k + d_v \hat{v}_k \right).$$

Here we used the fact that u_k and u_{k+1} are observed and thus deterministic, while all expectations act only on (v_k, v_{k+1}). Note that the posterior variance P_k enters $\mathbf{D}$ through $\langle v_k^2 \rangle = \hat{v}_k^2 + P_k$, so both the posterior mean and variance of the hidden process v_k explicitly contribute to the EM update (7.129).

The numerical experiment uses the dyad system with the parameters in (7.95). Only the coupling coefficient c and the forcing parameter F_v in the v equation are treated as unknown. The observed variable $u(t)$ exhibits a strongly nonlinear response to the hidden variable $v(t)$ through the term $(c\,v)u$ in the u equation, which triggers significant intermittency in the trajectory and causes extreme events. The hidden component $v(t)$ is influenced both by linear damping and the nonlinear term $-c\,u(t)^2$, making its evolution non-Gaussian in general. Although the marginal dynamics can be strongly non-Gaussian, CGNS ensures that the conditional distribution of $v(t)$ given the observed path $u_{0:T}$ remains Gaussian, enabling closed-form smoothing moments.

The results in Figure 7.13 show that the EM method converges rapidly and stably when applied to this partially observed dyad system. Starting from initial guesses ($c_0 - 0.2$ and $F_{v,0} = -1$), the parameter estimates quickly approach the true values $c_{\text{true}} = 1.2$ and $F_{v,\text{true}} = 1.0$. The time series plots highlight both the directly observed $u(t)$ and the recovery of the latent process $v(t)$, where the final smoother estimate closely matches the true hidden trajectory. This example illustrates that the EM algorithm can accurately estimate parameters and states for such nonlinear dynamical systems with only partial observations.

□

It is worth emphasizing that parameter constraints can be incorporated naturally into the EM framework without breaking the closed-form solvability of the M-step in either the linear Gaussian setting or the broader CGNS class [60]. Such constraints arise frequently in geophysical applications, where physical principles, such as conservation of energy, must be enforced during estimation. When these constraints are added, the maximization in the M-step becomes a constrained quadratic optimization problem whose solution satisfies the Karush-Kuhn-Tucker (KKT) conditions [45]. Importantly, the quadratic structure of the complete-data log-likelihood is preserved, so the constrained M-step still admits explicit analytic formulae. This allows EM-based parameter estimation to remain both computationally efficient and physically consistent, even in complex partially observed nonlinear systems.

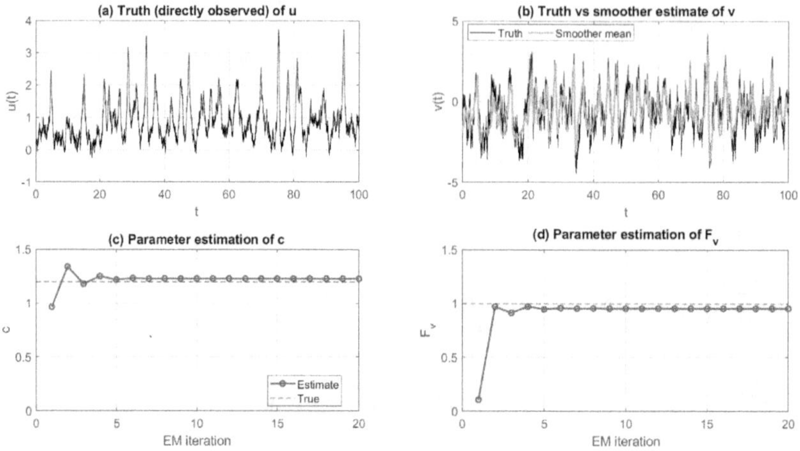

FIGURE 7.13: Parameter estimation for the dyad model (7.94) using the EM algorithm. Panel (a): Time series of the fully observed variable $u(t)$. Panel (b): True hidden variable $v(t)$ and its final smoother mean estimate. Panels (c)–(d): Convergence of the EM estimates of F_v and c, respectively.

7.9 Machine learning and data assimilation

With the rapid development of machine learning tools in recent years, they have become one of the dominant methods for improving the efficiency and accuracy of data assimilation [46, 20, 53].

Recall the two-step procedure of data assimilation: the forecast and the analysis. There are two typical ways to incorporate machine learning into data assimilation.

First, machine learning is widely used as a computationally efficient surrogate for complicated knowledge-based forecast models [116, 57, 115, 274, 207, 254, 46, 117, 53, 102, 38, 263, 191, 214, 212] or serves as a statistical correction to knowledge-based models that mitigates model error in the forecast step of data assimilation [108, 284, 222, 42, 100, 189, 101, 39]. It has also been used to optimize tuning parameters in ensemble data assimilation, such as the inflation rate of the covariance matrix [74, 266, 173]. Other studies focus on using machine learning to create specific ensemble members. For example, one method trains neural networks to generate ensembles with mass conservation and other constraints [231]. The work in [120, 290] utilizes analog ensemble data assimilation to create artificial ensemble members with variational autoencoders.

Second, machine learning has been applied more directly by building end-to-end learning schemes for the full data assimilation system [224, 43, 208, 190, 200, 44]. See a recent review paper [75] for more detailed examples of how machine learning facilitates data assimilation.

Recent studies have also integrated machine learning in stochastic systems. For example, a conditional Gaussian Koopman network has been developed [58, 59], which replaces the matrices and vectors in (7.56)–(7.55) with neural networks. The advantage is that the resulting neural stochastic differential equation can model many complex dynamical systems with flexibility, while the posterior solution remains analytically tractable.

Using machine learning to assist data assimilation is still a developing area. Combining machine learning with other tools continues to facilitate improvements in data assimilation.

7.10 Code listings for figures and examples

- `Chapter_7_Bayes.m`: generates Figure 7.1 in Example 7.2, which illustrates the posterior distributions under different prior and likelihood choices.
- `Chapter_7_Kalman_Filter.m`: generates Figure 7.2 in Example 7.5, which illustrates Kalman filter results under different observational noise levels.
- `Chapter_7_EKF.m`: generates Figure 7.3 in Example 7.8, which demonstrates the success and failure of the extended Kalman filter under different initial conditions.
- `Chapter_7_ETKF.m`: generates Figure 7.4 in Example 7.9, which illustrates the advantage of the ensemble Kalman filter over the extended Kalman filter for data assimilation in nonlinear dynamical systems.
- `Chapter_7_EnKBF.m`: generates Figure 7.5 in Example 7.10, which illustrates continuous-time data assimilation via the ensemble Kalman-Bucy filter.
- `Chapter_7_CGNF.m`: generates Figure 7.6 in Example 7.11, which applies a conditional Gaussian nonlinear filter to estimate the states of the Lorenz 63 model.
- `Chapter_7_Multi_Model_DA.m`: generates Figure 7.7 in Example 7.15, which illustrates multimodel data assimilation.
- `Chapter_7_CGNS_DA.m`: generates Figure 7.8 in Example 7.16, which presents the results of filtering, smoothing, and sampling based on a nonlinear dyad model.
- `Chapter_7_Dynamical_Interpolation.m`: generates Figure 7.9 in Example 7.17, which compares dynamical interpolation (via the Kalman smoother) with linear and spline interpolation in a noisy oscillatory system.
- `Chapter_7_Parameter_Estimation_Direct_Method.m`: generates Figure 7.10 in Example 7.19, which illustrates parameter estimation via standard state augmentation.
- `Chapter_7_Parameter_Estimation_Prior_Information.m`: generates Figure 7.11 in Example 7.21, which illustrates parameter estimation via state augmentation with prior information.
- `Chapter_7_Parameter_Estimation_EM_Linear.m`: generates Figure 7.12 in Example 7.23, which applies the expectation-maximization algorithm to estimate parameters in a partially observed linear Gaussian system.
- `Chapter_7_Parameter_Estimation_EM_CGNS.m`: generates Figure 7.13 in Example 7.24, which applies the expectation-maximization algorithm to estimate parameters in a partially observed nonlinear dyad model with conditional Gaussian structure.

8

Information Theory and Uncertainty Quantification

8.1 Information theory and measuring uncertainty

When uncertainty is present, the output of a system is no longer a single fixed value but can take different values with different probabilities. It is therefore natural to model the output as a random variable. The PDF of this random variable provides a quantitative description of how likely different outcomes are, and thus offers a natural way to characterize uncertainty.

Example 8.1 (Uncertainty due to inaccurate measurements). Consider a thermometer whose readings always exhibit some fluctuations. If we use it to measure room temperature multiple times, we will obtain slightly different results each time. By collecting these repeated measurements, we can numerically construct the PDF of the temperature readings. □

Qualitatively, the spread of a set of samples provides an intuitive measure of uncertainty. A wider spread indicates greater uncertainty. Quantitatively, for Gaussian distributions, the variance serves as a natural and complete metric for uncertainty, since it fully characterizes the spread of the PDF. More generally, uncertainty is encoded in the full PDF. Therefore, the measures of uncertainty should be defined in terms of the properties of the associated PDF of the samples. This finding motivates the use of information-theoretic quantities to systematically quantify the uncertainty.

8.1.1 Shannon entropy

Shannon entropy is widely used in uncertainty quantification [21, 203, 242]. It originates from communication theory, where a "word" is represented as a sequence of binary digits of length n. The set of all such words has $2^n = N$ elements [244]. Therefore, the amount of information needed to characterize one element, which is the number of digits, is $n = \log_2 N$. Now assume the full set is partitioned into disjoint subsets, where the i-th subset contains N_i elements. The probability that a randomly chosen element belongs to the i-th subset is $p_i = N_i/N$. Once the subset is known, the additional information needed to identify the element is $\log_2 N_i$. The average information required to determine an element is then

$$\begin{aligned}\sum_i \frac{N_i}{N}\log_2 N_i &= \sum_i \frac{N_i}{N}\log_2\left(\frac{N_i}{N}\cdot N\right)\\ &= \sum_i p_i \log_2 p_i + \log_2 N.\end{aligned} \tag{8.1}$$

Since $\log_2 N$ is the information required when no prior knowledge is available, the quantity $-\sum_i p_i \log_2 p_i$ represents the average lack of information, that is, the uncertainty. This

DOI: 10.1201/9781003738459-8

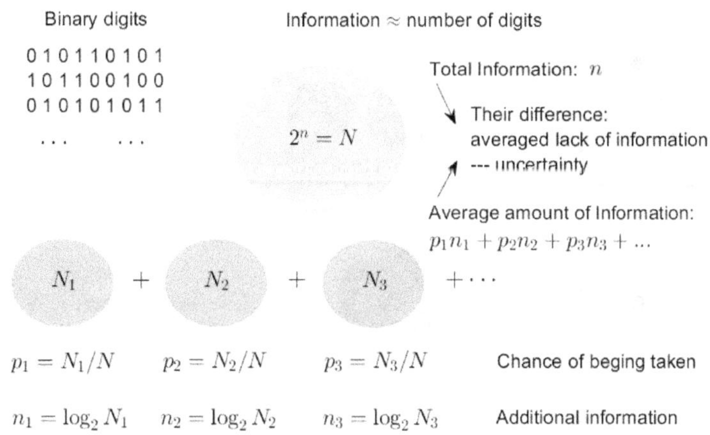

FIGURE 8.1: Illustration of Shannon entropy in the context of communication theory.

quantity is known as Shannon entropy. See Figure 8.1 for an illustration. The formal definition of Shannon entropy generalizes this idea by assigning a lack of information $-\ln(p(\mathbf{x}))$,to each possible outcome $\mathbf{x}$ and averaging over all outcomes. For discrete random variables, this average is a summation over probabilities, while for continuous random variables it becomes an integral over the probability density function. The choice of logarithm base only affects the units of entropy (e.g., bits for base 2 and nats for the natural logarithm).

Let $p(\mathbf{x})$ denote the probability mass function or probability density function of a random variable $\mathbf{X}$. Shannon entropy is defined as [119]:

$$\text{Continuous:} \qquad \mathcal{S}(p) = -\int p(\mathbf{x}) \ln p(\mathbf{x}), \mathrm{d}\mathbf{x}, \tag{8.2a}$$

$$\text{Discrete:} \qquad \mathcal{S}(p) = -\sum_i p_i \ln p_i. \tag{8.2b}$$

See [244, 3] for a more rigorous derivation of Shannon entropy and its uniqueness.

Example 8.2 (Quantifying the uncertainties via entropy). Figure 8.2 presents six probability distributions that demonstrate how Shannon entropy quantifies uncertainty. Each case shows how the predictability of outcomes varies, from complete certainty to maximal randomness, revealing how entropy measures the effectiveness of random guesses in matching actual outcomes. Shannon entropy (8.2b) for these cases can be computed by its definition:

$$\begin{aligned}
&\text{(a) } \mathcal{S}(p) = -1 \ln 1 && = 0\\
&\text{(b) } \mathcal{S}(p) = -\left(0.5 \ln(0.5) + 0.5 \ln(0.5)\right) && = 0.6931\\
&\text{(c) } \mathcal{S}(p) = -\left(1/3 \ln(1/3) + 1/3 \ln(1/3) + 1/3 \ln(1/3)\right) && = 1.0986\\
&\text{(d) } \mathcal{S}(p) = -\left(0.8 \ln(0.8) + 0.2 \ln(0.2)\right) && = 0.5004\\
&\text{(e) } \mathcal{S}(p) = -\left(0.9 \ln(0.9) + 0.1 \ln(0.1)\right) && = 0.3251\\
&\text{(f) } \mathcal{S}(p) = -\lim_{n\to\infty} n\left(\frac{1}{n}\ln\left(\frac{1}{n}\right)\right) = \lim_{n\to\infty} \ln n && = \infty
\end{aligned}$$

Shannon entropy captures uncertainty across distinct probability distributions. A deterministic case in Case (a) yields zero entropy, representing complete certainty. An equal-partition binary distribution in Case (b) introduces uncertainty. It maximizes entropy for two outcomes, which will be explained by the maximum entropy principle below. With three equally

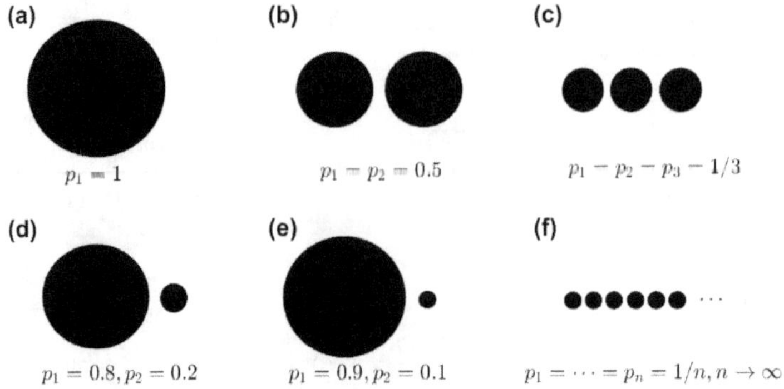

FIGURE 8.2: Computing Shannon entropy for systems with different probability distributions across partitions.

likely outcomes in Case (c), entropy increases further compared to (b) due to the more possible choices. A skewed binary distribution in Case (d) reduces entropy compared to (b), reflecting higher predictability. Extreme skewed Case (e) further lowers entropy, as it is closer to the deterministic case. Finally, the limiting case of uniform probabilities in Case (f) shows entropy diverging as uncertainty grows unbounded, corresponding to an infinite uncertainty. □

Gaussian distributions are widely used in practice. Therefore, it is natural to examine Shannon entropy when the underlying distribution is Gaussian.

Example 8.3 (Shannon entropy of Gaussian distributions). When $p \sim \mathcal{N}(\boldsymbol{\mu}, \mathbf{R})$ is a m-dimensional Gaussian distribution, where $\boldsymbol{\mu}$ and $\mathbf{R}$ are the mean and covariance, then Shannon entropy has the following form:

$$\mathcal{S}(p) = \frac{m}{2}(1 + \ln 2\pi) + \frac{1}{2}\ln\det(\mathbf{R}). \tag{8.3}$$

Intuitively, in the m-dimensional case, the entropy scales with the logarithm of the "volume" spanned by the covariance matrix $\mathbf{R}$, capturing how spread out the distribution is in all directions. In the one-dimensional situation, (8.3) implies that the uncertainty is uniquely determined by the covariance and is invariant under translations of the mean, consistent with the intuition. Specifically, the entropy increases with the variance, as a larger variance reflects greater dispersion and thus higher uncertainty.

The derivation of (8.3) is straightforward. Recall the m-dimensional Gaussian distribution

$$p(\mathbf{x}) = \frac{1}{(2\pi)^{m/2}\det(\mathbf{R})^{1/2}}\exp\left(-\frac{1}{2}(\mathbf{x}-\boldsymbol{\mu})^\top\mathbf{R}^{-1}(\mathbf{x}-\boldsymbol{\mu})\right). \tag{8.4}$$

Taking the logarithm of the above Gaussian distribution yields

$$\ln p(\mathbf{x}) = -\frac{m}{2}\ln(2\pi) - \frac{1}{2}\ln\det(\mathbf{R}) - \frac{1}{2}(\mathbf{x}-\boldsymbol{\mu})^\top\mathbf{R}^{-1}(\mathbf{x}-\boldsymbol{\mu}). \tag{8.5}$$

Taking the expectation with respect to $p(\mathbf{x})$, we obtain

$$\mathcal{S}(p) = -\mathbb{E}[\ln p(\mathbf{X})] = \frac{m}{2}\ln(2\pi) + \frac{1}{2}\ln\det(\mathbf{R}) + \frac{1}{2}\mathbb{E}\left[(\mathbf{X}-\boldsymbol{\mu})^\top\mathbf{R}^{-1}(\mathbf{X}-\boldsymbol{\mu})\right], \tag{8.6}$$

where we have used the fact that $\frac{m}{2}\ln(2\pi)$ and $\frac{1}{2}\ln\det(\mathbf{R})$ are constants and thus remain unchanged under expectation. For a Gaussian random variable, in light of the definition of covariance, the quadratic form (the last term on the right-hand side of (8.6)) satisfies

$$\mathbb{E}\left[(\mathbf{X}-\boldsymbol{\mu})^{\mathrm{T}}\mathbf{R}^{-1}(\mathbf{X}-\boldsymbol{\mu})\right] = \mathrm{tr}(\mathbf{R}^{-1}\mathbf{R}) = m. \tag{8.7}$$

Plugging (8.7) into (8.6) leads to the expression of Shannon entropy of a m-dimensional Gaussian distribution (8.3). □

For non-Gaussian PDFs, uncertainty quantification becomes more complicated. In such cases, features, such as outliers and extreme events, are better characterized by higher-order moments (e.g., kurtosis) rather than variance alone. Therefore, it is essential to use the full PDF for accurate uncertainty assessment. Shannon entropy serves as a particularly useful uncertainty quantification tool, providing an objective metric to quantify the uncertainty in a given PDF.

Example 8.4 (Shannon entropy of a Student's t-distribution). Consider a Student's t-distribution [251],

$$f(x;\nu) = \frac{\Gamma\left(\frac{\nu+1}{2}\right)}{\sqrt{\nu\pi}\,\Gamma\left(\frac{\nu}{2}\right)}\left(1+\frac{x^2}{\nu}\right)^{-\frac{\nu+1}{2}}, \tag{8.8}$$

where ν is the number of degrees of freedom, and Γ is the Gamma function defined in (5.118). For $\nu = 3$, the Student's t-distribution is symmetric about zero and exhibits characteristic heavy tails. Figure 8.3 compares this distribution with its Gaussian counterpart, where both distributions share identical mean and variance but differ markedly in their tail behavior. The entropies for the Student's t-distribution and its Gaussian counterpart are 1.773 and 1.946, respectively. Although the t-distribution exhibits heavier tails and therefore assigns higher probability to extreme events, entropy depends on the full shape of the distribution rather than tail behavior alone. The Gaussian distribution is known to maximize entropy among all distributions with a fixed mean and variance. Therefore, any other distribution with the same first two moments, including the t-distribution, must have equal or lower entropy. The reduction in entropy arises because the t-distribution redistributes probability mass away from the intermediate "shoulder" regions toward both the peak and the tails, resulting in a more structured (less maximally random) distribution overall. □

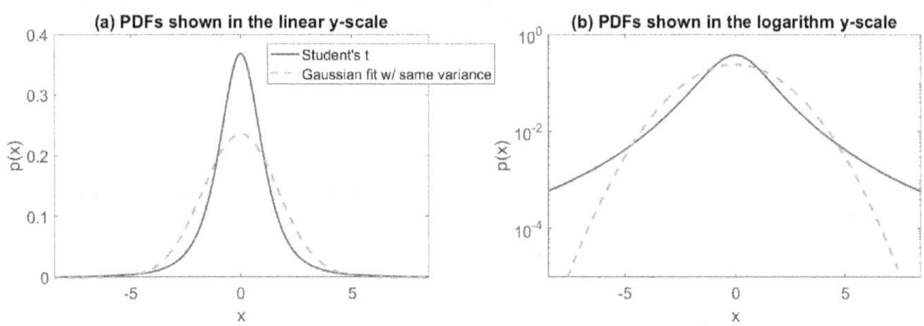

FIGURE 8.3: Comparison of the Student's t-distribution and a Gaussian distribution with matched variance. Panel (a): PDFs plotted on a linear scale, highlighting the heavier tails of the t-distribution. Panel (b): PDFs on a logarithmic scale, emphasizing tail behavior and the enhanced probability of extreme events in the t-distribution.

8.1.2 Principle of maximum entropy

Entropy is a fundamental concept not only in communication theory but also in thermodynamics, where it quantifies the degree of disorder in a system [109]. The second law of thermodynamics (Clausius' formulation) states that the entropy of an isolated system always increases toward a maximum. This physical principle parallels information-theoretic entropy, which measures uncertainty and decreases as additional information (constraints) about the system becomes available. These perspectives establish entropy maximization as a powerful principle for modeling complex systems with incomplete knowledge, that is, the most plausible statistical state is the one that satisfies all physical constraints while maximizing entropy [142].

The maximum entropy principle provides a natural framework for determining the optimal PDF subject to given constraints. It leads to a constrained optimization problem with entropy as the objective function to be maximized.

Example 8.5 (Least biased discrete distribution with no additional constraint). Consider determining the least biased discrete probability distribution p over the finite set $\{a_1, \ldots, a_n\}$, where no additional constraints are imposed.

While we refer to this as an "unconstrained" problem (meaning that there is no additional constraints imposed), the maximum entropy principle inherently requires two fundamental conditions:

$$\sum_{i=1}^{n} p_i = 1 \quad \text{(normalization)} \quad \text{and} \quad p_i \geq 0 \quad \text{(non-negativity)}. \tag{8.9}$$

This leads to the following constrained optimization problem:

$$\text{Maximize} \quad \mathcal{S}(p) = -\sum_{i=1}^{n} p_i \ln p_i, \qquad \text{Subject to} \quad \sum_{i=1}^{n} p_i = 1 \quad \text{and} \quad p_i \geq 0.$$

We solve this using the method of Lagrange multipliers. The Lagrangian function is:

$$\mathcal{L}(p, \lambda) = -\sum_{i=1}^{n} p_i \ln p_i + \lambda \left(\sum_{i=1}^{n} p_i - 1 \right),$$

where λ is the Lagrange multiplier associated with the normalization constraint. The optimality conditions are obtained by setting the partial derivatives to zero:

$$\frac{\partial \mathcal{L}}{\partial p_i} = 0 \quad \Rightarrow \quad -\ln p_i^* - 1 + \lambda = 0 \quad \Rightarrow \quad p_i^* = e^{\lambda - 1}.$$

By symmetry, all optimal probabilities p_i^* must be equal. Enforcing normalization yields the solution:

$$p_i^* = \frac{1}{n} \quad \text{(uniform distribution)}.$$

This uniform distribution represents the fairest possible assignment when no prior information is available, a direct consequence of the maximum entropy principle. □

Recall the results in Example 8.2, we notice that uniform distributions maximize entropy for systems with two and three outcomes (Cases (b) and (c), respectively). These findings are consistent with Example 8.5, confirming that uniform distributions maximize entropy when only normalization is imposed.

Example 8.6 (Least biased density given mean and variance). The least biased density p with the following constraints,

$$\mu = \int x p(x) \, \mathrm{d}x, \qquad R = \int (x - \mu)^2 p(x) \, \mathrm{d}x.$$

is the following Gaussian distribution:

$$p^*(x) = \frac{1}{\sqrt{2\pi R}} \exp\left(-\frac{(x-\mu)^2}{2R}\right). \tag{8.10}$$

The result in (8.10) complements the central limit theorem, which explains the frequent emergence of Gaussian distributions in nature. This is also one of the reasons that the Gaussian distribution is widely used in practice due to its maximum entropy property, making it the least informative (most unbiased) distribution for given moment constraints, while naturally emerging as the limiting distribution in many settings.

To derive the result in (8.10), we start by introducing Lagrange multipliers $\lambda_0, \lambda_1, \lambda_2$ and define the constrained functional

$$\begin{aligned} \mathcal{J}(p) = &-\int p(x)\ln p(x)\,\mathrm{d}x - \lambda_0\Big(\int p(x)\,\mathrm{d}x - 1\Big) \\ &- \lambda_1\Big(\int xp(x)\,\mathrm{d}x - \mu\Big) - \lambda_2\Big(\int (x-\mu)^2 p(x)\,\mathrm{d}x - R\Big). \end{aligned} \tag{8.11}$$

Unlike the finite-dimensional Lagrange multiplier method used in Example 8.5, which applies to functions of finitely many variables, the optimization problem here is posed over the space of probability density functions. Since both the entropy and the constraints depend on the entire function $p(x)$, they are functionals rather than ordinary functions. Consequently, optimality must be tested against all admissible infinitesimal perturbations of $p(x)$, which motivates the use of variational derivatives. Consider a variation $p \mapsto p + \varepsilon\eta$, where $\eta(x)$ is an arbitrary smooth function satisfying

$$\int \eta(x)\,\mathrm{d}x = 0. \tag{8.12}$$

The derivative with respect to ε therefore measures the first-order change of the functional $\mathcal{J}$ under perturbations of $p(x)$ and is known as the variational derivative. The first-order variation of $\mathcal{J}$ is given by

$$\frac{\mathrm{d}}{\mathrm{d}\varepsilon}\mathcal{J}(p+\varepsilon\eta)\bigg|_{\varepsilon=0} = -\int \eta(x)\big(\ln p(x)+1\big)\,\mathrm{d}x - \lambda_0\int \eta(x)\,\mathrm{d}x - \lambda_1\int x\eta(x)\,\mathrm{d}x - \lambda_2\int (x-\mu)^2\eta(x)\,\mathrm{d}x. \tag{8.13}$$

Since this variation must vanish for all admissible perturbations $\eta(x)$, we obtain the pointwise condition

$$\ln p(x) + 1 + \lambda_0 + \lambda_1 x + \lambda_2(x-\mu)^2 = 0. \tag{8.14}$$

Equivalently,

$$p(x) = \exp\big(-1 - \lambda_0 - \lambda_1 x - \lambda_2(x-\mu)^2\big). \tag{8.15}$$

This expression is Gaussian in x. Enforcing the mean constraint implies $\lambda_1 = 0$ in the centered coordinate $(x-\mu)$, while enforcing the variance constraint yields $\lambda_2 = 1/(2R)$. Finally, the normalization constraint determines the prefactor, giving the result in (8.10).

□

Example 8.6 shows that, for a fixed variance, the Gaussian distribution achieves higher entropy than any other distribution. This justifies the lower entropy in the fat-tailed Student's t-distribution in Example 8.4 compared to its Gaussian fit.

8.1.3 Relative entropy

In many practical applications, we seek to quantify the information loss when using an approximate distribution $p^M(\mathbf{x})$ instead of the true distribution $p(\mathbf{x})$. Typically, $p(\mathbf{x})$ represents a complete probabilistic model, while $p^M(\mathbf{x})$ corresponds to a reduced-order approximation that sacrifices some accuracy for computational efficiency.

Recall that, building upon Shannon entropy, we quantify information loss as follows. According to (8.1), the ignorance (information deficit) about $\mathbf{x}$ under the true distribution $p(\mathbf{x})$ is $\mathcal{I}(p(\mathbf{x})) = -\ln(p(\mathbf{x}))$. Its expected value yields Shannon entropy:

$$\langle \mathcal{I}(p(\mathbf{x})) \rangle = -\int p(\mathbf{x}) \ln(p(\mathbf{x}))\, \mathrm{d}\mathbf{x}. \tag{8.16}$$

For an approximate model $p^M(\mathbf{x})$, while the local ignorance measure remains $\mathcal{I}(p^M(\mathbf{x})) = -\ln\big(p^M(\mathbf{x})\big)$, its expectation must be computed under the true distribution:

$$\langle \mathcal{I}(p^M(\mathbf{x})) \rangle = -\int p(\mathbf{x}) \ln\big(p^M(\mathbf{x})\big)\, \mathrm{d}\mathbf{x}. \tag{8.17}$$

This expectation is taken under the true distribution because actual outcomes are generated by the true system $p(\mathbf{x})$, not by the model.

The relative entropy between distributions emerges from their difference [161, 160]:

$$\mathcal{P}(p, p^M) := \langle \mathcal{I}(p^M(\mathbf{x})) \rangle - \langle \mathcal{I}(p(\mathbf{x})) \rangle = \int p(\mathbf{x}) \ln\left(\frac{p(\mathbf{x})}{p^M(\mathbf{x})}\right) \mathrm{d}\mathbf{x}. \tag{8.18}$$

The relative entropy $\mathcal{P}(p, p^M)$ is also known as the Kullback-Leibler (KL) divergence and I-divergence [85]. It serves as an objective metric for model error and measures the expected increase in ignorance when using p^M instead of p to describe outcomes actually generated by p. Note that, if we instead used $p^M(\mathbf{x}) \ln\big(p^M(\mathbf{x})\big)$ to compute $\langle \mathcal{I}(p^M(\mathbf{x})) \rangle$ in (8.18), we would obtain the Shannon entropy difference, which is a quantity with limited utility for information loss quantification despite having other applications [288]. Analogous to (8.18), a corresponding definition also exists for discrete probability measures.

Example 8.7 (Computing the relative entropy between two simple distributions). Consider two simple probability distributions for a coin toss: (a) the true distribution p (fair coin): $p(\text{Heads}) = 0.5$, $p(\text{Tails}) = 0.5$, and (b) the model distribution q (biased coin): $q(\text{Heads}) = 0.9$, $q(\text{Tails}) = 0.1$. The relative entropy $\mathcal{P}(p, q)$ measures how much information is lost when using the biased model q to approximate the fair coin p:

$$\mathcal{P}(p, q) = \sum_{x \in \{\mathrm{H}, \mathrm{T}\}} p(x) \ln\left(\frac{p(x)}{q(x)}\right) = 0.5 \ln\left(\frac{0.5}{0.9}\right) + 0.5 \ln\left(\frac{0.5}{0.1}\right) \approx 0.5108 \text{ nats.}$$

This positive value indicates that the model q loses information by incorrectly assuming the coin is biased. If q matched p exactly, the relative entropy would be zero. □

Example 8.8 (Difference between relative entropy and Shannon entropy difference). Consider predicting outcomes for a soccer match between Alpha FC and Beta United. A well-informed model p might assign Alpha FC a 20% chance to win, 20% to draw, and 60% to lose. In contrast, an uninformed model p^M could incorrectly estimate an 60% chance for Alpha FC to win, 20% to draw, and 20% to lose. The Shannon entropy of both distributions is identical. This can be seen by computing the values using the definition in (8.2b):

$$\begin{aligned} \mathcal{S}(p) &= -\sum_i p_i \ln p_i = -(0.2 \ln 0.2 + 0.2 \ln 0.2 + 0.6 \ln 0.6) = 0.950, \\ \mathcal{S}(p^M) &= -\sum_i p_i^M \ln p_i^M = -(0.6 \ln 0.6 + 0.2 \ln 0.2 + 0.2 \ln 0.2) = 0.950, \end{aligned} \tag{8.19}$$

which leads to a misleading entropy difference of zero, i.e., $\mathcal{S}(p) - \mathcal{S}(p^M) = 0$. In contrast, the relative entropy correctly captures the biased probability assignments, quantifying the true information loss of the flawed model p^M:

$$\mathcal{P}(p, p^M) = \sum_i p_i \ln\left(\frac{p_i}{p_i^M}\right) = 0.2\ln\left(\frac{0.2}{0.6}\right) + 0.2\ln\left(\frac{0.2}{0.2}\right) + 0.6\ln\left(\frac{0.6}{0.2}\right) = 0.439. \quad (8.20)$$

□

Two crucial properties make the relative entropy particularly useful. First, $\mathcal{P}(p, p^M)$ is strictly positive unless $p = p^M$, and and increases as the discrepancy between the distributions grows. Second, it remains invariant under nonlinear coordinate transformations, making it a dimensionless measurement that's consistent across different units and coordinate systems used by various scientific communities.

Example 8.9 (Non-negativity of relative entropy). To show the non-negativity of relative entropy, we use Jensen's inequality. If f is a convex function and X is a real-valued random variable, then

$$\mathbb{E}[f(X)] \geq f(\mathbb{E}[X]). \quad (8.21)$$

In particular, the function $-\ln(\cdot)$ is convex. We denote by $\langle g\rangle$ the expectation of $g(x)$ with respect to $p(x)$, as was defined in Section 6.3.1: $\langle g\rangle \equiv \int g(x)\,p(x)\,\mathrm{d}x$. Assume $p(x) > 0$ whenever $p^M(x) > 0$, so that the ratio $p^M(x)/p(x)$ is well-defined. Then

$$\begin{aligned}\mathcal{P}(p, p^M) &= \int p(x)\ln\frac{p(x)}{p^M(x)}\,\mathrm{d}x = \left\langle -\ln\frac{p^M(x)}{p(x)}\right\rangle \\ &\geq -\ln\left\langle\frac{p^M(x)}{p(x)}\right\rangle = -\ln\int p(x)\frac{p^M(x)}{p(x)}\,\mathrm{d}x \\ &= -\ln\int p^M(x)\,\mathrm{d}x = -\ln 1 = 0,\end{aligned} \quad (8.22)$$

where the inequality in the second row comes from Jensen's inequality. □

Example 8.10 (Invariance of relative entropy under a change of variable). Let $y = T(x)$ be a one-to-one differentiable change of variable with $\frac{\mathrm{d}y}{\mathrm{d}x} \neq 0$. Define the corresponding densities in the y-variable by

$$\tilde{p}(y) = p(x)\left|\frac{\mathrm{d}x}{\mathrm{d}y}\right|, \qquad \tilde{q}(y) = q(x)\left|\frac{\mathrm{d}x}{\mathrm{d}y}\right|, \qquad x = T^{-1}(y). \quad (8.23)$$

Then

$$\mathcal{P}(p, q) = \int p(x)\ln\frac{p(x)}{q(x)}\,\mathrm{d}x = \int \tilde{p}(y)\ln\frac{\tilde{p}(y)\left|\frac{\mathrm{d}x}{\mathrm{d}y}\right|}{\tilde{q}(y)\left|\frac{\mathrm{d}x}{\mathrm{d}y}\right|}\,\mathrm{d}y = \int \tilde{p}(y)\ln\frac{\tilde{p}(y)}{\tilde{q}(y)}\,\mathrm{d}y = \mathcal{P}(\tilde{p}, \tilde{q}). \quad (8.24)$$

□

Similar to Shannon entropy, the relative entropy has more explicit expressions when both p and p^M are Gaussian distributions.

Example 8.11 (Relative entropy between two multivariate Gaussians). Let $p \sim \mathcal{N}(\boldsymbol{\mu}, \mathbf{R})$ and $p^M \sim \mathcal{N}(\boldsymbol{\mu}^M, \mathbf{R}^M)$ be m-dimensional Gaussian distributions with symmetric positive definite covariance matrices $\mathbf{R}$ and $\mathbf{R}^M$. The relative entropy has the explicit expression [180]:

$$\mathcal{P}(p, p^M) = \underbrace{\frac{1}{2}(\boldsymbol{\mu} - \boldsymbol{\mu}^M)^\top (\mathbf{R}^M)^{-1}(\boldsymbol{\mu} - \boldsymbol{\mu}^M)}_{\text{Signal}} + \underbrace{\frac{1}{2}\left[\mathrm{tr}(\mathbf{R}(\mathbf{R}^M)^{-1}) - m - \ln\det\big(\mathbf{R}(\mathbf{R}^M)^{-1}\big)\right]}_{\text{Dispersion}}. \tag{8.25}$$

The first terms is called the signal, which measures the mismatch between the means, weighted by the model precision $(\mathbf{R}^M)^{-1}$, while the second term is called the dispersion, which quantifies the discrepancy between covariance structures.

To derive (8.25), let us write the relative entropy as:

$$\mathcal{P}(p, p^M) \equiv \int p(\mathbf{x}) \ln \frac{p(\mathbf{x})}{p^M(\mathbf{x})}\, \mathrm{d}\mathbf{x} = \mathbb{E}_p\left[\ln p(\mathbf{X}) - \ln p^M(\mathbf{X})\right]. \tag{8.26}$$

Using the Gaussian log-density,

$$\ln p(\mathbf{x}) = -\frac{m}{2}\ln(2\pi) - \frac{1}{2}\ln\det(\mathbf{R}) - \frac{1}{2}(\mathbf{x} - \boldsymbol{\mu})^\top \mathbf{R}^{-1}(\mathbf{x} - \boldsymbol{\mu}), \tag{8.27}$$

and the analogous expression for $\ln p^M(\mathbf{x})$, where $\det(\cdot)$ denotes the determinant, we obtain

$$\ln \frac{p(\mathbf{x})}{p^M(\mathbf{x})} = \frac{1}{2}\ln\frac{\det(\mathbf{R}^M)}{\det(\mathbf{R})} - \frac{1}{2}(\mathbf{x}-\boldsymbol{\mu})^\top \mathbf{R}^{-1}(\mathbf{x}-\boldsymbol{\mu}) + \frac{1}{2}(\mathbf{x}-\boldsymbol{\mu}^M)^\top (\mathbf{R}^M)^{-1}(\mathbf{x}-\boldsymbol{\mu}^M). \tag{8.28}$$

Taking expectation with respect to p in (8.28) yields

$$\begin{aligned}\mathcal{P}(p, p^M) = \frac{1}{2}\ln\frac{\det(\mathbf{R}^M)}{\det(\mathbf{R})} &- \frac{1}{2}\mathbb{E}_p\left[(\mathbf{X} - \boldsymbol{\mu})^\top \mathbf{R}^{-1}(\mathbf{X} - \boldsymbol{\mu})\right] \\ &+ \frac{1}{2}\mathbb{E}_p\left[(\mathbf{X} - \boldsymbol{\mu}^M)^\top (\mathbf{R}^M)^{-1}(\mathbf{X} - \boldsymbol{\mu}^M)\right].\end{aligned} \tag{8.29}$$

For the second term, since $\mathbf{X} - \boldsymbol{\mu} \sim \mathcal{N}(\mathbf{0}, \mathbf{R})$,

$$\mathbb{E}_p\left[(\mathbf{X} - \boldsymbol{\mu})^\top \mathbf{R}^{-1}(\mathbf{X} - \boldsymbol{\mu})\right] = \mathrm{tr}(\mathbf{R}^{-1}\mathbf{R}) = m, \tag{8.30}$$

where $\mathrm{tr}(\cdot)$ denotes the matrix trace. For the third term, write $\mathbf{X} - \boldsymbol{\mu}^M = (\mathbf{X} - \boldsymbol{\mu}) + (\boldsymbol{\mu} - \boldsymbol{\mu}^M)$. Expanding the quadratic form and using $\mathbb{E}_p[\mathbf{X} - \boldsymbol{\mu}] = \mathbf{0}$ gives

$$\begin{aligned}&\mathbb{E}_p\left[(\mathbf{X} - \boldsymbol{\mu}^M)^\top (\mathbf{R}^M)^{-1}(\mathbf{X} - \boldsymbol{\mu}^M)\right] \\ &= \mathbb{E}_p\left[(\mathbf{X} - \boldsymbol{\mu})^\top (\mathbf{R}^M)^{-1}(\mathbf{X} - \boldsymbol{\mu})\right] + (\boldsymbol{\mu} - \boldsymbol{\mu}^M)^\top (\mathbf{R}^M)^{-1}(\boldsymbol{\mu} - \boldsymbol{\mu}^M) \\ &= \mathrm{tr}\big((\mathbf{R}^M)^{-1}\mathbf{R}\big) + (\boldsymbol{\mu} - \boldsymbol{\mu}^M)^\top (\mathbf{R}^M)^{-1}(\boldsymbol{\mu} - \boldsymbol{\mu}^M),\end{aligned} \tag{8.31}$$

where we used the identity $\mathbb{E}[(\mathbf{X} - \boldsymbol{\mu})^\top \mathbf{A}(\mathbf{X} - \boldsymbol{\mu})] = \mathrm{tr}(\mathbf{A}\mathbf{R})$. Substituting these results into (8.29) and noting that

$$\ln\frac{\det(\mathbf{R}^M)}{\det(\mathbf{R})} = -\ln\det\big(\mathbf{R}(\mathbf{R}^M)^{-1}\big), \tag{8.32}$$

we arrive at the closed-form expression (8.25). □

Beyond quantifying model error, relative entropy can also naturally be used to measure information gain in Bayesian inference [181]. For example, in data assimilation, we can compute the additional information in the posterior distribution $p(x|y)$ compared to the prior distribution $p(x)$ through $\mathcal{P}(p(x|y), p(x))$. Note that, different from traditional distance metrics such as the L2 norm, relative entropy is asymmetric between the two inputs and does not satisfy the triangle inequality. This property reflects its fundamental nature as a directional measure of information difference.

8.1.4 Mutual information

Mutual information (also called transinformation) quantifies how much information two random vectors share. Intuitively, it measures how much knowing one random vector reduces the uncertainty about the other. If two random vectors are completely independent, then knowing one provides no information about the other, and the mutual information is zero.

Formally, the mutual information between two random vectors $\mathbf{X}$ and $\mathbf{Y}$ is defined as the relative entropy between the joint distribution $p(\mathbf{x}, \mathbf{y})$ and the product of the marginal distributions $p(\mathbf{x})p(\mathbf{y})$ [79]:

$$\mathcal{M}(\mathbf{X}, \mathbf{Y}) = \mathcal{P}\big(p(\mathbf{x}, \mathbf{y}),\, p(\mathbf{x})p(\mathbf{y})\big) = \int_{\mathbf{X}}\int_{\mathbf{Y}} p(\mathbf{x}, \mathbf{y}) \ln \frac{p(\mathbf{x}, \mathbf{y})}{p(\mathbf{x})p(\mathbf{y})} \, d\mathbf{x}\, d\mathbf{y}. \tag{8.33}$$

This definition has a clear interpretation. The distribution $p(\mathbf{x})p(\mathbf{y})$ represents the situation in which $\mathbf{X}$ and $\mathbf{Y}$ are assumed to be independent. Thus, $\mathcal{M}(\mathbf{X}, \mathbf{Y})$ measures how different the true joint distribution $p(\mathbf{x}, \mathbf{y})$ is from what it would look like if no dependence existed. From the fundamental properties of relative entropy, it follows immediately that $\mathcal{M}(\mathbf{X}, \mathbf{Y}) \geq 0$, with equality if and only if $\mathbf{X}$ and $\mathbf{Y}$ are independent random vectors.

Mutual information is closely related to entropy and conditional entropy, which provides a more intuitive interpretation.

Example 8.12 (Mutual information and conditional entropy). Using the identity $p(\mathbf{x}, \mathbf{y}) = p(\mathbf{x} \mid \mathbf{y})p(\mathbf{y})$, the mutual information (8.33) can be rewritten as

$$\begin{aligned}
\mathcal{M}(\mathbf{X}, \mathbf{Y}) &= \int_{\mathbf{X}}\int_{\mathbf{Y}} p(\mathbf{x}, \mathbf{y}) \ln \frac{p(\mathbf{x} \mid \mathbf{y})}{p(\mathbf{x})} \, d\mathbf{x}\, d\mathbf{y} \\
&= -\int_{\mathbf{X}} p(\mathbf{x}) \ln p(\mathbf{x})\, d\mathbf{x} + \int_{\mathbf{X}}\int_{\mathbf{Y}} p(\mathbf{x}, \mathbf{y}) \ln p(\mathbf{x} \mid \mathbf{y})\, d\mathbf{x}\, d\mathbf{y}.
\end{aligned}$$

Recall that the Shannon entropy of $\mathbf{X}$, defined in (8.2a), measures the total uncertainty associated with $\mathbf{X}$. The conditional entropy of $\mathbf{X}$ given $\mathbf{Y}$ is defined as

$$\mathcal{S}(\mathbf{X} \mid \mathbf{Y}) := -\int_{\mathbf{X}}\int_{\mathbf{Y}} p(\mathbf{x}, \mathbf{y}) \ln p(\mathbf{x} \mid \mathbf{y})\, d\mathbf{x}\, d\mathbf{y}, \tag{8.34}$$

and represents the remaining uncertainty in $\mathbf{X}$ after $\mathbf{Y}$ is known. With these definitions, the mutual information admits the compact form

$$\mathcal{M}(\mathbf{X}, \mathbf{Y}) = \mathcal{S}(\mathbf{X}) - \mathcal{S}(\mathbf{X} \mid \mathbf{Y}). \tag{8.35}$$

Similarly, defining the joint entropy

$$\mathcal{S}(\mathbf{X}, \mathbf{Y}) := -\int_{\mathbf{X}}\int_{\mathbf{Y}} p(\mathbf{x}, \mathbf{y}) \ln p(\mathbf{x}, \mathbf{y})\, d\mathbf{x}\, d\mathbf{y}, \tag{8.36}$$

one can derive the following equivalent relationships: $\mathcal{M}(\mathbf{X}, \mathbf{Y}) = \mathcal{M}(\mathbf{Y}, \mathbf{X})$, $\mathcal{M}(\mathbf{X}, \mathbf{X}) = \mathcal{S}(\mathbf{X})$, $\mathcal{M}(\mathbf{X}, \mathbf{Y}) = \mathcal{S}(\mathbf{X}) - \mathcal{S}(\mathbf{X} \mid \mathbf{Y})$, $\mathcal{M}(\mathbf{X}, \mathbf{Y}) = \mathcal{S}(\mathbf{Y}) - \mathcal{S}(\mathbf{Y} \mid \mathbf{X})$, and $\mathcal{M}(\mathbf{X}, \mathbf{Y}) = \mathcal{S}(\mathbf{X}) + \mathcal{S}(\mathbf{Y}) - \mathcal{S}(\mathbf{X}, \mathbf{Y})$. These identities emphasize that mutual information is symmetric and can be viewed as the amount of information shared by $\mathbf{X}$ and $\mathbf{Y}$, or equivalently, the overlap between their individual uncertainties. □

Example 8.13 (Mutual information in the Gaussian framework). Consider a $2n$-dimensional jointly Gaussian random vector $(\mathbf{X}, \mathbf{Y})^{\mathtt{T}}$ with mean $(\boldsymbol{\mu}_x, \boldsymbol{\mu}_y)^{\mathtt{T}}$ and covariance matrix

$$\mathbf{R} = \begin{pmatrix} \mathbf{R}_x & \mathbf{R}_{xy} \\ \mathbf{R}_{yx} & \mathbf{R}_y \end{pmatrix}, \qquad \mathbf{R}_{yx} = \mathbf{R}_{xy}^{\mathtt{T}}, \tag{8.37}$$

where $\mathbf{X}, \mathbf{Y} \in \mathbb{R}^n$. The mutual information is given by

$$\mathcal{M}(\mathbf{X}, \mathbf{Y}) = \frac{1}{2} \ln\left(\frac{\det(\mathbf{R}_x) \det(\mathbf{R}_y)}{\det(\mathbf{R})} \right). \tag{8.38}$$

To derive the result in (8.38), we start by rewriting the mutual information as

$$\mathcal{M}(\mathbf{X}, \mathbf{Y}) = \mathcal{S}(\mathbf{X}) + \mathcal{S}(\mathbf{Y}) - \mathcal{S}(\mathbf{X}, \mathbf{Y}). \tag{8.39}$$

Using the expression of entropy for Gaussian distributions (8.3), we obtain

$$\mathcal{S}(\mathbf{X}) = \frac{1}{2} \ln\big((2\pi e)^n \det(\mathbf{R}_x)\big), \qquad \mathcal{S}(\mathbf{Y}) = \frac{1}{2} \ln\big((2\pi e)^n \det(\mathbf{R}_y)\big), \tag{8.40}$$

and

$$\mathcal{S}(\mathbf{X}, \mathbf{Y}) = \frac{1}{2} \ln\big((2\pi e)^{2n} \det(\mathbf{R})\big). \tag{8.41}$$

Substituting these expressions into (8.39) yields

$$\begin{aligned}\mathcal{M}(\mathbf{X}, \mathbf{Y}) &= \frac{1}{2} \ln\big((2\pi e)^n \det(\mathbf{R}_x)\big) + \frac{1}{2} \ln\big((2\pi e)^n \det(\mathbf{R}_y)\big) - \frac{1}{2} \ln\big((2\pi e)^{2n} \det(\mathbf{R})\big) \\ &= \frac{1}{2} \ln\left(\frac{\det(\mathbf{R}_x) \det(\mathbf{R}_y)}{\det(\mathbf{R})} \right).\end{aligned}$$

The mutual information depends only on the covariance structure of the joint Gaussian distribution and vanishes if and only if $\mathbf{R}_{xy} = \mathbf{0}$. In this case, $\mathbf{X}$ and $\mathbf{Y}$ are independent.□

Example 8.14 (Scalar Gaussian mutual information). Consider two jointly Gaussian scalar random variables (X, Y) with variances σ_X^2, σ_Y^2, and correlation coefficient ρ. The covariance matrix is

$$\mathbf{R} = \begin{pmatrix} \sigma_X^2 & \rho\sigma_X\sigma_Y \\ \rho\sigma_X\sigma_Y & \sigma_Y^2 \end{pmatrix}. \tag{8.42}$$

Using (8.38), we compute $\det(\mathbf{R}) = \sigma_X^2\sigma_Y^2(1-\rho^2)$. Therefore, the mutual information reduces to

$$\mathcal{M}(X, Y) = \frac{1}{2} \ln\left(\frac{\sigma_X^2\sigma_Y^2}{\sigma_X^2\sigma_Y^2(1-\rho^2)} \right) = -\frac{1}{2} \ln(1-\rho^2). \tag{8.43}$$

This expression shows that, in the scalar Gaussian case, the mutual information depends only on the squared correlation coefficient ρ^2. It vanishes when $\rho = 0$ (independence) and increases monotonically as $|\rho| \to 1$, diverging in the limit of perfect linear dependence. □

8.2 Uncertainty reduction

Since uncertainty is ubiquitous across scientific and engineering disciplines, uncertainty reduction becomes an important topic that advances the improvement of accuracy and confidence in state estimation and forecast.

8.2.1 General strategies for uncertainty reduction

Strategies for effectively reducing uncertainty in complex systems can often be classified into three broad and complementary categories: model improvement, observation optimization, and advanced data-model integration.

First, most models used in practice inevitably contain errors and uncertainties. One fundamental source of such uncertainty is the incomplete understanding of the underlying physical processes governing the system. Approximate errors in model development account for a large portion of the uncertainty. Note that even when the underlying physics is well understood, practical limitations on computational resources often prevent the simulation of the full system. For example, resolving all relevant spatial and temporal scales in complex Earth systems often demands substantial storage and computational resources. These challenges make it difficult to utilize a complete model to provide real-time forecasts. As a result, models used for practical purposes are typically constructed with finite resolution, and small-scale processes must be truncated, parameterized, or otherwise approximated. However, in highly nonlinear and turbulent systems, these unresolved small-scale features can have a profound impact on large-scale dynamics and statistics. Without a suitable description of these small-scale features, the model will introduce inaccurate forecasts and misrepresented uncertainty. To improve model fidelity and reduce uncertainty, it is therefore essential to develop more accurate reduced or approximate models. This may be achieved through improved physical theory, higher-order numerical schemes, or the incorporation of stochastic parameterizations and machine-learning-based closures that efficiently approximate unresolved processes while keeping the computational cost manageable.

Second, in addition to the models, observational data often provide supplementary information for reducing uncertainty through state estimation, as discussed in Chapter 7. By combining observations with model forecasts, data assimilation methods can substantially reduce uncertainty in state estimation. However, the effectiveness of uncertainty reduction depends critically on the quality, quantity, and spatiotemporal distribution of the observations. Poorly placed or noisy observations may contribute little useful information, while well-designed observation networks can significantly enhance estimation accuracy. Consequently, optimal observational design, such as determining where, when, and how frequently to collect data, plays a crucial role in uncertainty reduction and has become an essential topic in practice.

Third, uncertainty reduction can be further enhanced by employing more sophisticated methodologies for integrating data with models. Advanced data-model integration techniques facilitate integrating diverse information sources while accounting for their respective uncertainties. For example, optimally combining multiple models with different fidelities and resolutions can be an effective and generalized way to improve the standard data assimilation. In addition, since calculating exact information measurements can be computationally expensive for high-dimensional systems, it is of practical importance to develop fast yet effective approximations to these measurements that facilitate quantification of the uncertainty reduction. Notably, these methods not only improve understanding of uncertainty variation in state estimation but can also inform model calibration, parameter inference, and predictive uncertainty quantification, thereby providing valuable guidance for systematically reducing uncertainty in complex dynamical systems.

These topics will be discussed in this and the subsequent chapters.

8.2.2 Using the information distance as an optimization criterion for uncertainty reduction

Information-theoretic quantities provide a principled way to quantify uncertainty in probabilistic models. By treating information distances as objective functions, uncertainty reduction can be formulated as an optimization problem over model parameters or model classes.

Example 8.15 (Determining the optimal distribution via minimizing relative entropy). We seek the Gaussian distribution that best approximates the Student's t-distribution in (8.8) with $\nu = 3$ (Figure 8.3). Here, "best" is defined in the information-theoretic sense, namely as the Gaussian distribution that minimizes the relative entropy from the reference distribution.

Specifically, we minimize the relative entropy $\mathcal{P}(p, p^M)$, where p denotes the Student's t-distribution and p^M is a Gaussian distribution. Since a Gaussian distribution is uniquely determined by its mean μ and variance R, the optimization is carried out over these two parameters.

Figure 8.4 shows heat maps of the logarithmic relative entropy values over the (μ, R) parameter space. Panel (a) displays the exact relative entropy computations, while Panel (b) shows the signal–dispersion approximation based solely on the mean and variance of the Student's t-distribution. Both approaches consistently identify the optimal Gaussian distribution as having zero mean and variance $R = 3$, which exactly matches the first two moments of the reference distribution.

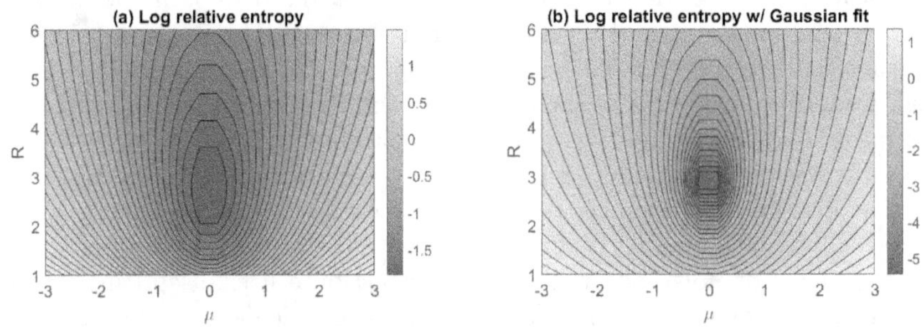

FIGURE 8.4: Heat maps of the logarithmic relative entropy values between the t-distribution and a candidate Gaussian distribution across the Gaussian mean and variance (μ, R). Panel (a): the exact relative entropy computations. Panel (b): using the signal-dispersion approximation based solely on the mean and variance of the t-distribution.

The agreement between the two panels in Figure 8.4 reflects a fundamental information-theoretic decomposition [180]:

$$\mathcal{P}(p, p^M) = \mathcal{P}(p, p_L) + \mathcal{P}(p_L, p^M), \tag{8.44}$$

where p is the reference distribution, p^M is the approximating Gaussian distribution, and p_L is the L-moment reconstruction of p (with $L = 2$ in this example). This identity holds because p_L is the information projection of p onto the moment-constrained Gaussian family, yielding a Pythagorean relation in relative entropy [180].

Here, $\mathcal{P}(p, p_L)$ represents the intrinsic information loss associated with retaining only the first L moments, and thus provides a fundamental lower bound on the achievable approximation accuracy. In contrast, $\mathcal{P}(p_L, p^M)$ measures the approximation error in matching the moment-reconstructed distribution, which is the component that can be reduced through optimization. Equation (8.44) shows that the minimization procedure effectively targets this reducible error while leaving the intrinsic limitation unchanged.

This result is also consistent with the maximum entropy principle. Among all distributions with a given mean and variance, the Gaussian distribution maximizes Shannon entropy. Consequently, the optimal Gaussian approximation under the relative entropy can be interpreted as the maximum entropy distribution that is consistent with the first two moments of the Student's t-distribution. □

8.2.3 Uncertainty reduction via data assimilation: Role of multiple independent measurements and information barrier

The posterior distribution in data assimilation reduces uncertainty compared to the prior distribution by incorporating additional information from observations. A particularly interesting aspect to examine is how various measurements contribute to uncertainty reduction. A comprehensive investigation of uncertainty reduction in practical data assimilation would need to address numerous additional factors, including optimal observation placement and sensor types. Here, we use a simple yet analytically tractable example to quantify the reduction in uncertainty as a function of the number of independent observations. Using the information measurements, we can quantitatively assess the reduction in uncertainty [71].

Example 8.16 (Uncertainty reduction via multiple independent measurements). To illustrate uncertainty reduction in data assimilation through multiple independent measurements, consider a scalar state variable u and noisy observations $\mathbf{v} = \mathbf{g}u + \boldsymbol{\epsilon}$, where $\mathbf{g} = (g, \ldots, g)^{\mathtt{T}}$ is an $L \times 1$ observation operator, $\boldsymbol{\epsilon} = (\epsilon_1, \ldots, \epsilon_L)^{\mathtt{T}}$ represents $L \times 1$ independent random measurement noise, and $\mathbf{v}$ itself is an $L \times 1$ vector comprising L repeated measurements of the system. Assume the prior distribution $p(u) \sim \mathcal{N}(\mu_f, R_f)$ and the observational noise with a diagonal covariance $\mathbf{R}^o$ (of size $L \times L$) are both Gaussian. According to the Kalman filter presented in Chapter 7, the posterior distribution $p(u|\mathbf{v}) \sim \mathcal{N}(\mu_a, R_a)$ is also Gaussian. The posterior mean μ_a and covariance R_a can be written down explicitly

$$\mu_a = (1 - \mathbf{K}\mathbf{g})\mu_f + \mathbf{K}\mathbf{v}, \quad R_a = (1 - \mathbf{K}\mathbf{g})R_f, \tag{8.45}$$

where $\mathbf{K}$ is of size $1 \times L$ and is given by $\mathbf{K} = R_f \mathbf{g}^{\mathtt{T}}(\mathbf{g}R_f\mathbf{g}^{\mathtt{T}} + \mathbf{R}^o)^{-1}$. Figure 8.5 presents numerical simulations demonstrating the dependence of assimilation results on the number of observations L. Here $g = 1$, $R_f = 1$, and $\mathbf{R}^o = \mathbf{I}$ (an $L \times L$ identity matrix) are used. When $L \to \infty$, the posterior mean asymptotically converges to the true state and the posterior variance decays to zero, indicating perfect certainty in the limit. These are consistent with intuitions.

An interesting finding is the distinct asymptotic behavior of the two components in the relative entropy. While the signal part converges to a constant since the posterior mean converges to the truth, the dispersion component grows unboundedly at a logarithmic rate. This behavior comes from the mathematical structure of relative entropy, where the dispersion term involves the ratio of prior to posterior variances, which can diverge as the posterior variance shrinks with increasing size of observations. These results have important implications for data assimilation. While additional observations always provide new information, the information gain exhibits diminishing returns. When the number of existing observations is small, including additional observations dramatically reduces uncertainty. However, the marginal information gain from additional observations diminishes as L increases, reflecting a logarithmic growth in total information. This is called a practical information barrier, indicating that achieving a fixed amount of uncertainty reduction requires exponentially more data at later stages. Such a finding naturally raises the question of determining the optimal number of observations in practical applications, which involves balancing information gain and deployment costs. Despite being extremely simple, this example provides an intuition of the information barrier in the uncertainty reduction of the Lagrangian data assimilation [68] using the setup in Example 7.12. The mathematical derivation below also explains why the information reduction is a function of the logarithm of the tracer number.

While numerical results provide intuitive understanding, this extremely simple example allows rigorous derivation of such conclusions. The analytical derivation not only reveals the underlying logical framework but also serves as a paradigmatic case study for examining uncertainty reduction mechanisms. Below is the mathematical derivation.

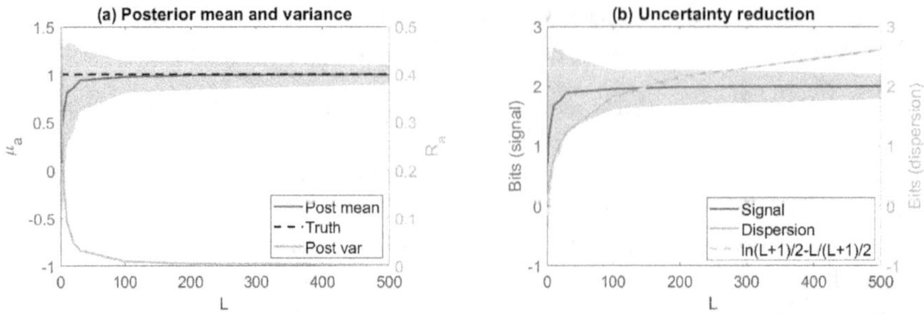

FIGURE 8.5: Posterior state estimation (Panel (a)) and uncertainty reduction (Panel (b)) as a function of the measurement number L. Results are based on 50 independent experimental trials. The posterior mean (and thus signal part in the uncertainty reduction) varies across trials (shaded confidence band) due to dependence on observational noise realizations, consistent with Kalman filter theory, while the posterior variance (and thus the dispersion part) remains identical across all trials as it is noise-realization-independent.

For further notational simplicity, let us assume that the prior variance is $R_f = 1$ and that the observational noise is $\mathbf{R}^o = \mathbf{I}$, meaning that the noise of each measurement is of unit variance and that the measurements are independent of each other. The observational operator $\mathbf{g}$ is an $L \times 1$ column vector. The dimension of the outer product $\mathbf{g}\mathbf{g}^{\mathrm{T}}$, which will be used below, is $L \times L$. With these, the Kalman gain $\mathbf{K}$ is necessarily a $1 \times L$ row vector, since the dimensions of $\mathbf{g}^{\mathrm{T}}$ and $(\mathbf{g}\mathbf{g}^{\mathrm{T}} + \mathbf{I})^{-1}$ are $1 \times L$ and $L \times L$, respectively, and so we denote their product by $\mathbf{k}^{\mathrm{T}}$ for the remainder of this subsection, and its expression can be simplified to,

$$\mathbf{k}^{\mathrm{T}} = \mathbf{g}^{\mathrm{T}}(\mathbf{g}\mathbf{g}^{\mathrm{T}} + \mathbf{I})^{-1}. \tag{8.46}$$

In light of (8.46), the corresponding expressions for the posterior mean and posterior variance in (8.45) are,

$$\mu_a = \frac{1}{\mathbf{g}^{\mathrm{T}}\mathbf{g} + 1}\mu_f + \frac{\mathbf{g}^{\mathrm{T}}}{\mathbf{g}^{\mathrm{T}}\mathbf{g} + 1}\mathbf{v}, \tag{8.47}$$

and,

$$R_a = \frac{1}{\mathbf{g}^{\mathrm{T}}\mathbf{g} + 1}. \tag{8.48}$$

Note that the $L \times L$ matrix $\mathbf{g}\mathbf{g}^{\mathrm{T}}$ appears in the parentheses in (8.46) while the denominators in (8.47) and (8.48) contain the scalar inner product $\mathbf{g}^{\mathrm{T}}\mathbf{g}$, which is precisely the squared Euclidean norm of $\mathbf{g}$. Both μ_a and R_a are scalars. The first term on the right-hand side of (8.47) is a product of two scalars, while the second term is a product of a $1 \times L$ row vector and an $L \times 1$ column vector.

The results in (8.47)–(8.48) can be derived using the following two matrix identities (given by the well-known Woodbury matrix identity [133]):

$$(\mathbf{A} + \mathbf{BCD})^{-1} = \mathbf{A}^{-1} - \mathbf{A}^{-1}\mathbf{B}(\mathbf{C}^{-1} + \mathbf{D}\mathbf{A}^{-1}\mathbf{B})^{-1}\mathbf{D}\mathbf{A}^{-1}, \tag{8.49}$$

$$(\mathbf{I} - (\mathbf{A} + \mathbf{E})^{-1}\mathbf{A})\mathbf{E}^{-1} = (\mathbf{A} + \mathbf{E})^{-1}, \tag{8.50}$$

where $\mathbf{A}$, $\mathbf{B}$, $\mathbf{C}$, $\mathbf{D}$, and $\mathbf{E}$ are matrices of sizes $m \times m$, $m \times L$, $L \times L$, $L \times m$, and $m \times m$, respectively. According to (8.46),

$$1 - \mathbf{k}^{\mathrm{T}}\mathbf{g} = 1 - \mathbf{g}^{\mathrm{T}}(\mathbf{g}\mathbf{g}^{\mathrm{T}} + \mathbf{I})^{-1}\mathbf{g}. \tag{8.51}$$

Denote by $\mathbf{A} = 1$, $\mathbf{B} = \mathbf{g}^{\mathrm{T}}$, $\mathbf{C} = \mathbf{I}$ and $\mathbf{D} = \mathbf{g}$ (which essentially sets $m = 1$ in the above matrix dimensions). The relationship in (8.49) simplifies (8.51),

$$1 - \mathbf{k}^{\mathrm{T}}\mathbf{g} = \frac{1}{1 + \mathbf{g}^{\mathrm{T}}\mathbf{g}}, \tag{8.52}$$

which is a scalar. Likewise, denote by $\mathbf{A} = \mathbf{g}\mathbf{g}^{\mathrm{T}}$ and $\mathbf{E} = \mathbf{I}$, both of which are of size $L \times L$. The relationship in (8.50) leads to the $1 \times L$ vector,

$$\begin{aligned}
\mathbf{k}^{\mathrm{T}} &= \mathbf{g}^{\mathrm{T}}(\mathbf{g}\mathbf{g}^{\mathrm{T}} + \mathbf{I})^{-1} = \mathbf{g}^{\mathrm{T}}(\mathbf{A} + \mathbf{E})^{-1} \\
&= \mathbf{g}^{\mathrm{T}}(\mathbf{I} - (\mathbf{A} + \mathbf{E})^{-1}\mathbf{A})\mathbf{E}^{-1} \\
&= \mathbf{g}^{\mathrm{T}}(\mathbf{I} - (\mathbf{g}\mathbf{g}^{\mathrm{T}} + \mathbf{I})^{-1}\mathbf{g}\mathbf{g}^{\mathrm{T}}) \\
&= (1 - \mathbf{g}^{\mathrm{T}}(\mathbf{g}\mathbf{g}^{\mathrm{T}} + \mathbf{I})^{-1}\mathbf{g})\mathbf{g}^{\mathrm{T}} \\
&= \frac{\mathbf{g}^{\mathrm{T}}}{1 + \mathbf{g}^{\mathrm{T}}\mathbf{g}},
\end{aligned} \tag{8.53}$$

where the last equality is due to (8.51) and (8.52). Therefore, (8.52) and (8.53) give the posterior mean expression (8.47) while (8.52) leads to the expression of the posterior variance (8.48).

Next, further assume $\mathbf{g} = (1, \ldots, 1)^{\mathrm{T}}$ for simplicity. The L observations are given by,

$$\mathbf{v} = \mathbf{g}u + \boldsymbol{\epsilon} \Leftrightarrow \begin{pmatrix} v_1 \\ \vdots \\ v_L \end{pmatrix} = \begin{pmatrix} 1 \\ \vdots \\ 1 \end{pmatrix} u + \begin{pmatrix} \epsilon_1 \\ \vdots \\ \epsilon_L \end{pmatrix}, \tag{8.54}$$

where $\epsilon_1, \ldots \epsilon_L$ are L independent standard Gaussian white noises, each having a variance of 1. Thus, (8.47) and (8.48) can further be simplified to

$$\mu_a = \frac{1}{L+1}\mu_f + \frac{1}{L+1}\sum_{l=1}^{L} v_l, \tag{8.55}$$

and

$$R_a = \frac{1}{L+1}. \tag{8.56}$$

The result in (8.55) indicates that the observational noise can cancel off when multiple observations appear. This is consistent with the law of large numbers. In particular, when more and more observations are available, which means the noise is canceled off to a more significant degree, the weight toward the prior information becomes less and less. Another way to express this observation is to notice that the posterior mean in (8.47) can be expressed as a convex linear combination of the prior mean and the maximum likelihood estimator (MLE). Specifically, observe that for,

$$w = \frac{1}{\mathbf{g}^{\mathrm{T}}\mathbf{g} + 1}, \tag{8.57}$$

we have from (8.47) and (8.54) that,

$$\mu_a = \frac{1}{\mathbf{g}^{\mathrm{T}}\mathbf{g} + 1}\mu_f + \frac{\mathbf{g}^{\mathrm{T}}}{\mathbf{g}^{\mathrm{T}}\mathbf{g} + 1}\mathbf{v} = w\mu_f + (1 - w)\frac{\mathbf{g}^{\mathrm{T}}\mathbf{v}}{\mathbf{g}^{\mathrm{T}}\mathbf{g}} = \frac{1}{L+1}\mu_f + \frac{L}{L+1}\frac{\sum_{l=1}^{L} v_l}{L}. \tag{8.58}$$

The result in (8.58) highlights a general and intuitive feature of data assimilation. The posterior mean is a weighted average of the prior mean and the estimate based solely on observations. When only a few observations are available, the prior information plays an important role. As the number of observations increases, the influence of observational noise is reduced through averaging, and the posterior mean becomes increasingly dominated by the information provided by the data. In the limit as $L \to \infty$, the weight placed on the prior mean vanishes, and the posterior mean converges to the sample mean of the observations. This reflects a fundamental principle of uncertainty reduction: with sufficient independent measurements, observational evidence overwhelms prior uncertainty, leading to increasingly confident and accurate state estimates.

These multiple observations also help reduce the uncertainty reflected by the posterior variance. By plugging (8.56) into the dispersion part of the relative entropy expression, it is shown that,

$$\begin{aligned}\text{Dispersion} &= \frac{1}{2}\Big[R_a(R_f)^{-1} - 1 - \ln\big(R_a(R_f)^{-1}\big)\Big]\\ &= \frac{1}{2}\left(R_a - 1 - \ln R_a\right) = \frac{1}{2}\left(-\frac{L}{L+1} + \ln(L+1)\right),\end{aligned} \tag{8.59}$$

which asymptotically, as $L \to \infty$, behaves like $\ln L$. Thus, although variance decreases at rate $1/L$, the accumulated information grows only logarithmically with the number of observations. □

8.2.4 Optimal experimental design in data assimilation based on uncertainty reduction

Uncertainty reduction is central to many optimal experimental design problems. A central theme in practice is that observations are costly, limited, and often insufficient to fully recover the underlying dynamical properties. This leads to one crucial optimal design problem, in which the observer seeks to determine where, when, and how to collect data to maximize the information gained about the system state or parameters [154, 40, 148]. Depending on specific goals, optimal design problems can focus on sensor placement, adaptive observation scheduling, multi-step decision strategies, optimal forcing for system identification, multi-fidelity model selection, or design criteria that balance accuracy and computational cost. In many practical settings, such as weather forecasting, ocean monitoring, environmental surveillance, and engineering digital twins, optimal design has become an essential component of modern predictive systems.

One of the most classical and widely used optimal design problems in data assimilation is optimal observation placement. The goal is to choose sensor locations (or, equivalently, which components of the state to observe) so that posterior uncertainty is minimized. Let $\mathbf{x} \in \mathbb{R}^n$ denote the state, $\mathbf{H}$ the observation operator, and $\mathbf{R}$ the observation-error covariance. In the linear Gaussian case, the posterior covariance satisfies

$$\mathbf{P}_a^{-1} = \mathbf{P}_f^{-1} + \mathbf{H}^{\mathrm{T}}\mathbf{R}^{-1}\mathbf{H}, \tag{8.60}$$

where $\mathbf{P}_f$ is the forecast covariance. Optimal placement can then be formulated as selecting $\mathbf{H}$ to minimize a scalar criterion of $\mathbf{P}_a$, such as $\mathrm{tr}(\mathbf{P}_a)$ (A-optimal), $\log\det(\mathbf{P}_a)$ (D-optimal), or $\lambda_{\max}(\mathbf{P}_a)$ (E-optimal) [287].

Example 8.17 (Location selection with a sinusoidal observation operator). A simple way to encode location-dependent information content is through a space-dependent observation operator. Consider a scalar unknown parameter a (e.g., the amplitude of a mode) with prior

$$a \sim \mathcal{N}(0, P_0), \qquad P_0 = 1. \tag{8.61}$$

Suppose we can place a single sensor at a spatial location x on a one-dimensional domain, and the measurement at that location is

$$y = \sin(x)\, a + \varepsilon, \qquad \varepsilon \sim \mathcal{N}(0, R), \quad R = 0.25. \tag{8.62}$$

Here $\sin(x)$ plays the role of the observation operator; at locations where $\sin(x)$ is small, the observation is only weakly sensitive to a and thus contains little information about it.

Conditioned on an observation at location x, the posterior variance of a is

$$P_a(x) = \left(P_0^{-1} + \sin^2(x)\, R^{-1}\right)^{-1}. \tag{8.63}$$

To see how location affects information content, consider three candidate sensor locations:

$$x_1 = 0.1\pi, \qquad x_2 = 0.5\pi, \qquad x_3 = 0.9\pi, \tag{8.64}$$

which give

$$\sin^2(x_1) \approx 0.095, \qquad \sin^2(x_2) = 1, \qquad \sin^2(x_3) \approx 0.095. \tag{8.65}$$

Substituting (8.65) into the posterior variance formula (8.63) with $P_0 = 1$ and $R = 0.25$ gives

$$P_a(x_1) = P_a(x_3) = \left(1 + 0.095/0.25\right)^{-1} = 0.724, \qquad P_a(x_2) = \left(1 + 1/0.25\right)^{-1} = 0.200. \tag{8.66}$$

Thus, placing the sensor at $x_2 = \pi/2$, where $|\sin(x)|$ is maximal, reduces the variance of a from 1 to 0.2, while placing it near x_1 or x_3, where $|\sin(x)|$ is relatively small, yields only a modest variance reduction. In other words, the sensitivity of the observation to the unknown parameter, measured here by $|\sin(x)|$, directly controls the information gain. Locations where the observation operator has a large magnitude are far more informative for estimating a. This simple example highlights such a crucial principle of location selection, which can be generalized to more complex data assimilation settings. □

Example 8.18 (Selecting one variable to observe in a correlated two-dimensional system). Consider a two-dimensional state where we are allowed to observe only one component. Let the forecast covariance and the observational error variance be

$$\mathbf{P}_f = \begin{pmatrix} 4 & 1.5 \\ 1.5 & 1 \end{pmatrix}, \qquad R = 0.25, \tag{8.67}$$

respectively. We need to choose whether to observe the first or second component of the state. If we observe x_1, the observation operator is $\mathbf{H} = (1\ 0)$; if we observe x_2, it is $\mathbf{H} = (0\ 1)$. Making use of the Kalman posterior covariance in (8.60), a direct computation yields

$$\text{trace}(\mathbf{P}_a^{(1)}) \approx 0.71, \qquad \text{trace}(\mathbf{P}_a^{(2)}) \approx 2.40. \tag{8.68}$$

Thus, observing x_1 leads to substantially lower posterior uncertainty.

This example highlights several essential aspects of optimal observation placement. Since x_1 has the largest marginal variance, it provides intuitive evidence that x_1 is likely to be selected as the observational variable. As a result, observing x_1 yields a substantially smaller posterior trace than observing x_2. However, the marginal variance alone does not determine the best sensor location. The correlation structure of the forecast covariance plays a crucial role: since x_1 and x_2 are positively correlated, measuring x_1 reduces uncertainty not only in that component but also in x_2, where the information propagates across variables depending on the strength of the cross-covariance term. In more complex situations, optimal design can be nonintuitive. The optimal sensor placement depends on how information spreads through the covariance structure rather than on local variances alone. □

Example 8.19 (Optimal sensor not at the largest-variance variable). Consider a three-dimensional state with forecast covariance and the observational error variance

$$\mathbf{P}_f = \begin{pmatrix} 3.0 & 1.3 & 1.8 \\ 1.3 & 3.5 & -0.2 \\ 1.8 & -0.2 & 3.0 \end{pmatrix}, \qquad R = 0.5. \tag{8.69}$$

The marginal variances are

$$\mathrm{Var}(x_1) = 3.0, \qquad \mathrm{Var}(x_2) = 3.5, \qquad \mathrm{Var}(x_3) = 3.0, \tag{8.70}$$

so x_2 has the largest prior variance. We are allowed to observe exactly one component x_i, using $\mathbf{H}_i = \mathbf{e}_i^{\mathsf{T}}$ with $\mathbf{e}_i$ the i-th coordinate vector. Making use of the standard Kalman formula for the posterior covariance in (8.60), a direct computation yields

$$\mathrm{trace}(\mathbf{P}_a^{(1)}) \approx 5.52, \qquad \mathrm{trace}(\mathbf{P}_a^{(2)}) \approx 6.01, \qquad \mathrm{trace}(\mathbf{P}_a^{(3)}) \approx 5.99. \tag{8.71}$$

Thus, the optimal choice (in the sense of minimizing $\mathrm{trace}(\mathbf{P}_a^{(i)})$) is to observe x_1, even though x_2 has slightly larger variance.

The reason lies in the correlation structure encoded in $\mathbf{P}_f$. The first component is moderately or strongly correlated with the others (correlations $\rho_{12} \approx 0.40$ and $\rho_{13} \approx 0.60$), so observing x_1 reduces uncertainty not only in x_1 itself but also in x_2 and x_3. In contrast, x_2 is only weakly correlated with x_3 ($\rho_{23} \approx -0.06$). Therefore, although observing x_2 reduces its own uncertainty, it has a smaller impact on the global state. This example demonstrates that, in multivariate systems, the optimal observation location need not coincide with the variable with the largest variance. The optimal sensor is the one that most effectively reduces overall uncertainty by leveraging the covariance structure rather than just the local variance.
□

8.3 The role of nonlinearity in uncertainty quantification

A fundamental challenge in uncertainty quantification for many practical problems comes from nonlinear effects. These nonlinearities may appear in different forms, including nonlinear dynamics in physical systems, nonlinear mappings between variables, and nonlinear diagnostic criteria. Such nonlinearities dramatically alter how uncertainty propagates from inputs to outputs. They often produce complex features that differ significantly from those of linear systems. A typical example involves systems with Gaussian inputs. It is well known that Gaussianity is preserved under linear transformations; namely, any linear operation on a Gaussian variable yields another Gaussian distribution. However, nonlinear transformations fundamentally break this preservation property. Nonlinear mappings or diagnostic criteria typically convert Gaussian inputs into outputs with non-Gaussian statistics, often with heavy tails.

8.3.1 Uncertainty propagation in linear and nonlinear dynamics

To understand how uncertainty propagates in dynamical systems, we begin with the mean-fluctuation decomposition of a random variable x [201] introduced in Section 5.4:

$$x = \langle x \rangle + x', \qquad \text{with } \langle x' \rangle = 0, \tag{8.72}$$

where $\langle\cdot\rangle$ denotes the ensemble average. Here, $\langle x\rangle$ represents the mean (or average) state, while x' represents the random deviation from the mean.

In many physical systems, uncertainty arises from random fluctuations, which may be due to unresolved processes, external disturbances, or measurement noise. In general, the uncertainty of a random variable is described by its full probability distribution, or equivalently by all of its moments. Sometimes, the variance $\langle(x')^2\rangle$ provides a simple and commonly used measure of uncertainty, as it quantifies the typical amplitude of fluctuations around the mean. In special cases, such as Gaussian distributions, the variance alone is sufficient to fully characterize the uncertainty, as illustrated in Example 8.3.

We now examine how such uncertainty evolves in a simple linear system, which will serve as a reference case before considering nonlinear dynamics.

Example 8.20 (Uncertainty propagation in a linear system). Consider the one-dimensional linear stochastic system

$$\frac{\mathrm{d}x}{\mathrm{d}t} = -ax + f + \sigma\dot{W}, \tag{8.73}$$

where $a > 0$, f is a constant forcing, and $\dot{W}$ denotes Gaussian white noise. The random forcing term $\sigma\dot{W}$ continuously perturbs the system and acts as a source of uncertainty.

Taking the ensemble average of (8.73), and using the fact that the noise has zero mean, we obtain the mean equation

$$\frac{\mathrm{d}\langle x\rangle}{\mathrm{d}t} = -a\langle x\rangle + f. \tag{8.74}$$

This equation shows that the evolution of the mean state is deterministic and is not affected by the uncertainty described by the fluctuations.

Subtracting (8.74) from (8.73), we obtain the equation governing the fluctuation part:

$$\frac{\mathrm{d}x'}{\mathrm{d}t} = -ax' + \sigma\dot{W}. \tag{8.75}$$

This equation shows that the fluctuations evolve independently of the mean and are controlled by a balance between linear damping and stochastic forcing.

Because the system is linear and the noise is Gaussian, any Gaussian (or deterministic) initial condition will remain Gaussian at all later times. As a result, the uncertainty of the system is fully characterized by the variance $\langle(x')^2\rangle$, which evolves independently of the mean. □

However, the situation becomes very different if nonlinear terms appear in the dynamics.

Example 8.21 (Uncertainty propagation in a simple nonlinear system). Consider a quadratic nonlinear term on the right-hand side of the dynamics,

$$\frac{\mathrm{d}x}{\mathrm{d}t} = bx^2 + ... \tag{8.76}$$

Since

$$\langle x^2\rangle = \langle(\langle x\rangle + x')^2\rangle = \langle(\langle x\rangle^2 + (x')^2 + 2\langle x\rangle x')\rangle = \langle x\rangle^2 + \langle(x')^2\rangle, \tag{8.77}$$

taking the ensemble average of (8.76) leads to

$$\frac{\mathrm{d}\langle x\rangle}{\mathrm{d}t} = b\langle x\rangle^2 + b\langle(x')^2\rangle + ... \tag{8.78}$$

This result demonstrates two important aspects of nonlinear systems. First, higher-order moments (such as the variance $\langle(x')^2\rangle$) directly influence the evolution of lower-order moments (like the mean $\langle x\rangle$), leading to more complex mean dynamics. Second, and more fundamentally, the time evolution of any given moment depends on higher-order moments, creating an infinite hierarchy of unclosed equations. Consequently, uncertainty quantification in nonlinear systems typically requires approximation methods to truncate this hierarchy, presenting significant additional challenges compared to linear systems. □

Example 8.22 (Numerical illustration of the uncertainty propagation in simple scalar linear and nonlinear equations). Let us compare the uncertainty in the simple linear model (8.73) with a simple scalar nonlinear model:

$$\frac{\mathrm{d}x}{\mathrm{d}t} = ax + bx^2 - cx^3 + f + \sigma\dot{W}. \tag{8.79}$$

The parameter values are as follows:

$$\begin{aligned} \text{Linear model}: &\quad a = 2, \quad f = 4, \\ \text{Nonlinear model}: &\quad a = 2, \quad b = -1, \quad c = 2, \quad f = 0.5 \quad \sigma = 1. \end{aligned} \tag{8.80}$$

Figure 8.6 illustrates key differences between linear and nonlinear systems. Panel (a) displays time series from both the linear model (8.73) and its deterministic counterpart obtained by removing the stochastic noise. In this linear case, the deterministic solution exactly matches the mean dynamics (8.74), demonstrating complete separability between uncertainty and mean evolution.

The situation becomes fundamentally different for nonlinear systems. The cubic nonlinear model (8.79) exhibits more complex behavior when the stochastic forcing is removed. The deterministic system possesses three equilibrium points (corresponding to the roots of the cubic function), with two stable attractors and one unstable equilibrium. Consequently, the final state depends critically on the initial condition. Solutions may converge to either a stable attractor or remain at the unstable equilibrium if it starts from there. See Panel (b).

Panel (c) compares time series from both deterministic and stochastic versions of the nonlinear model using identical initial conditions, where the deterministic version means the noise term is dropped in (8.79) while other terms on the right-hand side are all unchanged. Notably, the trajectory from such a deterministic model diverges from the stochastic system's mean state due to the ignorance of the nonlinear terms associated with the fluctuation components in the actual mean dynamics, as discussed in Example 8.21. When the full stochastic model is used (Panel (d)), the solution intermittently transitions between neighborhoods of both stable attractors, resulting in a characteristic bimodal probability distribution.

□

8.3.2 Uncertainty propagation in nonlinear diagnostics

The impact of nonlinearity on uncertainty quantification can be demonstrated in even simpler scenarios without dynamical systems. In particular, nonlinear transformations of random variables can distort probability distributions and couple higher-order moments, even when the input uncertainty is simple or Gaussian.

Example 8.23 (Parameter estimation in a simple linear algebraic equation). Consider a basic parameter estimation problem for the linear algebraic system

$$w = \theta u. \tag{8.81}$$

For known (deterministic) input u and output w, the parameter θ admits the exact closed-form solution $\theta = w/u$.

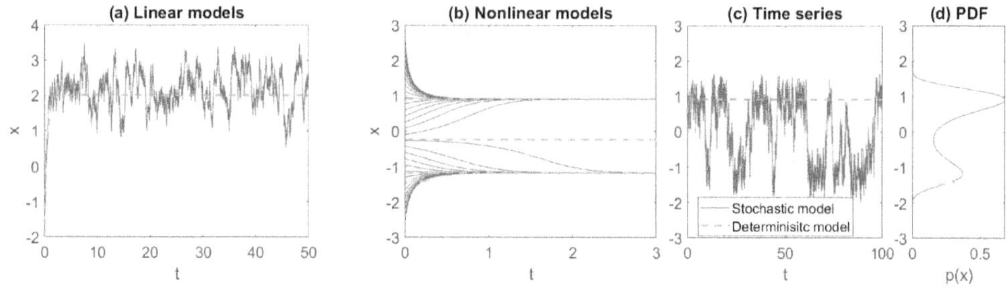

FIGURE 8.6: Comparison between the linear model (8.73) and the nonlinear model (8.79). Panel (a): a single realization of the linear stochastic model alongside its deterministic counterpart (which in this linear case is the mean equation). Panel (b): time series of the nonlinear deterministic model (stochastic term removed) from various initial conditions. The thick magenta lines indicate the three fixed points (two stable and one unstable). Panel (c): a single realization of the nonlinear stochastic model and its deterministic counterpart, starting from the same initial condition. Panel (d): PDF of the nonlinear stochastic model.

However, when u is uncertain and described by a probability distribution $p(u)$, the estimated parameter θ becomes a random variable. For simplicity, instead of computing the full distribution of θ, we focus on its expectation. A key observation is that

$$\mathbb{E}[\theta] = \mathbb{E}\left[\frac{w}{u}\right] \neq \frac{w}{\mathbb{E}[u]}, \tag{8.82}$$

demonstrating that expectation and inversion do not commute.

For illustration, suppose $u \sim \mathrm{Unif}[a, b]$ with $0 < a < b$. Then

$$\mathbb{E}[u] = \frac{a+b}{2}.$$

Defining the transformed variable $v = 1/u$, its probability density function is

$$p(v) = \frac{1}{(b-a)v^2}, \qquad v \in \left[\frac{1}{b}, \frac{1}{a}\right],$$

with mean

$$\mathbb{E}[v] = \frac{\ln b - \ln a}{b-a} \neq \frac{1}{\mathbb{E}[u]}.$$

This example reveals an important insight: although the original relation (8.81) is linear in the random variable u, the parameter estimation $\theta = w/u$ involves a nonlinear transformation of the uncertainty through inversion. As a result, a uniform distribution of u induces a non-uniform distribution for θ, and higher-order statistical effects influence the mean. This demonstrates that linear intuition for statistical moments can fail when nonlinear diagnostics are involved. □

Example 8.24 (Parameter estimation in a linear system via least squares). Consider the two-dimensional linear system

$$\frac{\mathrm{d}x}{\mathrm{d}t} = ay, \qquad \frac{\mathrm{d}y}{\mathrm{d}t} = bx, \tag{8.83}$$

which is linear in both the parameter vector $\boldsymbol{\theta} = (a, b)^\mathrm{T}$ and the state vector $(x, y)^\mathrm{T}$. Let us first assume observations of x, y, and their time derivatives $\dot{x} := \mathrm{d}x/\,\mathrm{d}t$ and $\dot{y} := \mathrm{d}y/\,\mathrm{d}t$ are available at discrete times t_i, $i = 1, \ldots, I$.

To estimate the two parameters, define, at each time t_i, the regression matrix:

$$\mathbf{M}_i = \begin{pmatrix} y_i & 0 \\ 0 & x_i \end{pmatrix} \in \mathbb{R}^{2\times 2}, \qquad \mathbf{z}_i = \begin{pmatrix} \dot{x}_i \\ \dot{y}_i \end{pmatrix}. \tag{8.84}$$

Stacking all observations, the parameters can be estimated via least squares, yielding

$$\boldsymbol{\theta}_{\mathrm{LS}} = \left(\sum_{i=1}^{I} \mathbf{M}_i^{\mathrm{T}}\mathbf{M}_i\right)^{-1} \left(\sum_{i=1}^{I} \mathbf{M}_i^{\mathrm{T}}\mathbf{z}_i\right). \tag{8.85}$$

In component form, the estimator for a reads

$$a_{\mathrm{LS}} = \frac{\sum_i y_i \dot{x}_i}{\sum_i y_i^2}. \tag{8.86}$$

Now consider the practically relevant case where only x_i and $\dot{x}_i$ are observed. Since estimating a requires y_i through (8.86), uncertainty arises when y_i must be inferred. Assume y_i is modeled as a Gaussian random variable and written in Reynolds decomposition form,

$$y_i = \langle y_i \rangle + y_i', \qquad \langle y_i' \rangle = 0. \tag{8.87}$$

In this case, a_{LS} becomes a random variable. Rather than computing the exact distribution of the estimator, which is complicated due to the nonlinear dependence on y_i, we adopt a simple approximation by replacing the random quadratic terms in the least-squares formula with their expected values. This approximation is sufficient for our purpose here, which is to illustrate how uncertainty in an unobserved variable enters parameter estimation through nonlinear diagnostic relationships and systematically alters the estimated parameters, even when the underlying dynamics are linear. Using

$$\langle y_i^2 \rangle = \langle y_i \rangle^2 + \langle (y_i')^2 \rangle, \tag{8.88}$$

the approximate estimator becomes

$$a \approx \frac{\sum_i \langle y_i \rangle \dot{x}_i}{\sum_i \left(\langle y_i \rangle^2 + \langle (y_i')^2 \rangle\right)}. \tag{8.89}$$

Here we treat $\dot{x}_i$ as deterministic (or independent of y_i) for simplicity, so that the expectation only affects the quadratic terms in y_i. This expression highlights one important feature. Although the underlying dynamical system (8.83) is linear, the least-squares diagnostic involves quadratic and reciprocal operations. As a result, uncertainty in y_i enters the estimator through the variance term $\langle (y_i')^2 \rangle$, leading to systematic bias.

This can be understood even at a single time instant. If y_i is deterministic and nonzero, then $a = \dot{x}_i / y_i$. However, if y_i is stochastic, $\langle \dot{x}_i / y_i \rangle \neq \dot{x}_i / \langle y_i \rangle$, because inversion is nonlinear, as shown in the previous example. Thus, uncertainty in y_i fundamentally alters parameter estimation through the nonlinear diagnostic criterion, even for linear dynamics. □

Let us now explore how nonlinear diagnostics alter uncertainty quantification in a more practical setting. Oceanic eddies are rotating flow structures that play a crucial role in transporting momentum, heat, and tracers in the ocean. Because of their complex spatial structure, there is no universal criterion for eddy identification. One widely used diagnostic is the Okubo-Weiss (OW) parameter [206, 278], defined by

$$\Lambda = s_{\mathrm{n}}^2 + s_{\mathrm{s}}^2 - \omega^2, \tag{8.90}$$

where the normal strain, shear strain, and relative vorticity are

$$s_\mathrm{n} = u_x - v_y, \qquad s_\mathrm{s} = v_x + u_y, \qquad \omega = v_x - u_y, \tag{8.91}$$

with (u, v) denoting the two-dimensional velocity field. Negative values of Λ indicate regions dominated by rotation and are commonly associated with eddies. Because Λ depends quadratically on velocity gradients, it is a nonlinear diagnostic of the flow field.

Example 8.25 (Uncertainty propagation in eddy identification via the OW parameter). To illustrate how uncertainty propagates through a nonlinear diagnostic, consider a simple two-dimensional incompressible flow described by a stream function $\psi(\mathbf{x}, t)$, where $\mathbf{x} = (x, y)$. In two dimensions, defining the velocity by $u = -\partial_y \psi$ and $v = \partial_x \psi$, automatically ensures incompressibility ($\nabla \cdot \mathbf{u} = 0$).

Assume that ψ consists of only four Fourier modes,

$$\psi(\mathbf{x}, t) = \sum_{\mathbf{k} \in \mathcal{K}} \hat{\psi}_\mathbf{k}(t) e^{i\mathbf{k}\cdot\mathbf{x}}, \qquad \mathcal{K} = \{(1,0), (-1,0), (0,1), (0,-1)\}. \tag{8.92}$$

The complex coefficients $\hat{\psi}_\mathbf{k}(t)$ are assumed to be independent Gaussian random variables with zero mean and variance σ^2. The reality condition $\hat{\psi}_{-\mathbf{k}} = \hat{\psi}^*_\mathbf{k}$ ensures that ψ is real-valued. Using the definition of the velocity, the nonzero velocity gradients are

$$\partial_y u = \hat{\psi}_{(0,1)} e^{iy} + \hat{\psi}_{(0,-1)} e^{-iy} := X, \tag{8.93}$$

$$\partial_x v = -\left(\hat{\psi}_{(1,0)} e^{ix} + \hat{\psi}_{(-1,0)} e^{-ix}\right) := Y, \tag{8.94}$$

while $\partial_x u = \partial_y v = 0$. Here, X and Y are real-valued random variables.

Because X and Y are linear combinations of Gaussian random variables, they are Gaussian as well. Moreover, since the two Fourier modes contributing to X (and similarly Y) are complex conjugates with equal variance, their contributions add together. A direct calculation gives

$$\langle X \rangle = \langle Y \rangle = 0, \qquad \langle X^2 \rangle = \langle Y^2 \rangle = 4\sigma^2, \qquad \langle XY \rangle = 0. \tag{8.95}$$

Substituting into the OW definition yields

$$\Lambda = X^2 + 6XY + Y^2, \tag{8.96}$$

which can be written as the quadratic form

$$\Lambda = \begin{pmatrix} X & Y \end{pmatrix} \begin{pmatrix} 1 & 3 \\ 3 & 1 \end{pmatrix} \begin{pmatrix} X \\ Y \end{pmatrix}. \tag{8.97}$$

Diagonalizing the matrix leads to

$$\Lambda = 4Z_1^2 - 2Z_2^2, \tag{8.98}$$

where Z_1 and Z_2 are independent Gaussian random variables with variance $4\sigma^2$. Since the square of a Gaussian random variable follows a χ^2 distribution (with one degree of freedom, up to a scaling factor), each term in Λ is χ^2-distributed. As a result, Λ is given by the difference of two scaled χ^2 random variables and therefore does not follow a Gaussian distribution. Although the velocity field itself is Gaussian, the OW parameter depends quadratically on velocity gradients. This quadratic nonlinearity produces an asymmetric distribution with heavy tails, demonstrating how nonlinear diagnostics can fundamentally

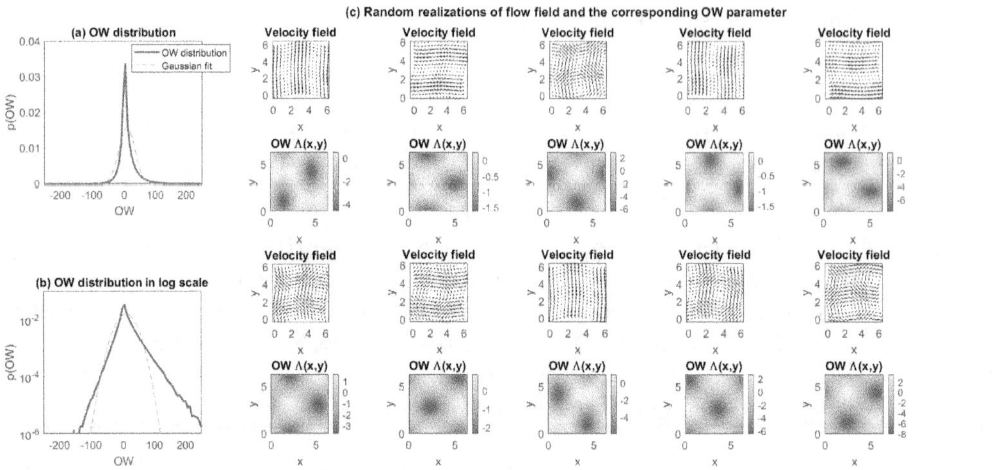

FIGURE 8.7: Eddy identification test results. Panels (a): PDF of the OW parameter and its Gaussian fit. Panel (b): similar to Panel (a), but the y-axis is plotted on a logarithmic scale. Panel (c): random realizations of the flow field and the corresponding OW parameter. The parameter $\sigma = 1$ is used in this numerical test.

alter the uncertainty characteristics of derived quantities, even when the underlying flow field is statistically simple [198].

Figure 8.7 show a numerical experiment by setting $\sigma = 1$. Note that, in this setup, Λ has the same distribution at different spatial grid point.

This example provides a clear illustration of non-Gaussianity arising from nonlinear turbulence diagnostics.

□

8.4 Code listings for figures and examples

- `Chapter_8_Gaussian_t_Distributions.m`: generates Figure 8.3 in Example 8.4, which compares a Student's t-distribution with the Gaussian distribution and their corresponding Shannon entropies.

- `Chapter_8_Optimal_Distribution_Relative_Entropy.m`: generates Figure 8.4 in Example 8.15, which determines the optimal distribution by minimizing relative entropy.

- `Chapter_8_UQ_Multiple_Obs.m`: generates Figure 8.5 in Example 8.16, which illustrates uncertainty reduction as a function of the number of observations.

- `Chapter_8_UQ_Linear_Nonlinear_Models.m`: generates Figure 8.6 in Example 8.22, which illustrates uncertainty propagation in simple scalar linear and nonlinear equations.

- `Chapter_8_UQ_Eddy.m`: generates Figure 8.7 in Example 8.25, which illustrates uncertainty propagation in eddy identification via the Okubo-Weiss parameter.

9

Causal Inference

9.1 Fundamentals of causal inference

Causal inference determines cause-and-effect relationships between variables [213, 132]. It has found wide application across disciplines such as atmospheric and ocean sciences, economics, and neuroscience. In addition to discovering the interactions between variables, causal inference plays a significant role in model identification, policy evaluation, and decision-making.

9.1.1 Correlation vs. causality

The difference between correlation and causality is one of the most fundamental concepts in statistics and data analysis. Correlation describes a statistical relationship in which two variables tend to vary together in a systematic way. It is commonly measured through correlation coefficients,

$$\mathrm{Corr}(X, Y) = \frac{\mathbb{E}[(X - \mu_X)(Y - \mu_Y)]}{\sigma_X \sigma_Y}, \tag{9.1}$$

where μ_X and σ_X are the mean and standard deviation of X, and μ_Y and σ_Y are those of Y. The correlation coefficient $\mathrm{Corr}(X, Y)$ ranges from -1 to 1, indicating the strength and direction of a linear association. However, correlation alone does not indicate why the variables move together. It merely quantifies the degree to which these two variables are statistically related.

In contrast, causality refers to a genuine cause-and-effect relationship in which changes in one variable directly produce changes in another. Different from correlation, which is symmetric, causality has a clear direction. If variable X causes variable Y, then intervening on X will lead to changes in Y, but intervening on Y will not necessarily affect X. While causal relationships often induce correlation, the reverse is not true. Correlation does not necessarily imply causation.

9.1.2 An overview of causal inference

In the following subsections, we will introduce the concepts of (1) confounding variables, (2) interventions, (3) counterfactuals, and (4) reverse causation. They play crucial roles in the study of causal inference.

Causal inference usually starts with an observed correlation. However, correlation alone is insufficient because it may arise from confounding variables, which are third factors that influence both variables of interest. Therefore, identifying and accounting for confounders is essential for ruling out spurious associations. Next, to define causality precisely, we need to consider the so-called interventions. A causal effect exists if actively forcing a variable to change would alter the outcome. Interventions formalize this idea by separating passive

DOI: 10.1201/9781003738459-9

observation from active manipulation and by breaking the natural links between a variable and its usual causes. Yet, despite the central role of interventions, causal effects cannot be directly observed at the individual level. This leads to counterfactual reasoning. For any individual or system, we can observe only one outcome, which is the one that actually occurred. The outcome under an alternative intervention remains unobserved. Causal inference thus becomes a missing-data problem, requiring assumptions and models to estimate counterfactual outcomes. Finally, causal reasoning must establish the correct direction of influence. Reverse causation occurs when cause and effect are mistakenly changed. This is often because variables influence each other over time or because future expectations affect present behavior. Careful attention to temporal order, interventions, and plausible mechanisms is required to avoid such errors. Table 9.1 includes these concepts and the questions they aim to clarify.

TABLE 9.1: Key concepts in causal inference and the questions they clarify.

Concept	Core question	What it clarifies
Correlation	Do variables move together?	Describes statistical association but does not explain why it occurs
Confounder	Could a third factor explain the association?	Explains spurious correlations caused by common influences
Interventions	What happens if we force a change?	Defines causality by breaking natural dependencies and mimicking experiments
Counterfactuals	What would have happened under a different action?	Shows why causal effects are unobservable and must be inferred
Reverse causation	Is the direction of cause correctly identified?	Prevents mistakenly swapping cause and effect

Based on these concepts, the following roadmap may serve as guidance for studying causal inference. Each step corresponds to a key question that needs to be addressed before a causal claim can be justified. We begin with an observed correlation, which motivates the investigation. We then ask whether the correlation could be explained by confounding variables, that is, factors that influence both variables simultaneously. Next, we consider interventions to investigate what would happen if we could actively force one variable to change. Because such interventions cannot usually be performed for every individual, we must reason in terms of counterfactuals to ask what would have happened under an alternative action. Finally, we examine whether the assumed direction of causality is correct, guarding against reverse causation.

Observed correlation

↓

Could this be confounding?

↓

What if we intervene?

↓

What is the counterfactual?

↓

Is the causal direction correct?

Below, we introduce each of these concepts.

9.1.3 Confounding variables

Many correlated relationships are explained by confounding variables, which are third factors that influence both variables being studied.

Example 9.1 (Confounding variables). Consider the observed relationship between ice cream sales and drowning incidents in Table 9.2. We observe that over the summer months,

TABLE 9.2: The observed ice cream sales, drowning incidents, and temperature.

Month	Ice Cream Sales	Drowning Deaths	Temperature
June	\$10,000	15	75°F
July	\$15,000	25	85°F
August	\$12,000	18	80°F

both ice cream sales and drowning deaths increase significantly. Based on such a strong correlation, we may mistakenly conclude that eating ice cream causes drowning. However, the true explanation lies in the temperature, which is called a confounding variable. High temperatures cause more people to buy ice cream for refreshment and also make more people want to swim. The increased number of people who swim naturally leads to more drowning accidents, which creates the spurious correlation between ice cream and drowning without any direct causal link between them. □

9.1.4 Interventions

Interventions form the basis of causal reasoning by distinguishing between seeing and doing. When we observe that two variables are associated, we cannot know if this relationship reflects causation without understanding what happens when we actively change one variable. The mathematical tool that formalizes this is the *do*-operator. While conditional probability, $P(Y|X = x)$, describes what we see in the data when X takes a value x, the interventional probability, $P(Y|do(X = x))$, describes what happens when we actively force X to be x. This action of $do(X = x)$ is crucial because it simulates a randomized experiment, breaking all the natural causes that would normally influence X and allowing us to see its pure effect on Y.

Example 9.2 (Interventions).

To see the difference between the standard conditional distribution and the *do*-operator, consider a lighting system controlled by an automatic timer. This timer (T) is a common cause. It turns the switch (S) on at night and also powers the light (L), making it illuminate. During the day, the timer turns both the switch and the light off. If we only collect passive observational data, we will find a perfect correlation: whenever the switch is on, the light is on. This leads to a conditional probability of $P(\text{light on}|\text{switch} = 1) = 1$, which misleadingly suggests that flipping the switch causes the light to turn on.

The *do*-operator allows us to test this hypothesis. The expression $P(\text{light on}|do(\text{switch} = 1))$ represents the probability that the light is on after we physically intervene and flip the switch to the "on" position, irrespective of the timer. If we perform this intervention during the day, the timer is still set to "day mode" and will not send power to the light. Consequently, even with the switch forced on, the light remains off, and we find that $P(\text{light on}|do(\text{switch} = 1)) = 0$. This result reveals the true causal structure. The switch itself has no causal power over the light. Their correlation was entirely driven by the hidden common cause, the timer. This example demonstrates how the *do*-operator separates causal effects from spurious correlations, providing a foundation for reliable causal inference. □

9.1.5 Counterfactuals

Counterfactuals involve reasoning about "what might have been". It considers alternative scenarios that did not actually occur. While interventions look forward ("what if I do this?"), counterfactuals look backward ("what would have happened if I had done something different?"). This type of reasoning is fundamental to learning from past experiences and making better decisions in the future.

Example 9.3 (Counterfactuals).

Imagine Sarah takes an aspirin and her headache disappears within an hour. The factual outcome is that her headache went away. The counterfactual question is: "Would Sarah's headache have persisted if she hadn't taken the aspirin?" We can never know the definite answer for a single individual, but we can estimate probabilities using groups of similar people. If clinical trials show that 70% of people with headaches still have them after two hours without aspirin, we might estimate a 70% probability that Sarah's headache would have persisted without treatment. This counterfactual reasoning helps us understand whether the aspirin actually caused the headache relief. □

9.1.6 Reverse causation

Reverse causation arises when the direction of cause and effect is mistakenly reversed. In such cases, what we interpret as the "cause" is actually a consequence of the true causal factor. This issue often occurs when two variables influence each other over time or when data are collected in a way that obscures the true temporal order of events. Identifying reverse causation requires careful attention to timing, logical precedence, and plausible mechanisms linking the variables.

Example 9.4 (Reverse causation). A well-known illustration comes from studies examining the relationship between economic growth and stock market performance. At first glance, one might notice that strong stock market returns are followed by periods of economic growth and conclude that rising stock prices drive economic expansion. In reality, the opposite direction is more credible: when investors expect future economic growth, they bid up stock prices today. Thus, it is the anticipation of future growth that causes current stock price increases, not the other way around. This example highlights how confusing the direction of causality can lead to misleading conclusions. □

9.2 Time-series causality

9.2.1 Granger causality

Granger causality provides a statistical framework for evaluating whether one time series contains predictive information about another [118]. The basic idea is intuitive. If including past values of variable X significantly enhances forecasts of variable Y beyond what can be achieved using Y's own past, then X is said to Granger-cause Y. It is worth highlighting that Granger causality is based on predictive rather than true causality. Granger causality detects temporal precedence and incremental predictive power, not direct causal mechanisms. Nevertheless, it is a valuable tool in fields such as economics, neuroscience, and Earth science, where controlled experiments are often infeasible.

TABLE 9.3: Monthly data on consumer confidence (X) and retail sales (Y).

Month	t	X_t	Y_t	X_{t-1}	Y_{t-1}	$\hat{Y}_t$ (Model 1)
January	1	105	1000	–	–	–
February	2	108	1020	105	1000	1005
March	3	110	1040	108	1020	1035
April	4	107	1030	110	1040	1055
May	5	112	1060	107	1030	1045
June	6	115	1080	112	1060	1075

Mathematically, we compare two autoregressive models:

$$\begin{aligned}
\text{Model 1 (Restricted):}\quad & Y_t = \alpha_0 + \sum_{i=1}^{m} \alpha_i Y_{t-i} + \epsilon_t, \\
\text{Model 2 (Unrestricted):}\quad & Y_t = \beta_0 + \sum_{i=1}^{m} \beta_i Y_{t-i} + \sum_{j=1}^{n} \gamma_j X_{t-j} + \nu_t.
\end{aligned} \tag{9.2}$$

Model 1 uses only the past of Y to predict its current value, while Model 2 augments this with past values of X. If X truly provides predictive information, the residual (unexplained) variation from Model 2 should be notably smaller than that from Model 1. Although the variables presented here are scalars for notational simplicity, the method can easily be generalized to vector forms.

The null hypothesis of no Granger causality is:

$$H_0 : \gamma_1 = \gamma_2 = \cdots = \gamma_n = 0, \tag{9.3}$$

meaning that the lagged values of X do not help predict Y. The test statistic is based on an F-test comparing the two models [122]:

$$F = \frac{(RSS_1 - RSS_2)/n}{RSS_2/(T - m - n - 1)}, \tag{9.4}$$

where RSS_1 and RSS_2 are the residual sums of squares from the restricted and unrestricted models, T is the sample size, and n is the number of X lags added. The F-test evaluates whether the reduction in residual error achieved by adding X is statistically significant, accounting for the loss of degrees of freedom. Under H_0, the statistic follows an $F_{n,T-m-n-1}$ distribution.

Example 9.5 (Economic indicators). Consider monthly data on consumer confidence (X) and retail sales (Y) in Table 9.3. Suppose we use one lag for each variable ($m = n = 1$): The fitted models are:

$$\begin{aligned}
\text{Model 1 (Y only):}\quad & \hat{Y}_t = 50 + 0.95Y_{t-1}, \quad RSS_1 = 1225, \\
\text{Model 2 (Y + X):}\quad & \hat{Y}_t = 30 + 0.85Y_{t-1} + 1.5X_{t-1}, \quad RSS_2 = 73.
\end{aligned} \tag{9.5}$$

The F-statistic is computed as:

$$F = \frac{(1225 - 73)/1}{73/(6 - 1 - 1 - 1)} = 47.33. \tag{9.6}$$

The critical value for $F_{1,3}$ at $\alpha = 0.05$ is 10.13. Since $47.33 > 10.13$, we reject the null hypothesis that $\gamma_1 = 0$.

This result indicates that past consumer confidence provides statistically significant predictive power for current retail sales. Model 2's residual error ($RSS_2 = 73$) is much smaller than Model 1's ($RSS_1 = 1225$), confirming the improvement in forecast accuracy. However, it is essential to note that Granger causality reflects predictive rather than true causality. Both variables might still be jointly influenced by broader macroeconomic conditions. In practice, the test suggests that consumer confidence is a valuable leading indicator for forecasting retail sales. □

9.2.2 Transfer entropy

While Granger causality assesses directional predictability through linear regression, transfer entropy generalizes this idea using concepts from information theory [239, 136]. It measures the directed flow of information from one time series to another, capturing both linear and nonlinear dependencies. In essence, transfer entropy quantifies how much knowing the past of variable X reduces the uncertainty in the future of variable Y, beyond what can be explained by Y's own past.

Different from Granger causality, which assumes a specific linear model and Gaussian noise, transfer entropy does not assume a specific parametric model. Transfer entropy relies purely on observed probability distributions. This makes it especially useful in understanding complex dynamical systems with nonlinear and non-Gaussian features, such as those in neuroscience, atmosphere and ocean science, and ecological dynamics, where standard linear assumptions often fail. For linear Gaussian systems, transfer entropy reduces to a function of the prediction error variance and is therefore equivalent to Granger causality.

Mathematically, transfer entropy from X to Y is defined as:

$$\mathrm{TE}_{X\to Y} = \sum_{y_{t+1}, y_t^{(m)}, x_t^{(n)}} p(y_{t+1}, y_t^{(m)}, x_t^{(n)}) \log \frac{p(y_{t+1}|y_t^{(m)}, x_t^{(n)})}{p(y_{t+1}|y_t^{(m)})}, \tag{9.7}$$

where $y_t^{(m)}$ represents the m previous values of Y, and $x_t^{(n)}$ represents the n previous values of X. Even though these are vectors, we omit the bold font to keep the notation simple here. Intuitively, $p(y_{t+1}|y_t^{(m)})$ captures how well we can predict Y using its own past while $p(y_{t+1}|y_t^{(m)}, x_t^{(n)})$ captures how prediction improves when we also include X's past. Their ratio (inside the logarithm) measures the incremental information gain contributed by X. A positive value of $\mathrm{TE}_{X\to Y}$ indicates that knowledge of X's history provides additional predictive information about Y, which is analogous to rejecting the null hypothesis in a Granger causality test.

Example 9.6 (Comparing Granger causality and transfer entropy). Consider two time series $\{X_t\}$ and $\{Y_t\}$ generated by the following nonlinear process:

$$\begin{aligned} &X_t \sim \mathrm{Uniform}(0,1), \\ &Y_t = 0.2\,Y_{t-1} + F(X_{t-1}, Y_{t-1}) + \varepsilon_t, \quad \varepsilon_t \sim \mathcal{N}(0, 0.5^2), \end{aligned} \tag{9.8}$$

where

$$F(X_{t-1}, Y_{t-1}) = \begin{cases} +2, & X_{t-1} > 0.8 \text{ and } Y_{t-1} < 0, \\ -2, & X_{t-1} > 0.8 \text{ and } Y_{t-1} \geq 0, \\ 0, & \text{otherwise.} \end{cases} \tag{9.9}$$

Intuitively, when X_{t-1} is large, it pushes Y_t either upward or downward depending on Y_{t-1}'s sign. This generates a strong but nonlinear and state-dependent relationship. See Figure

9.1 for the time sequence of X_t and Y_t as well as their nonlinear relationship. However, the average linear correlation between X_{t-1} and Y_t may be close to zero even though X_{t-1} clearly influences Y_t, which can be easily seen from Panel (c) of Figure 9.1.

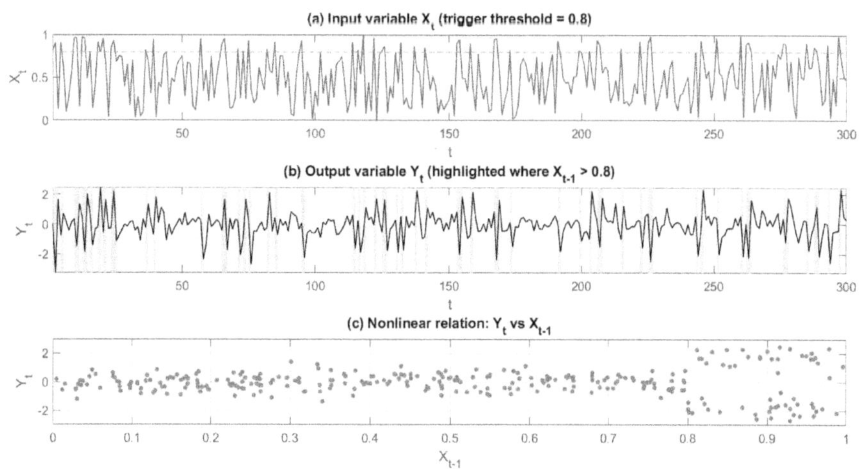

FIGURE 9.1: Comparison of the Granger causality and transfer entropy using the model in (9.8)–(9.9). Panels (a)–(b): time sequences of X_t and Y_t. Panel (c): nonlinear relation between Y_t and X_{t-1}.

To apply the Granger causality analysis, we test whether past values of X improve predictions of Y in a linear autoregressive model:

$$\begin{aligned} \text{Model 1 (Restricted): } Y_t &= \alpha_0 + \alpha_1 Y_{t-1} + \epsilon_t, \\ \text{Model 2 (Unrestricted): } Y_t &= \beta_0 + \beta_1 Y_{t-1} + \gamma_1 X_{t-1} + \nu_t. \end{aligned} \tag{9.10}$$

Let RSS_1 and RSS_2 denote the residual sum of squares from the restricted and unrestricted models, respectively. We set the length of the time sequence be $T = 300$. The two residuals are close to each other $\text{RSS}_1 = 279.26$ and $\text{RSS}_2 = 278.21$. The F-statistic in (9.4) is $F = 1.12$. This leads to a p-value of $p = 0.29 > \alpha = 0.05$. Therefore, the Granger test fails to detect causality. This happens because X_{t-1}'s positive and negative effects cancel out when modeled linearly.

In contrast, transfer entropy can provide a different perspective on this problem. For presentation purposes, we estimate a coarse-grained version of the transfer entropy by discretizing the continuous variables:

$$X_{t-1}^{(d)} = \begin{cases} 1, & X_{t-1} > 0.8, \\ 0, & \text{otherwise}, \end{cases} \quad Y_{t-1}^{(d)} = \begin{cases} 1, & Y_{t-1} > 0, \\ 0, & Y_{t-1} \le 0, \end{cases} \quad Y_t^{(d)} = \begin{cases} 1, & Y_t > 0, \\ 0, & Y_t \le 0, \end{cases} \tag{9.11}$$

which is nevertheless sufficient to demonstrate its advantages over the Granger causality. From the simulated data, we estimate joint frequencies for (y_t, y_{t-1}, x_{t-1}) and compute the conditional probabilities. A simplified summary of the main terms is shown in Table 9.4. Plugging these probabilities in (9.7) leads to

$$\begin{aligned} \text{TE}_{X\to Y} &= 0.10 \log_2 \frac{0.90}{0.55} + 0.10 \log_2 \frac{0.80}{0.45} + 0.20 \log_2 \frac{0.45}{0.55} + 0.20 \log_2 \frac{0.40}{0.45} + \cdots \\ &\approx 0.21 \text{ bits}. \end{aligned}$$

TABLE 9.4: Applying transfer entropy to detect the causality of the data generated by the nonlinear model (9.8). A coarse-grained version of the transfer entropy (9.4) is utilized here.

x_{t-1}	y_{t-1}	y_t	$p(y_t, y_{t-1}, x_{t-1})$	$p(y_t \vert y_{t-1})$	$p(y_t \vert y_{t-1}, x_{t-1})$
1	0	1	0.10	0.55	0.90
1	1	0	0.10	0.45	0.80
0	0	1	0.20	0.55	0.45
0	1	0	0.20	0.45	0.40

Because $\text{TE}_{X \to Y} > 0$, we conclude that knowledge of X_{t-1} improves the predictability of Y_t beyond the information contained in Y_{t-1} alone. This is consistent with the observations in Panel (c) of Figure 9.1.

This example highlights why transfer entropy can reveal relationships that Granger causality misses. The Granger test assumes a constant linear effect of X_{t-1} on Y_t, so opposite effects in different states cancel out. In contrast, transfer entropy measures information flow across the entire conditional distribution, capturing nonlinear and state-dependent dependencies. In practice, Granger causality serves as a useful first check for linear systems, while transfer entropy provides deeper insight when nonlinear interactions or complex dynamics are suspected.

□

9.3 Graph-based methods

Graph-based methods provide a powerful framework for representing and reasoning about causal relationships among variables. In this view, a system is modeled as a directed acyclic graph (DAG) [26], where each node represents a variable and each directed edge $X \to Y$ represents a direct causal influence of X on Y. These methods advance both qualitative reasoning (e.g., identifying whether causal effects can be estimated from data) and quantitative estimation (e.g., computing the average causal effect).

9.3.1 The structural causal model (SCM)

The structural causal model (SCM) is a graph-based causal model [213]. It is defined by a set of variables $\{X_1, X_2, \ldots, X_n\}$, a DAG $\mathcal{G}$ encoding causal dependencies, and structural equations of the form:

$$X_i = f_i(\text{Pa}(X_i), U_i), \tag{9.12}$$

where $\text{Pa}(X_i)$ are the parents of X_i in $\mathcal{G}$, and U_i are independent external noise terms. The setup allows us to reason about interventions using the $do(\cdot)$ operator, which conceptually "breaks" incoming edges into the variable being intervened on.

9.3.2 D-separation: Identifying conditional independence

Consider the causal graph:

$$X \to Z \to Y, \tag{9.13}$$

where X causes Z, and Z causes Y. Under the structural model assumptions, X and Y are conditionally independent given Z. That is,

$$X \perp Y \mid Z, \tag{9.14}$$

because once we control for the mediator Z, knowing X provides no additional information about Y. This property, called d-separation [130], is key to directly reading off independence relationships from the graph structure.

To make the conditioning explicit, we regress both X and Y on Z and examine their residuals. In the linear setting, this corresponds to the projections

$$X = \alpha_X Z + r_X, \qquad Y = \alpha_Y Z + r_Y, \tag{9.15}$$

where α_X and α_Y are the least-squares regression coefficients, and r_X and r_Y are the corresponding residuals. By construction, the residuals satisfy

$$\mathbb{E}[r_X Z] = 0, \qquad \mathbb{E}[r_Y Z] = 0. \tag{9.16}$$

In the linear Gaussian setting, conditioning on Z is equivalent to examining the statistical relationship between the residuals r_X and r_Y.

Example 9.7 (D-separation). Recall the DAG in (9.13), where Z mediates the causal flow from X to Y. Consider the following specific relationship:

$$X = \epsilon_X, \qquad Z = 0.9X + \epsilon_Z, \qquad Y = 2Z + \epsilon_Y, \tag{9.17}$$

where ϵ_X, ϵ_Y and ϵ_Z are independent standard Gaussian white noise.

Figure 9.2 shows the scatter plots of variables, aiming to demonstrate their correlations. Panel (a) shows that X and Y are correlated because information flows through Z. Similarly, the correlation between Y and Z is clearly shown in Panel (b). However, once we condition on Z, that is, by examining the residuals of X and Y after regressing out Z, their correlation vanishes.This is because, once Z is fixed, the remaining variability in X comes from ϵ_X, while the remaining variability in Y comes from ϵ_Y. Since ϵ_X and ϵ_Y are independent, the residuals r_X and r_Y are uncorrelated. Consequently,

$$\mathrm{Corr}(r_X, r_Y) = 0, \tag{9.18}$$

which indicates the conditional independence $X \perp Y \mid Z$. This implies the principle of d-separation. In a chain structure $X \to Z \to Y$, X and Y become independent once Z is known. Panel (c) shows an uncorrelated scatter plot of these two residual terms, which confirms the argument above.

□

9.3.3 Intervention and the do-operator

In causal inference, the main question is not merely whether two variables are associated, but whether changing one variable would cause a change in another. This difference becomes particularly important when confounding variables are present.

Consider the following causal graph:

$$X \to Y, \qquad Z \to X, \qquad Z \to Y. \tag{9.19}$$

Here, X has a direct causal effect on Y, which is the effect we would like to understand. However, the variable Z is a common cause of both X and Y. Such a variable is called a confounder.

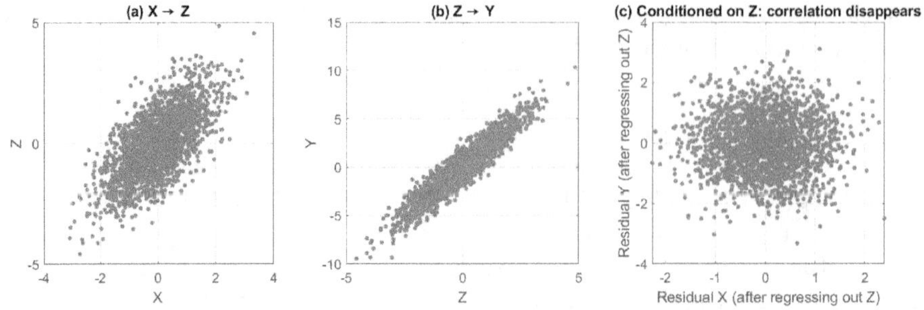

FIGURE 9.2: Scatter plots of variables X, Y, and Z, corresponding to the DAG in (9.13), where the detailed relationship is given by (9.17).

The first thing we need to explicitly point out is why observational relationships can be misleading. Let us assume we compute the conditional distribution $P(Y \mid X = x)$ from observational data. This quantity answers the question:

"Among individuals for whom X happens to take the value x, what does Y look like?"

The problem is that individuals with $X = x$ are not selected at random. Because Z influences X, those with similar values of X tend to have similar values of Z. Since Z also affects Y, the observed association between X and Y reflects two effects mixed together: (1) the genuine causal effect of X on Y along the path $X \to Y$, and (2) a spurious association created by the backdoor path $X \leftarrow Z \to Y$. As a result, the quantity $P(Y \mid X = x)$ does not, in general, represent the causal effect of X on Y.

Therefore, to precisely define causality, let us introduce an intervention. An intervention means that we actively force the variable X to take a particular value, rather than letting it be determined by its usual causes. This is the idea behind the do-operator, denoted by $do(X = x)$.

Conceptually, applying $do(X = x)$ corresponds to performing an experiment in which everyone is assigned the value $X = x$, regardless of the value of Z. In the causal graph, this intervention removes the arrow $Z \to X$, while leaving all other causal relationships unchanged. The variable Z continues to influence Y, but it no longer influences how X is assigned.

To understand the difference between interventional and observational quantities, let us start with the interventional distribution

$$P(Y \mid do(X = x)), \tag{9.20}$$

which answers a fundamentally different question from $P(Y \mid X = x)$. It asks:

"If we were to force X to equal x for everyone, what would Y become?"

Because X is externally set, the confounding influence of Z on X is eliminated. This makes $P(Y \mid do(X = x))$ the correct quantity for describing the causal effect of X on Y.

Although we usually cannot perform interventions directly, the causal graph tells us how to compute the interventional distribution using observational data. In this setting, the causal effect of X on Y can be obtained by averaging over the confounder Z:

$$P(Y \mid do(X = x)) = \sum_{z} P(Y \mid X = x, Z = z)\, P(Z = z). \tag{9.21}$$

This formula is known as the backdoor adjustment. Note that under intervention, X is no longer influenced by Z.

Equation (9.21) mirrors the logic of a randomized experiment. Individuals differ in their values of Z, but because X is assigned externally, these differences are no longer related to how X is chosen. By averaging over all possible values of Z, we recover the causal effect of X on Y while removing the spurious association caused by confounding.

To summarize, in the presence of confounding variables, observational associations do not generally reflect causal effects. The do-operator provides a formal way to describe interventions that isolate causality by breaking natural dependencies. The distinction between $P(Y \mid X)$ and $P(Y \mid do(X))$ captures the essential difference between observing what happens naturally and actively changing the system [213]. This idea underlies randomized experiments and forms the foundation of modern causal inference.

Example 9.8 (Observational v.s. interventional effects). Figure 9.3 illustrates the difference between the observational conditional expectation $E[Y \mid X]$ and the interventional expectation $E[Y \mid do(X)]$ in a system with confounding using the causal graph (9.19). The system is given as follows:

$$Z = \epsilon_Z, \qquad X = 0.8Z + \epsilon_X, \qquad Y = c_x X + 2Z + \epsilon_Y, \tag{9.22}$$

where ϵ_X, ϵ_Y and ϵ_Z are independent standard Gaussian white noise. Two regimes are studies here,

$$\begin{aligned} \text{Regime I:} \qquad & c_x = 1.5, \\ \text{Regime II:} \qquad & c_x = 0. \end{aligned} \tag{9.23}$$

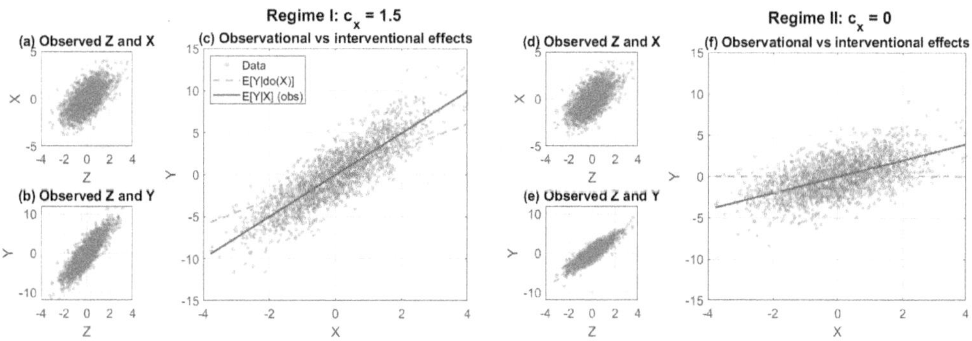

FIGURE 9.3: Comparison of observational and interventional effects in a confounded system. Panels (a)–(b): scatter points show simulated data between Z and X and between Z and Y in Regime I. Panel (c): scatter points between the simulated X and Y. The solid line represents the naive observational regression $E[Y \mid X]$, which includes both the causal effect of X on Y and the spurious association induced by Z. The dashed line represents the interventional expectation $E[Y \mid do(X)]$, computed by averaging over all values of Z via the backdoor adjustment formula. Panels (d)–(f) are similar but for Regime II.

The scatter points in Panels (a)–(b) show that the observed Z and X and the observed Z and Y directly from the simulation using (9.22) are both highly correlated. The scatter points in Panel (c) also illustrate that the raw simulated data for X and Y are correlated, where Z acts as a confounder influencing both variables. The solid line represents the naive regression of Y on X, i.e., $E[Y \mid X]$, which captures not only the causal effect of X on Y but also the spurious association induced by Z. In contrast, the dashed line corresponds to $E[Y \mid do(X)]$, computed by averaging over all values of Z according to the backdoor adjustment formula. Specifically, for a given x value, we simulate $P(Y \mid X = x, Z)$ for all observed values of Z (with added noise) and then average the results to obtain $E[Y \mid do(X = x)]$. This procedure effectively removes the confounding influence of Z and isolates the true causal effect of X on Y. As expected, the slope of the interventional line is lower than the observational regression. This leads to an important finding: naive conditioning can overestimate causal effects when confounders are present. Panels (d)–(f) show the case in Regime II, where Y does not directly rely on X according to (9.22) when $c_x = 0$. However, since X depends on Z and Z affects Y, a strong correlation between X and Y still exists. Yet, when applying the do-operator, the dashed line clearly indicates that there is no direct causal effect of X on Y. This example provides a concrete illustration of the do-operator: by externally setting X and averaging over Z, we simulate an intervention that reveals the genuine causal relationship between X and Y.

□

9.4 Causation entropy

9.4.1 Basic concept

Causation entropy provides an information-theoretic framework for quantifying directed causal influence in complex dynamical systems [253]. It specifically captures the directional information flow from one variable to another while conditioning on other relevant variables. This makes it particularly powerful for identifying genuine causal relationships in the presence of confounding factors and indirect couplings.

The causation entropy from variable $\mathbf{Z}$ to variable $\mathbf{X}$ conditioned on a set of variables $\mathbf{Y}$ is defined as:

$$C_{\mathbf{Z}\to\mathbf{X}|\mathbf{Y}} = \mathcal{S}(\mathbf{X} \mid \mathbf{Y}) - \mathcal{S}(\mathbf{X} \mid \mathbf{Y}, \mathbf{Z}) \tag{9.24}$$

where $\mathcal{S}(\cdot|\cdot)$ denotes conditional entropy defined in (8.34). This formulation measures the reduction in uncertainty about $\mathbf{X}$ when incorporating knowledge of $\mathbf{Z}$, beyond what is already provided by $\mathbf{Y}$. A positive causation entropy indicates that $\mathbf{Z}$ contains unique information about $\mathbf{X}$ that is not captured by the conditioning set $\mathbf{Y}$.

9.4.2 Model identification

One of the main applications of the causation entropy is data-driven model identification, particularly for discovering sparse dynamical structures [149]. Assume we have a n-dimensional time series of state variables $\mathbf{u} = [u_1, \ldots, u_n]^\mathtt{T}$. Let us assume a function library $\mathbf{f}(\mathbf{u}(t), t)$ which contains linear and nonlinear combinations of these variables. To learn the underlying dynamics of the observed time series, we start by assuming a governing equation that includes all functions in the library for each component of $\mathbf{u}$:

$$\frac{\mathrm{d}\mathbf{u}}{\mathrm{d}t} = \boldsymbol{\Theta} \times \mathbf{f}(\mathbf{u}(t), t), \tag{9.25}$$

with component-wise form:

$$\begin{bmatrix} \frac{du_1}{dt} \\ \frac{du_2}{dt} \\ \vdots \\ \frac{du_n}{dt} \end{bmatrix} = \begin{bmatrix} \xi_{1,1} & \cdots & \xi_{1,r} \\ \xi_{2,1} & \cdots & \xi_{2,r} \\ \vdots & \ddots & \vdots \\ \xi_{n,1} & \cdots & \xi_{n,r} \end{bmatrix} \begin{bmatrix} f_1\left(u_1(t), \ldots, u_n(t), t\right) \\ f_2\left(u_1(t), \ldots, u_n(t), t\right) \\ \vdots \\ f_r\left(u_1(t), \ldots, u_n(t), t\right) \end{bmatrix} \tag{9.26}$$

The goal is to find a sparse model structure, which corresponds to a sparse matrix $\boldsymbol{\Theta}$ and estimate its non-zero elements.

The causation entropy $C_{f_i \to \dot{u}_j | [\mathbf{f} \setminus f_i]}$ specifically quantifies the influence of the basis function f_i on the time derivative $\dot{u}_j := du_j/dt$ while accounting for all other basis functions. This directed measure helps distinguish between spurious correlations and genuine causal drivers in the system dynamics. Unlike Least Absolute Shrinkage and Selection Operator (LASSO) regression, which enforces sparsity through penalization, causation entropy evaluates conditional information transfer to determine structural relevance before parameter estimation.

The model identification procedure proceeds as follows. For each pair (i, j) where $i = 1, \ldots, r$ (basis functions) and $j = 1, \ldots, n$ (state variables), compute the causation entropy $C_{f_i \to \dot{u}_j | [\mathbf{f} \setminus f_i]}$. If this quantity is zero (or practically negligible), then f_i does not contribute meaningful information to $\dot{u}_j$ and the corresponding parameter $\xi_{i,j}$ is set to zero. This process yields a sparse structure for the coefficient matrix $\boldsymbol{\Theta}$, after which least squares or maximum likelihood estimation via quadratic optimization can often be used to determine the non-zero parameter values.

In practice, by approximating all joint and marginal distributions as Gaussians, the causation entropy can be computed analytically [72]:

$$
\begin{aligned}
C_{\mathbf{Z} \to \mathbf{X} | \mathbf{Y}} &= \mathcal{S}(\mathbf{X} \mid \mathbf{Y}) - \mathcal{S}(\mathbf{X} \mid \mathbf{Y}, \mathbf{Z}) \\
&= \mathcal{S}(\mathbf{X}, \mathbf{Y}) - \mathcal{S}(\mathbf{Y}) - \mathcal{S}(\mathbf{X}, \mathbf{Y}, \mathbf{Z}) + \mathcal{S}(\mathbf{Y}, \mathbf{Z}) \\
&= \frac{1}{2}\ln(\det(\mathbf{R}_{\mathbf{XY}})) - \frac{1}{2}\ln(\det(\mathbf{R}_{\mathbf{Y}})) - \frac{1}{2}\ln(\det(\mathbf{R}_{\mathbf{XYZ}})) + \frac{1}{2}\ln(\det(\mathbf{R}_{\mathbf{YZ}}))
\end{aligned}
\tag{9.27}
$$

where $\mathbf{R}_{\mathbf{XYZ}}$ denotes the covariance matrix of $(\mathbf{X}, \mathbf{Y}, \mathbf{Z})^{\mathtt{T}}$, with similar notation for other covariances. The Gaussian approximation enables practical computation of causation entropy even in high-dimensional settings. While exact probability density function reconstruction from time series data suffers from the curse of dimensionality, the Gaussian approach provides closed-form expressions that remain computationally tractable. Crucially, for model structure identification, we only need to determine whether the causation entropy is zero while the exact value is secondary. Although nonlinear dependencies may not always be captured fully by second-order statistics, in many practical dynamical systems strong causal influences manifest in covariance structure, making the Gaussian approximation effective for structural screening. This allows efficient determination of the sparse structure of $\boldsymbol{\Theta}$ while maintaining the nonlinear character of the underlying system dynamics.

Example 9.9 (Model identification using causation entropy). Consider the noisy Lorenz 63 model:

$$
\begin{aligned}
\frac{\mathrm{d}x}{\mathrm{d}t} &= \sigma(y - x) + \sigma_x \dot{W}_x, \\
\frac{\mathrm{d}y}{\mathrm{d}t} &= x(\rho - z) - y + \sigma_y \dot{W}_y, \\
\frac{\mathrm{d}z}{\mathrm{d}t} &= xy - \beta z + \sigma_z \dot{W}_z,
\end{aligned}
\tag{9.28}
$$

with the standard parameters $\sigma = 10$, $\rho = 28$ and $\beta = 8/3$ and small stochastic noise levels $\sigma_x = \sigma_y = \sigma_z = 1$. Figure 9.4 illustrates the application of causation entropy for identifying sparse dynamics of the noisy Lorenz 63 model in (9.28) from time series data with a total length of $t = 100$. The candidate functions include linear and quadratic terms: x, y, z, xy, yz, zx, x^2, y^2, z^2. The derivatives $\dot{x}$, $\dot{y}$, and $\dot{z}$ are computed using finite differences with a tiny numerical integration time step $\Delta t = 0.001$.

Figure 9.4 visualizes the original causation entropy matrix, which quantifies the causal influence of candidate functions on state variable derivatives. The seven terms with the highest causation entropy correspond precisely to the terms in the true dynamics. By applying a threshold (here 0.0004), the method recovers a sparse model that matches the true system. Notably, the y term in the dy/dt equation has a significantly lower entropy value. Ignoring this term has a negligible impact, as the system's statistics, including autocorrelation and equilibrium distributions, remain accurate. Such a finding indicates the ability of using the causation entropy to rank the relative importance of dynamical terms.

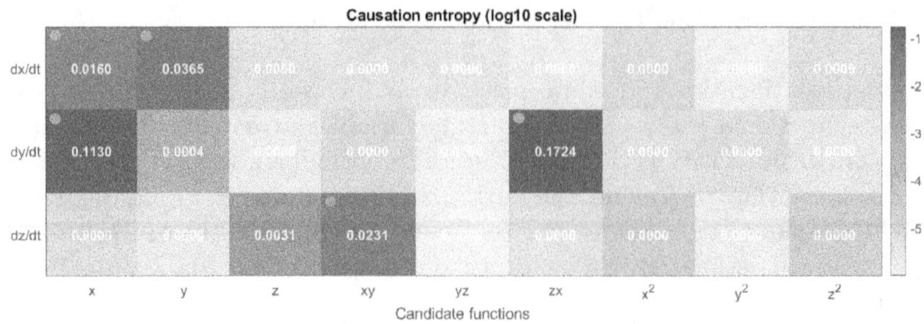

FIGURE 9.4: Causation entropy matrix for model selection. The matrix elements represent the contribution strength between model components. A logarithmic scale for the shade is adopted to enhance the contrast between dominant terms (selected for the model) and negligible terms (discarded). The true dynamical terms are indicated by a dot in each corresponding cell. The numerical value of each entropy, computed from the Gaussian approximation, is displayed in bold white.

□

9.5 Model-based methods

In this subsection, we briefly introduce two dynamical, model-based methods that facilitate the discovery of causal relationships within complex systems.

9.5.1 Detecting causal relationships via linear response theory

Linear response theory provides a systematic framework for inferring dynamical causal relationships in complex systems, particularly turbulent systems, by quantifying how statistical observables respond to small external perturbations [177, 25]. If a small perturbation applied to variable X induces a measurable change in a statistical observable of variable Y, then X exerts a dynamical causal influence on Y within the governing model. This notion of causality is defined relative to the assumed dynamical equations.

Mathematically, this relationship is captured by the response function:

$$\chi_{YX}(t) = \frac{\delta}{\delta f_X(0)} \langle g(Y(t)) \rangle_f \Big|_{f=0}, \tag{9.29}$$

where f_X denotes a small perturbing force applied to X, $g(Y(t))$ is a chosen observable of Y, and the expectation $\langle \cdot \rangle_f$ is taken with respect to the invariant measure of the perturbed system. The functional derivative is evaluated at zero forcing. Unlike purely correlational analyses, linear response theory explicitly exploits the temporal evolution of the dynamical system. Because the perturbation breaks equilibrium correlations and propagates forward in time, the resulting response function reveals directional dynamical influence. In practice, response functions can often be estimated from unforced simulations using fluctuation-dissipation relations, providing an efficient way to uncover the direction and strength of dynamical couplings between variables.

9.5.2 Information transfer

Another model-based method for detecting directional influence is information transfer. Consider a two-dimensional dynamical system

$$\frac{dx}{dt} = F_1(x, y), \tag{9.30a}$$

$$\frac{dy}{dt} = F_2(x, y). \tag{9.30b}$$

The information transfer from y to x is defined as the difference in the time evolution of the marginal entropy of x under two different conditions [168, 184]:

$$T_{y \to x} = \frac{\mathrm{d}}{\mathrm{d}t} H_1(p(x)) - \frac{\mathrm{d}}{\mathrm{d}t} H_1^*(p(x)). \tag{9.31}$$

Here, $H_1(p(x))$ denotes the marginal entropy of x derived from the full coupled system (9.30), where both x and y evolve dynamically. In contrast, $H_1^*(p(x))$ denotes the marginal entropy of x computed under a modified system in which y is treated as an externally fixed parameter when evaluating the evolution of x. In stochastic systems, the entropy evolution is governed by the corresponding Fokker-Planck equation. The difference between these two entropy rates isolates the contribution of y to the uncertainty evolution of x. A nonzero value of $T_{y \to x}$ indicates directed information flow from y to x under the system dynamics, meaning that the evolution of y actively influences the uncertainty and evolution of x. Unlike purely statistical measures such as Granger causality or transfer entropy, this method relies explicitly on the governing dynamical equations and characterizes causality within the specified model.

9.6 Instantaneous causal inference from an inverse problem viewpoint

In addition to purely data-driven or model-driven methods, incorporating partial observations into a dynamical model via data assimilation also provides a valuable framework for detecting cause-and-effect relationships. Notably, it has the unique advantage of revealing instantaneous causal relationships by examining whether observations of a future state of a time series $\mathbf{x}$ facilitate the reduction of the current uncertainty of another time series $\mathbf{y}$, through which a causal link can be established. Importantly, it also provides the instantaneous causal inference range, which is particularly important for revealing the predictability and attribution scale of events in complex dynamical systems. The method uniquely identifies dynamic causal interactions without requiring observations of candidate causes, accommodates short datasets, and scales efficiently to high dimensions. Crucially, it provides online tracking of causal roles, which may reverse intermittently.

9.6.1 Assimilative causal inference

Recall the Granger's predictive causality or its nonlinear extension, transfer entropy, presented in Section 9.2. A variable $\mathbf{y}$ is said to be the cause of $\mathbf{x}$ at time t if the knowledge of $\mathbf{y}(s \leq t)$ improves the prediction or reduces the uncertainty of $\mathbf{x}(t)$. Using information theory, this can be expressed as:

$$\mathcal{S}\big(p(\mathbf{x}(t)|\mathbf{y}(s \leq t))\big) < \mathcal{S}\big(p(\mathbf{x}(t))\big), \tag{9.32}$$

where $\mathcal{S}(\cdot)$ is the Shannon entropy and the history of the target variable in the conditioning, i.e., $\mathbf{x}(s < t)$, is omitted in (9.32) for notation simplicity.

The above argument can be interpreted in a reverse way, from the perspective of statistical inference through an inverse problem in uncertainty quantification. If $\mathbf{x}$ is the subsequent effect of $\mathbf{y}(t)$, then knowing future information about $\mathbf{x}$ reduces the uncertainty in inferring the current state of $\mathbf{y}$:

$$\mathcal{S}\big(p(\mathbf{y}(t)|\mathbf{x}(s > t))\big) < \mathcal{S}\big(p(\mathbf{y}(t))\big). \tag{9.33}$$

Equation (9.33) expresses the same causal mechanism as (9.32), but viewed from the opposite direction in time. If $\mathbf{y}(t)$ genuinely influences $\mathbf{x}$, then future realizations of $\mathbf{x}$ necessarily carry information about the present state of $\mathbf{y}$. In other words, one can infer the cause by observing its effects downstream in time. This reversal in perspective is standard in inverse problems: rather than predicting outputs from inputs, we infer the inputs from observed outputs using a dynamical model. Here, causality is revealed through how much uncertainty about $\mathbf{y}(t)$ can be reduced by observing future states of $\mathbf{x}$. Consider a time interval $[0, T]$ and a fixed time instant $t \in [0, T]$. Conditioning both sides of (9.33) on the past history of $\mathbf{x}$ yields

$$\mathcal{S}\big(p(\mathbf{y}(t) \mid \mathbf{x}(s \leq T))\big) < \mathcal{S}\big(p(\mathbf{y}(t) \mid \mathbf{x}(s \leq t))\big), \tag{9.34}$$

which preserves the causal implication in (9.33).

A key observation is that (9.34) has a direct interpretation in Bayesian data assimilation. The right-hand side corresponds to a filtering distribution, which estimates $\mathbf{y}(t)$ using information available up to time t. The left-hand side corresponds to a smoothing distribution, which additionally incorporates future observations of $\mathbf{x}$. Any reduction in uncertainty between these two distributions must therefore arise from information propagated forward from $\mathbf{y}(t)$ to $\mathbf{x}(s > t)$ through the system dynamics. For notation simplicity, they are denoted as $p_t^{\mathrm{s}} := p(\mathbf{y}(t)|\mathbf{x}(s \leq T))$ and $p_t^{\mathrm{f}} := p(\mathbf{y}(t)|\mathbf{x}(s \leq t))$, respectively.

The relative entropy provides a natural tool to quantify the uncertainty reduction in p_t^{s} related to p_t^{f}. This difference is crucial for reflecting the additional information gained by incorporating future observations. As such, if the relative entropy between the smoother and filter distributions is nonzero, i.e.,

$$\mathcal{P}(p_t^{\mathrm{s}}, p_t^{\mathrm{f}}) = \int p_t^{\mathrm{s}} \ln\big(p_t^{\mathrm{s}}/p_t^{\mathrm{f}}\big) > 0, \tag{9.35}$$

then $\mathbf{y}$ is identified as the cause of $\mathbf{x}$ at time t. A nonzero relative entropy in (9.35) indicates that future observations of $\mathbf{x}$ contain information about $\mathbf{y}(t)$ that is not present in the past alone. Since information cannot propagate backward in time without a dynamical connection, this asymmetry identifies $\mathbf{y}(t)$ as a causal contributor to $\mathbf{x}$ at time t. By computing (9.35) at different t, a time-dependent causal link is established. Since data assimilation is utilized during this process, the method is named assimilative causal inference (ACI) [17, 16].

Under (9.35), ACI provides a unique formulation for identifying causal relationships. In traditional approaches, causal links are inferred by running a model forward to quantify the informational response on the target variables' entropy when conditioned on the history of the candidate causes. ACI instead solves an inverse problem, thus tracing causes back from the data of the observed subsequent effects. By interpolating causality from effects to causes instead of extrapolating the latter forward-in-time, ACI offers an inherently more robust framework compared to classical predictive causality.

9.6.2 Causal inference range

While (9.35) detects whether a causal link exists at time t, it does not specify how long the influence of $\mathbf{y}(t)$ persists into the future of $\mathbf{x}$. In realistic turbulent and nonlinear systems, such influence is neither instantaneous nor permanent since information propagates forward in time but gradually dissipates. The causal inference range (CIR) quantifies this finite temporal footprint. This raises two key questions: How can we determine the CIR and, more importantly, how can we do so objectively, i.e., without relying on empirical cutoff thresholds?

Recall from (9.35) that the relative entropy $\mathcal{P}(p_t^{\mathrm{s}}, p_t^{\mathrm{f}})$ between the filter and smoother estimates quantifies the causal influence from $\mathbf{y}(t)$ to $\mathbf{x}$. Here, $p_t^{\mathrm{s}} = p(\mathbf{y}(t)|\mathbf{x}(s \leq T))$ incorporates the complete future information of $\mathbf{x}$. Due to the finite effective influence length, the complete smoother solution p_t^{s} achieves a substantial uncertainty reduction compared to the lagged posterior distributions $p(\mathbf{y}(t)|\mathbf{x}(s \leq T'))$, which only utilize future information up to $t \leq T' \leq T$, for a short time period after t. The difference between these distributions vanishes as $T' \to T$. As additional future observations of $\mathbf{x}$ are incorporated, the smoother estimate of $\mathbf{y}(t)$ becomes progressively more accurate, up to the point where further future data no longer provide meaningful new information. The CIR is defined as the time interval over which this additional information gain remains significant. We therefore define the CIR through the relative entropy criterion: For a predetermined threshold $\varepsilon \geq 0$, the maximum lag T'_ε satisfying

$$\mathcal{P}\big(p(\mathbf{y}(t)|\mathbf{x}(s \leq T)), p(\mathbf{y}(t)|\mathbf{x}(s \leq T'))\big) > \varepsilon, \tag{9.36}$$

yields the innate CIR measure $\tau_\varepsilon(t) = T'_\varepsilon - t$. This adaptive time variable $\tau_\varepsilon(t)$ defines the effective causal period $[t, t + \tau_\varepsilon(t)]$ during which $\mathbf{y}(t)$ exerts measurable influence on $\mathbf{x}$.

While the CIR computed via (9.36) is intuitive, it remains inherently subjective due to its dependence on the threshold parameter ε. In the following, an objective CIR formulation is developed, eliminating the need for a prescribed threshold. This threshold-independent measure is subsequently integrated into the ACI framework. For simplicity, assume that the relative entropy between the aforementioned complete and lagged smoother distributions decreases monotonically as more future information of $\mathbf{x}$ is incorporated. This commonly holds in practice, though the theory of the objective CIR doesn't require this condition. At a given time instant t, let $M = \mathcal{P}(p_t^{\mathrm{s}}, p_t^{\mathrm{f}})$ denote the maximum value of the relative entropy between the complete and lagged smoother distributions over $t \leq T' \leq T$, which is the difference between the smoother estimate and the filter estimate (where the latter can be viewed as a degenerate smoother using no future information). Given a threshold $\varepsilon \in [0, M]$, (9.36) yields a subjective CIR as $\tau_\varepsilon(t) = T'_\varepsilon - t$. The objective CIR is then naturally defined by

$$\tau(t) := \frac{1}{M}\int_0^M \tau_\varepsilon(t)d\varepsilon. \tag{9.37}$$

Dividing M guarantees the unit of the objective CIR is "time". The objective CIR $\tau(t)$ can be interpreted as an information-weighted average causal duration, measuring how long future observations meaningfully contribute to reducing uncertainty about $\mathbf{y}(t)$. By integrating over all possible thresholds, this definition removes subjective tuning and yields a robust, intrinsic causal timescale. Note that numerically evaluating the integral in (9.37) requires repeated smoother computations, leading to a computational complexity that scales

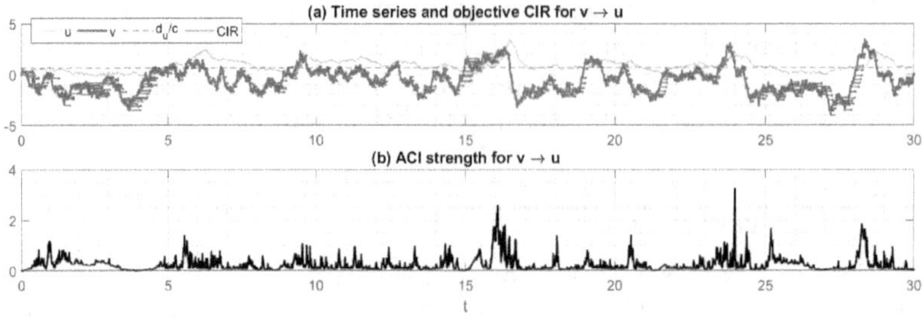

FIGURE 9.5: Assimilative causal inference. Panel (a): observed time series of $u(t)$ (thin curve), the stochastic damping $v(t)$ (thick curve), and the threshold d_u/c (dashed line). The horizontal lines show the CIR from v to u. Panel (b): ACI strength from v to u.

quadratically with the number of discretization points. An efficient computational method of (9.37) has been developed in [17].

The subjective and objective CIR can be regarded as analogs of the autocorrelation function and decorrelation time, respectively. The autocorrelation function measures the memory of a turbulent system based on the duration of its values that remain above a predetermined threshold. However, determining the memory using the autocorrelation function is subjective and can vary significantly depending on the threshold value. In contrast, the decorrelation time, which integrates the autocorrelation function, provides an objective way to define the memory length and is free from any predetermined cutoff threshold.

Example 9.10 (Assimilative causal inference for extreme events). Consider the dyad model and parameters introduced in (7.94)–(7.95). This system produces extreme events in the observable component u when the auxiliary variable v exceeds the threshold $v \geq d_u/c$, at which point the effective linear damping in the u equation becomes negative and u undergoes transient growth. Figure 9.5 illustrates this mechanism. Panel (a) shows the time series of u and v together with the CIR. Panel (b) shows the ACI strength, which attains local maxima when v enters the triggering regime and before the corresponding bursts in u fully develop. Notably, the CIR rises earlier when conditioning on v, reflecting the fact that the buildup of an extreme event in u takes time after v activates the anti-damping. Therefore, monitoring the triggering variable v improves the predictability of extreme events observed in u.

□

ACI and CIR have broad applications, including detecting noise- or bifurcation-induced tipping points, identifying the key variables that trigger extreme events, and leveraging causal structure to construct minimal yet accurate reduced-order models [16].

9.7 Code listings for figures and examples

- `Chapter_9_Granger_Transfer_Entropy.m`: generates Figure 9.1 in Example 9.6, which compares Granger causality and transfer entropy in analyzing nonlinear time series.
- `Chapter_9_DAG.m`: generates Figure 9.2 in Example 9.7, which illustrates confounding and d-separation in graph-based causal inference.

- `Chapter_9_Confounded_System.m`: generates Figure 9.3 in Example 9.8, which compares observational and interventional effects in a confounded system.
- `Chapter_9_CEM.m`: generates Figure 9.4 in Example 9.9, which applies causation entropy for model identification.
- `Chapter_9_ACI.m`: generates Figure 9.5 in Example 9.10, which applies assimilative causal inference to study the mechanisms of extreme events.

10

Reduced-Order Models

10.1 Galerkin reduced-order models

Reduced-order models play a crucial role in many science and engineering disciplines. While accuracy remains a fundamental requirement in modeling complex phenomena, computational efficiency is equally important, particularly for applications that require online prediction, real-time control, or rapid decision-making. Therefore, developing appropriate reduced-order models becomes essential to preserve the crucial dynamical and statistical features of the original system while significantly reducing computational cost.

Galerkin reduced-order modeling is a widely used method and serves as a building block for many related reduced-order modeling techniques. It is a projection-based method for constructing low-dimensional dynamical systems that approximate the behavior of the original high-dimensional models. The key idea is to restrict the dynamics to a low-dimensional subspace and determine the reduced dynamics by enforcing orthogonality of the residual with respect to this subspace.

10.1.1 The Galerkin method

Consider a deterministic dynamical system

$$\frac{\mathrm{d}\mathbf{x}}{\mathrm{d}t} = \mathbf{F}(\mathbf{x}), \qquad \mathbf{x}(t) \in \mathbb{R}^N. \tag{10.1}$$

Assume that the solution can be well approximated in a low-dimensional subspace spanned by orthonormal basis vectors $\{\boldsymbol{\phi}_1, \ldots, \boldsymbol{\phi}_r\}$, with $r \ll N$. Here, orthonormal means that the vectors are mutually perpendicular and normalized, and therefore provide independent directions for representing the dynamics. The reduced representation can be written as

$$\mathbf{x}(t) \approx \sum_{k=1}^{r} a_k(t)\boldsymbol{\phi}_k := \boldsymbol{\Phi}\mathbf{a}(t), \tag{10.2}$$

where $\boldsymbol{\Phi} \in \mathbb{R}^{N\times r}$ and $\mathbf{a}(t) \in \mathbb{R}^r$ denote the collections of the basis vectors and the reduced coordinates, respectively. Substituting the ansatz (10.2) into the governing equation (10.1) gives the residual

$$\mathbf{r} = \boldsymbol{\Phi}\dot{\mathbf{a}} - \mathbf{F}(\boldsymbol{\Phi}\mathbf{a}), \tag{10.3}$$

where $\dot{\mathbf{a}} := \mathrm{d}\mathbf{a}/\,\mathrm{d}t$. The residual measures the mismatch between the time derivative implied by the reduced approximation and that required by the full dynamical system. Since the reduced solution is constrained to remain in the subspace spanned by $\{\boldsymbol{\phi}_1, \ldots, \boldsymbol{\phi}_r\}$, it is generally impossible to make this residual identically zero.

Therefore, instead of requiring the residual to be exactly zero, the Galerkin condition imposes the simpler requirement that the residual be orthogonal to the trial subspace,

$$\boldsymbol{\Phi}^{\mathtt{T}}\mathbf{r} = \mathbf{0}. \tag{10.4}$$

DOI: 10.1201/9781003738459-10

This condition ensures that, among all possible time derivatives that remain within the reduced subspace, the chosen reduced dynamics minimizes the residual error in a least-squares sense. Geometrically, this corresponds to projecting the full vector field $\mathbf{F}(\mathbf{x})$ onto the reduced subspace spanned by $\mathbf{\Phi}$. The procedure is qualitatively similar to PCA, which projects data onto a low-dimensional subspace to minimize reconstruction error. In contrast, the Galerkin method assumes a given subspace and projects the dynamics onto it, so that the remaining error is orthogonal to that subspace and cannot be further reduced by adjusting the reduced coordinates $\mathbf{a}(t)$.

Multiplying $\mathbf{\Phi}^{\mathrm{T}}$ to both sides of (10.3) and utilizing the orthogonality condition in (10.4), the Galerkin projection leads to the reduced-order model [27, 156, 229]:

$$\dot{\mathbf{a}} = \mathbf{\Phi}^{\mathrm{T}}\mathbf{F}(\mathbf{\Phi}\mathbf{a}), \tag{10.5}$$

which gives a closed system of r ordinary differential equations governing the reduced dynamics.

The above procedure for developing Galerkin reduced-order models can naturally be separated into two steps. First, a low-dimensional subspace is chosen to represent the dominant behavior of the system. Second, the governing equations are projected onto this subspace to obtain the reduced dynamics. Notably, the Galerkin method itself only specifies the projection step and does not prescribe how the basis functions should be chosen.

In many classical applications, the basis functions are selected a priori based on the structure of the model or physical insight. Common examples include Fourier modes, eigenfunctions of linear operators, polynomial bases, or other problem-specific functions. In such cases, no data analysis is involved, and the resulting reduced-order model is a standard Galerkin model.

Alternatively, the basis functions can be learned from data. Data decomposition methods, such as PCA, can provide a systematic way to construct an orthonormal basis from solution snapshots of the state variable $\mathbf{x}(t)$, rather than from the dynamical operator itself. When the basis vectors $\{\boldsymbol{\phi}_k\}$ are chosen as the leading principal components of simulated or observed snapshots, the resulting reduced-order model is commonly referred to as a Proper Orthogonal Decomposition (POD)-Galerkin model [220, 234, 134, 157]. POD-Galerkin modeling can be viewed as a data-driven variant of the Galerkin framework.

Example 10.1 (Galerkin projection of a linear system). Consider the linear system

$$\frac{\mathrm{d}\mathbf{x}}{\mathrm{d}t} = \mathbf{A}\mathbf{x}, \qquad \mathbf{A} = \begin{pmatrix} 0 & 1 \\ -2 & -3 \end{pmatrix}, \qquad \mathbf{x}(t) \in \mathbb{R}^2. \tag{10.6}$$

We choose a one-dimensional subspace spanned by

$$\boldsymbol{\phi} = \frac{1}{\sqrt{2}}\begin{pmatrix} 1 \\ 1 \end{pmatrix}, \qquad \mathbf{x}(t) \approx a(t)\boldsymbol{\phi}. \tag{10.7}$$

Since $\mathbf{F}(\mathbf{x}) = \mathbf{A}\mathbf{x}$ in this example, the Galerkin projection (10.5) gives

$$\dot{a} = \boldsymbol{\phi}^{\mathrm{T}}\mathbf{A}\boldsymbol{\phi}\, a. \tag{10.8}$$

A direct computation yields $\boldsymbol{\phi}^{\mathrm{T}}\mathbf{A}\boldsymbol{\phi} = \frac{1}{2}(1,1)\cdot(1,-5) = -2$, and therefore the reduced-order model is

$$\dot{a} = -2a. \tag{10.9}$$

The reconstructed state is given by $\mathbf{x}_{\mathrm{ROM}}(t) = a(t)\boldsymbol{\phi}$. □

Example 10.2 (Nonlinearity vanishing under Galerkin projection). Consider the following three-dimensional nonlinear system

$$\dot{\mathbf{x}} = \mathbf{A}\mathbf{x} + \mathbf{N}(\mathbf{x}), \qquad \mathbf{x} \in \mathbb{R}^3, \tag{10.10}$$

where

$$\mathbf{A} = \begin{pmatrix} -1 & 0 & 0 \\ 0 & -2 & 0 \\ 0 & 0 & -3 \end{pmatrix}, \qquad \mathbf{N}(\mathbf{x}) = \begin{pmatrix} 0 \\ x_1 x_3 \\ -x_1 x_2 \end{pmatrix}. \tag{10.11}$$

We again select a one-dimensional subspace,

$$\boldsymbol{\phi} = \frac{1}{\sqrt{3}} \begin{pmatrix} 1 \\ 1 \\ 1 \end{pmatrix}, \qquad \mathbf{x}(t) \approx a(t)\boldsymbol{\phi}. \tag{10.12}$$

Following (10.5), the Galerkin reduced-order model takes the form

$$\dot{a} = \boldsymbol{\phi}^{\mathrm{T}}\mathbf{A}\boldsymbol{\phi}\, a + \boldsymbol{\phi}^{\mathrm{T}}\mathbf{N}(a\boldsymbol{\phi}). \tag{10.13}$$

The linear contribution is

$$\boldsymbol{\phi}^{\mathrm{T}}\mathbf{A}\boldsymbol{\phi} = \frac{1}{3}(-1-2-3) = -2. \tag{10.14}$$

For the nonlinear term, note that $x_1 = x_2 = x_3 = a/\sqrt{3}$ from (10.12), which implies

$$\mathbf{N}(a\boldsymbol{\phi}) = \begin{pmatrix} 0 \\ a^2/3 \\ -a^2/3 \end{pmatrix}, \qquad \boldsymbol{\phi}^{\mathrm{T}}\mathbf{N}(a\boldsymbol{\phi}) = 0. \tag{10.15}$$

Hence, the reduced-order model becomes

$$\dot{a} = -2a. \tag{10.16}$$

Although the full system is nonlinear, the nonlinear terms vanish under projection onto this particular subspace. This example illustrates how the structure of the reduced-order model depends on the chosen basis. □

Galerkin reduced-order models are widely used for PDEs, especially when the energy spectrum decays rapidly, so that the large-scale modes capture most of the features of the underlying dynamics. In such cases, the solution can be well approximated using only a small number of basis functions. As a result, the reduced-order model has a much lower dimension than the original system, which gives significant gains in computational efficiency.

Among various choices of basis functions, Fourier bases are especially useful for problems with periodic boundary conditions, as they provide a simple and physically interpretable representation of scale interactions.

Example 10.3 (Fourier-Galerkin method for a 1D PDE: viscous Burgers' equation). Consider the viscous Burgers' equation on a periodic domain $x \in [0, 2\pi]$:

$$\partial_t u(x,t) + u(x,t)\,\partial_x u(x,t) = \nu\,\partial_{xx} u(x,t), \qquad u(x+2\pi,t) = u(x,t), \tag{10.17}$$

where $\nu > 0$ is the viscosity. The Galerkin reduced-order model based on the Fourier bases can be developed utilizing the following three-step procedure.

Step 1: Fourier expansion and truncation.

We approximate $u(x,t)$ by a truncated Fourier series,

$$u_r(x,t) = \sum_{k=-r}^{r} \hat{u}_k(t)\, e^{ikx}, \tag{10.18}$$

where $r \ll \infty$ is the truncation level. For real-valued u in physical space, the Fourier coefficients must satisfy the conjugate condition $\hat{u}_{-k} = \hat{u}_k^*$.

Step 2: Galerkin projection.

Define the residual of (10.17) under the ansatz (10.18):

$$\mathcal{R}(x,t) := \partial_t u_r + u_r\, \partial_x u_r - \nu\, \partial_{xx} u_r. \tag{10.19}$$

The Fourier-Galerkin condition requires the residual be orthogonal to each basis function e^{ikx} for $|k| \le r$, i.e., using the standard L^2 inner product on $[0, 2\pi]$,

$$\frac{1}{2\pi}\int_0^{2\pi} \mathcal{R}(x,t) e^{-ikx}\,\mathrm{d}x = 0, \qquad |k| \le r. \tag{10.20}$$

Step 3: Reduced ODE system for Fourier coefficients.

Compute the needed derivatives:

$$\partial_t u_r = \sum_{k=-r}^{r} \dot{\hat{u}}_k e^{ikx}, \qquad \partial_x u_r = \sum_{k=-r}^{r} (ik)\hat{u}_k e^{ikx}, \qquad \partial_{xx} u_r = \sum_{k=-r}^{r} (-k^2)\hat{u}_k e^{ikx}. \tag{10.21}$$

The nonlinear term is a product of Fourier series:

$$u_r\, \partial_x u_r = \left(\sum_{p=-r}^{r} \hat{u}_p e^{ipx}\right)\left(\sum_{q=-r}^{r} (iq)\hat{u}_q e^{iqx}\right) = \sum_{p=-r}^{r}\sum_{q=-r}^{r} (iq)\hat{u}_p\hat{u}_q\, e^{i(p+q)x}. \tag{10.22}$$

Projecting onto the k-th Fourier mode using (10.20) isolates the contribution associated with e^{ikx}, yielding for each $|k| \le r$:

$$\dot{\hat{u}}_k = -\sum_{\substack{p+q=k\\ |p|\le r,\ |q|\le r}} (iq)\, \hat{u}_p\hat{u}_q - \nu k^2 \hat{u}_k, \qquad |k| \le r. \tag{10.23}$$

This is a closed deterministic system of $2r+1$ ordinary differential equations for the Fourier coefficients $\{\hat{u}_k(t)\}_{k=-r}^{r}$, known as a Fourier-Galerkin reduced-order model. Note that the quadratic nonlinearity can generate modes with $|k| > r$. Truncation discards their feedback onto the resolved modes, which can lead to aliasing and motivates dealiasing or closure modeling in practice.

Truncation removes all modes with $|k| > r$. The viscous term $-\nu k^2 \hat{u}_k$ provides deterministic damping, which increases with the wavenumber and therefore preferentially suppresses small-scale modes. In contrast, the nonlinear term in (10.23) couples the retained modes through a convolution sum ($p+q = k$), allowing energy to be transferred among modes. This mode interaction plays a central role in the accuracy and limitations of Fourier-Galerkin reduced-order models and will be discussed further in the next subsections. □

10.1.2 Truncation effects in nonlinear models

Recall that Galerkin projection reduces a PDE to a finite-dimensional ODE system by restricting the solution to a low-dimensional subspace. For linear problems, this procedure often leads to an accurate reduced-order model, since linear operators preserve subspaces. However, for nonlinear PDEs, Galerkin truncation generally introduces an additional approximation error, because nonlinear interactions couple resolved and unresolved modes.

To make this precise, consider a nonlinear PDE of the form

$$\partial_t u = \mathcal{L}u + \mathcal{N}(u), \tag{10.24}$$

where $\mathcal{L}$ is a linear operator and $\mathcal{N}$ is a nonlinear operator. Let P_r denote the orthogonal projection onto a finite-dimensional subspace (e.g., the span of Fourier modes with $|k| \leq r$), and define the decomposition

$$u = u_r + u_u, \qquad u_r := P_r u, \qquad u_u := (I - P_r)u, \tag{10.25}$$

where u_r and u_u represent the resolved and unresolved components of the solution, respectively.

Projecting (10.24) onto the resolved subspace yields the exact evolution equation

$$\partial_t u_r = \mathcal{L}u_r + P_r\mathcal{N}(u_r) + P_r\big(\mathcal{N}(u_r + u_u) - \mathcal{N}(u_r)\big). \tag{10.26}$$

If the subspace is invariant under $\mathcal{L}$ (as in Fourier-Galerkin methods for constant-coefficient PDEs), then $\mathcal{L}u_r$ depends only on the resolved variables. On the other hand, the nonlinear operator $\mathcal{N}$ generally mixes different modes. The term $P_r\mathcal{N}(u_r)$ represents nonlinear interactions among resolved modes that remain within the resolved subspace. In contrast, the term $P_r\big(\mathcal{N}(u_r + u_u) - \mathcal{N}(u_r)\big)$ collects all nonlinear contributions involving at least one unresolved component. Specifically, it includes interactions between resolved and unresolved modes, as well as interactions entirely among unresolved modes that nevertheless project back onto the resolved subspace. Although u_u itself does not appear in the reduced-order model of u_r, the evolution of u_r still depends on the nonlinear interactions among u_u through this projection. As u_u is not evolved explicitly, the contribution of this term cannot be computed directly in a truncated Galerkin system. This introduces additional approximate errors in the Galerkin projection.

Example 10.4 (Fourier interpretation of unresolved mode feedback). To clarify the mechanism by which unresolved modes influence the time evolution of resolved modes in nonlinear systems, consider a Fourier representation of the full solution:

$$u(x,t) = \sum_{k\in\mathbb{Z}} \hat{u}_k(t)\, e^{ikx}. \tag{10.27}$$

Let $\mathcal{N}(u)$ denote the nonlinear term in the PDE, and write its Fourier expansion as

$$\mathcal{N}(u)(x,t) = \sum_{k\in\mathbb{Z}} \widehat{\mathcal{N}}_k(t)\, e^{ikx}, \tag{10.28}$$

where $\widehat{\mathcal{N}}_k$ denotes the k-th Fourier coefficient of the nonlinear term.

For a quadratic nonlinearity, which appears widely in fluids and geophysics, these coefficients take the convolution form

$$\widehat{\mathcal{N}}_k = \sum_{p+q=k} C(p,q)\, \hat{u}_p \hat{u}_q, \tag{10.29}$$

where the coefficients $C(p,q)$ depend on the specific PDE.

If the resolved modes are defined by $|k| \leq r$, then the sum contributing to $\widehat{\mathcal{N}}_k$ for a resolved mode k includes three types of interactions: (a) resolved-resolved interactions, with $|p| \leq r$ and $|q| \leq r$, (b) resolved-unresolved interactions, with one of $|p|$ or $|q|$ exceeding r, and (c) unresolved-unresolved interactions, with $|p| > r$ and $|q| > r$. Importantly, two unresolved modes can interact to produce a resolved mode if their wavenumbers satisfy $p+q=k$ with $|k| \leq r$. For example,

$$p = 12, \quad q = -11 \quad \Rightarrow \quad k = 1. \tag{10.30}$$

Thus, discarding unresolved modes removes not only their direct contribution to the solution but also their indirect influence on the resolved dynamics through nonlinear interactions.□

The truncated Galerkin system neglects these unresolved feedback terms, which may lead to inaccuracies such as spurious energy accumulation or incorrect long-time behavior. Closure models aim to approximate the net effect of unresolved modes on the resolved dynamics using only information from the resolved variables.

10.1.3 Closure modeling

Closure models are often used to mitigate the missing information in the truncated Galerkin system. A closure model replaces the unresolved feedback term $P_r(\mathcal{N}(u_r+u_u) - \mathcal{N}(u_r))$ by an approximation depending only on the resolved variables u_r. The resulting closed reduced-order model takes the form [219, 7, 235]:

$$\partial_t u_r = \mathcal{L} u_r + P_r \mathcal{N}(u_r) + \mathcal{C}(u_r), \tag{10.31}$$

where $\mathcal{C}(u_r)$ is the closure term. The goal of closure modeling is to capture the net effect of unresolved scales on the resolved dynamics, rather than to reproduce the detailed behavior of the discarded modes.

Example 10.5 (A simple deterministic closure method). A simple and widely used deterministic closure is linear damping, often referred to as eddy viscosity in spectral methods. In such a method, additional dissipation is introduced to mimic the transfer of energy from resolved modes to unresolved ones:

$$\mathcal{C}(u_r) = -\gamma u_r, \qquad \gamma > 0. \tag{10.32}$$

Alternatively, in Fourier-Galerkin models, eddy viscosity is often implemented as scale-dependent damping in spectral space:

$$\widehat{\mathcal{C}}_k = -\nu_{\mathrm{cl}} k^2 \hat{u}_k, \qquad |k| \leq r, \tag{10.33}$$

where ν_{cl} is a closure viscosity parameter. Despite its simplicity, such closures often significantly improve stability and long-time accuracy of Galerkin reduced-order models. □

Example 10.6 (A low-dimensional closure improving Galerkin truncation). Consider the following three-dimensional nonlinear system with $\mathbf{x} \in \mathbb{R}^3$,

$$\begin{aligned}
\frac{\mathrm{d}x_1}{\mathrm{d}t} &= -x_2 - a x_1 x_3, \\
\frac{\mathrm{d}x_2}{\mathrm{d}t} &= \quad x_1 - a x_2 x_3, \\
\frac{\mathrm{d}x_3}{\mathrm{d}t} &= \frac{1}{\varepsilon}\left(x_1^2 + x_2^2 - x_3\right),
\end{aligned} \tag{10.34}$$

where $a > 0$ and $0 < \varepsilon \ll 1$. The variables (x_1, x_2) represent resolved modes, while x_3 is a fast unresolved variable that relaxes rapidly toward $x_1^2 + x_2^2$.

A naive Galerkin truncation discards the unresolved variable by setting $x_3 = 0$, leading to the reduced system

$$\begin{aligned} \frac{dx_1}{dt} &= -x_2, \\ \frac{dx_2}{dt} &= \quad x_1. \end{aligned} \tag{10.35}$$

This system describes an undamped linear oscillator, and therefore cannot reproduce the decay observed in the full system (10.34).

To improve the performance of the reduced-order model, let us consider adding a closure term. Since x_3 relaxes rapidly, a simple closure approximation is to replace it by its quasi-equilibrium value,

$$x_3 \approx \alpha(x_1^2 + x_2^2). \tag{10.36}$$

Substituting this approximation into the resolved equations yields the closed reduced-order model

$$\begin{aligned} \frac{dx_1}{dt} &= -x_2 - a(x_1^2 + x_2^2)x_1, \\ \frac{dx_2}{dt} &= \quad x_1 - a(x_1^2 + x_2^2)x_2. \end{aligned} \tag{10.37}$$

The additional nonlinear damping term acts as a deterministic closure that approximates the net influence of the unresolved variable.

Figure 10.1 presents a numerical comparison between the full system (10.34), the Galerkin reduced-order model without closure (10.35), and the reduced-order model with closure (10.36)–(10.37). The parameters used in the simulation are

$$a = 2.0, \qquad \varepsilon = 0.05, \qquad \text{and} \qquad \alpha = 1.0. \tag{10.38}$$

Although the unresolved variable x_3 decays rapidly and remains at a small amplitude, neglecting it entirely leads to a qualitative change in the system dynamics. In contrast, approximating the net effect of x_3 through a closure model depending only on the resolved variables x_1 and x_2 yields a reduced-order model that accurately reproduces the resolved trajectories.

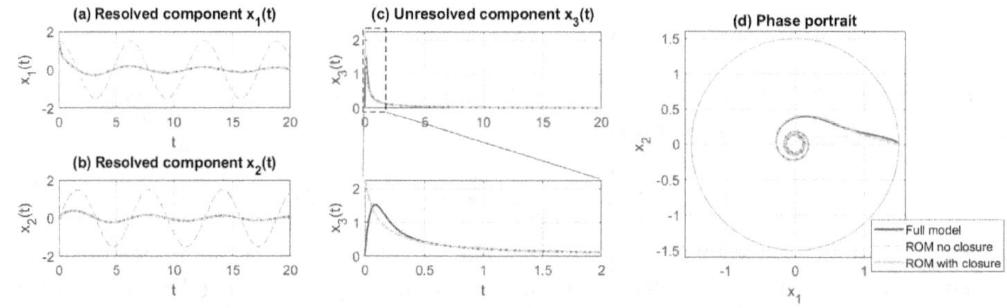

FIGURE 10.1: Comparison of the full system (10.34) (solid curves), the Galerkin reduced-order model without closure (10.35) (dashed curves), and the reduced-order model with closure (10.36)–(10.37) (dash-dot curves). Panels (a)–(b): time evolution of the resolved components x_1 and x_2. Panel (c): unresolved component x_3, together with a zoom-in view over $t \in [0, 2]$. Panel (d): phase portrait of the resolved dynamics.

The example here illustrates how a simple closure model can significantly improve the long-time accuracy of a reduced-order model by accounting for unresolved feedback. □

10.1.4 Challenges of Galerkin reduced-order models for chaotic systems

Galerkin reduced-order models are effective for many systems whose dynamics are dominated by a small number of coherent structures or weakly nonlinear interactions. However, for strongly chaotic systems, standard Galerkin truncation (even when augmented with simple closure models) often fails to reproduce the correct long-time behavior. This limitation reflects fundamental challenges associated with chaos rather than purely numerical or implementation issues.

Example 10.7 (Limitations of Galerkin reduced-order model for the Lorenz 63 system). Recall the Lorenz 63 model,

$$\begin{aligned}\frac{\mathrm{d}x}{\mathrm{d}t} &= \sigma(y-x),\\ \frac{\mathrm{d}y}{\mathrm{d}t} &= x(\rho-z)-y,\\ \frac{\mathrm{d}z}{\mathrm{d}t} &= xy-\beta z,\end{aligned}\tag{10.39}$$

with the standard parameters $\sigma = 10$, $\rho = 28$, $\beta = 8/3$ that generates chaotic behavior.

A standard Galerkin or POD-Galerkin reduced-order model may be constructed by projecting the dynamics onto a two-dimensional subspace learned from solution snapshots. Such a model often reproduces the full trajectory accurately over short time intervals. However, over long times, the reduced system typically exhibits significant discrepancies in its attractor geometry and statistics. Adding a simple closure term, such as linear damping, may improve numerical stability but does not robustly recover the correct long-time behavior. Figure 10.2 shows the gap between the true trajectory of x and its POD-Galerkin reduced-order models with and without closure. The latter two converge to a fixed point, which is significantly different from the butterfly attractor of the original Lorenz 63 model.

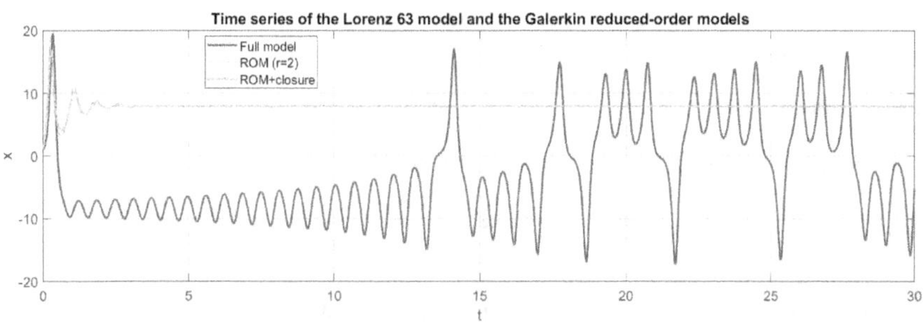

FIGURE 10.2: Comparison of the full system (10.39) (solid curves), the Galerkin reduced-order model without closure (dashed curves), and the reduced-order model with closure (dash-dot curves). Only the x component is shown but the same behavior is found in y and z.

□

Example 10.8 (Charney-DeVore model with regime transitions). The Charney-DeVore model is a low-dimensional nonlinear dynamical system originally introduced to study multiple equilibria and regime transitions in midlatitude atmospheric flow. Despite its simplicity, the Charney-DeVore system captures essential mechanisms underlying planetary-scale atmospheric variability, including the interaction between zonal flow and stationary Rossby waves. A six-dimensional formulation of this model can be found in [84]. For appropriate parameter regimes, the system exhibits strong chaotic behavior and intermittent transitions between distinct flow regimes [83].

A particularly interesting finding is that when this six-dimensional system is projected onto a standard five-dimensional subspace using the POD-Galerkin method, the retained modes capture 99.611% of the total energy, leaving a residual of only 0.389%. From a purely energetic perspective, the projection appears highly accurate. However, despite this small truncation error, the resulting five-mode POD-Galerkin reduced-order model fails to reproduce the regime transitions observed in the full system. Instead, its trajectories converge to a fixed point, indicating a qualitative breakdown of the long-term dynamics.□

The failure of Galerkin reduced-order models in chaotic systems is fundamentally linked to sensitivity to small structural perturbations in the governing equations. A small structural error in the reduced-order model can grow exponentially over time, leading to rapid divergence from the true dynamics. Galerkin truncation often removes unresolved degrees of freedom whose nonlinear interactions contribute essential feedback to the resolved variables. Simple closures typically model unresolved feedback as an instantaneous function of the resolved state. However, in chaotic systems, unresolved modes often exert history-dependent and state-dependent influences that cannot be captured by such approximations. Even if short-time trajectories are well approximated, reduced models may converge to an incorrect invariant measure, resulting in biased long-time statistics.

These observations indicate that, for highly chaotic systems, reduced-order modeling should not be evaluated solely on the basis of short-time trajectory accuracy. Instead, emphasis should be placed on reproducing long-term statistical properties and invariant measures, as well as on capturing feedback from the unresolved modes to the resolved ones. Addressing these challenges typically requires going beyond standard Galerkin projection and simple deterministic closures. Stochastic closures that represent unresolved feedback statistically, memory-based models that account for non-Markovian effects, and data-driven methods that infer effective reduced dynamics will play a crucial role in advancing reduced-order modeling. Some of these topics will be discussed in the subsequent subsections.

10.2 Stochastic reduced-order models

10.2.1 Stochastic approximations of nonlinear dynamics for turbulent statistics

The governing equations of many complex turbulent systems are nonlinear PDEs. Consider the following prototype nonlinear PDE with a scalar variable $u(x,t)$:

$$\partial_t u(x,t) = -\alpha\, u(x,t) - \beta\, u(x,t)\, \partial_x u(x,t) + \omega\, \partial_x u(x,t) + \nu\, \partial_{xx} u(x,t) + f(x,t), \quad (10.40)$$

where $\alpha > 0$ is a linear damping rate, β controls the strength of the quadratic nonlinear advection, $\omega \in \mathbb{R}$ represents dispersion (a phase-transport term), $\nu > 0$ is a diffusion/viscosity coefficient, and $f(x,t)$ is an external deterministic forcing. Such a form serves as a canonical model that combines (i) nonlinear energy transfer across scales through advection, (ii) linear decay through damping and diffusion, and (iii) wave-like phase propagation through the first-derivative term. Quadratic nonlinearities are commonly encountered in fluid and geophysical systems.

Among various numerical methods, spectral methods remain one of the most widely used approaches, especially for problems with periodic boundary conditions. For the simplicity of presenting the idea, consider the above model on a periodic domain $x \in [0, 2\pi]$. Nevertheless, the methodology of building the stochastic reduced-order model applies to other boundary conditions. Using Fourier basis functions $\exp(ikx)$, where k denotes the wavenumber and x the spatial coordinate, we can write the solution in terms of its spectral coefficients $\hat{u}_k(t)$. Expanding u in a Fourier series,

$$u(x,t) = \sum_{k\in\mathbb{Z}} \hat{u}_k(t)\, e^{ikx}, \qquad \hat{u}_k(t) = \frac{1}{2\pi}\int_0^{2\pi} u(x,t)e^{-ikx}\,\mathrm{d}x, \tag{10.41}$$

and using $\partial_x e^{ikx} = ike^{ikx}$ and $\partial_{xx}e^{ikx} = -k^2e^{ikx}$, the linear terms in (10.40) yield

$$-\alpha u \mapsto -\alpha\hat{u}_k, \qquad \omega\,\partial_x u \mapsto i(\omega k)\hat{u}_k, \qquad \nu\,\partial_{xx}u \mapsto -\nu k^2\hat{u}_k, \qquad f \mapsto \hat{f}_k. \tag{10.42}$$

The quadratic advection term produces a convolution sum. Specifically,

$$u\,\partial_x u = \Big(\sum_m \hat{u}_m e^{imx}\Big)\Big(\sum_n (in)\hat{u}_n e^{inx}\Big) = \sum_{m,n}(in)\hat{u}_m\hat{u}_n e^{i(m+n)x}, \tag{10.43}$$

so the coefficient of mode k is

$$\widehat{u\,\partial_x u}_k = \sum_{m\in\mathbb{Z}} i(k-m)\,\hat{u}_m\,\hat{u}_{k-m}. \tag{10.44}$$

Therefore, after truncating to $k \in [-K, K]$, the Fourier-mode dynamics take the form:

$$\frac{\mathrm{d}\hat{u}_k}{\mathrm{d}t} = \underbrace{\Big[-(\alpha+\nu k^2) + i(\omega k)\Big]}_{-d_k + i\omega_k}\,\hat{u}_k + \hat{f}_k(t) - \beta\mathcal{P}_K\left(\sum_{-K\le m\le K} i(k-m)\,\hat{u}_m\,\hat{u}_{k-m}\right), \tag{10.45}$$

where

$$d_k = \alpha + \nu k^2, \qquad \omega_k = \omega k, \qquad c_{m,k} = -\beta\, i(k-m), \tag{10.46}$$

and $\mathcal{P}_K$ project the solution back to the truncated space. In turbulent regimes, the nonlinear convolution term couples many interacting modes and is typically the dominant computational cost in direct numerical simulation. This motivates the replacement of this term with a calibrated stochastic surrogate when the goal is to reproduce forecast statistics rather than individual trajectories.

Due to the strongly turbulent nature of these systems, many practical tasks, such as statistical forecasting and data assimilation, require repeated integrations of the governing equations to obtain forecast distributions. Each run is already computationally expensive, as all $2K+1$ modes must be evolved and each involves numerous multiplications arising from nonlinear interactions. Consequently, ensemble forecasting quickly becomes computationally prohibitive. This motivates the development of much cheaper surrogate models that can significantly accelerate computations.

A key observation is that stochasticity can effectively mimic many essential features of turbulent dynamics. Since the objective in these applications is to reproduce forecast statistics rather than individual trajectories, an appropriately designed stochastic reduced-order model can capture the statistical behavior of the underlying deterministic system, provided suitable calibration criteria are imposed.

In many high-resolution turbulent systems, a substantial fraction of the total energy is concentrated in a relatively small number of Fourier modes. Nevertheless, in direct numerical simulations, the full set of modes must be evolved simultaneously to ensure numerical stability. The computational savings offered by stochastic reduced-order models, therefore, arise from three sources. First, the expensive nonlinear interaction terms in the equation of each mode are replaced by a small number of computationally inexpensive stochastic terms. Second, once nonlinear coupling is replaced by stochastic forcing, the equations for different modes become decoupled, allowing only the leading energetic modes to be retained. This dramatically reduces storage requirements and computational cost. Third, the absence of small-scale stiff dynamics allows the use of much larger numerical time steps.

A simple but commonly used stochastic reduced-order modeling strategy replaces the nonlinear interaction terms in each mode equation (10.45) with an additive white-noise forcing term. This approximation is particularly effective when the statistics of the mode are close to Gaussian. The resulting stochastic model takes the form [111]

$$\frac{\mathrm{d}\hat{u}_k(t)}{\mathrm{d}t} = \big(-d_k + i\omega_k\big)\hat{u}_k(t) + \hat{f}_k(t) + \sigma_k \dot{W}_k, \tag{10.47}$$

where $\dot{W}_k$ denotes Gaussian white noise. This is essentially a complex OU process. For systems exhibiting strongly non-Gaussian behavior, more sophisticated stochastic modeling strategies can be employed [61]. See the discussions in the next subsection.

In practice, the model parameters are optimized so that the surrogate (10.47) reproduces the forecast uncertainty of the original equation (10.45), which is essential for applications such as data assimilation and ensemble prediction. This is typically achieved by matching key statistical quantities, including equilibrium PDFs and decorrelation times. For the linear Gaussian stochastic model (10.47), closed-form analytic expressions are available for this calibration (see Section 6.5.1), which makes the model particularly useful for practical applications.

Example 10.9 (Statistical approximations to a two-layer QG model by linear Gaussian reduced-order models). Consider a two-layer quasi-geostrophic (QG) model governed by a system of nonlinear PDEs [265]. Each layer is discretized using a spatial resolution of 128×128. Examination of the energy spectrum indicates that the modes up to $|\mathbf{k}| = \sqrt{k_1^2 + k_2^2} \leq 11$ contain the majority of the system's energy. Based on this observation, we construct a reduced-order model by representing each of these modes with an independent linear stochastic equation of the form (10.47).

Although the Fourier coefficients in this reduced-order model evolve independently, the Fourier basis functions depend on the spatial coordinates (x, y). As a result, the reconstructed flow field in physical space still exhibits spatial correlations. The reduced-order model retains 380 modes ($|\mathbf{k}| = \sqrt{k_1^2 + k_2^2} \leq 11$), which is only about 1.16% of the total number of modes in the full model ($128 \times 128 \times 2 = 32768$). In addition to the drastic reduction in dimensionality, each mode equation in the surrogate model is far simpler than its nonlinear counterpart in the original system. The absence of small-scale modes also allows for much larger time steps in numerical integration. Consequently, the overall computational gain is substantial. It is worth emphasizing that directly running the original PDE at such a low resolution would lead to numerical instability.

Panels (a) and (e) of Figure 10.3 show the energy spectra, while Panel (b) shows a snapshot of the stream function from the full two-layer QG model. Panel (c) shows the truncated flow field using models $|\mathbf{k}| \leq 11$. Panel (d) shows the corresponding field reconstructed from the calibrated stochastic surrogate models [80]. Since the two simulations correspond to different realizations, pointwise agreement is not expected. Nevertheless, the two fields are qualitatively similar. Panels (g) and (h) compare the time series of Fourier mode (6, 6) from the two models. Despite being independent realizations, they exhibit similar probability density functions and temporal correlation structures. □

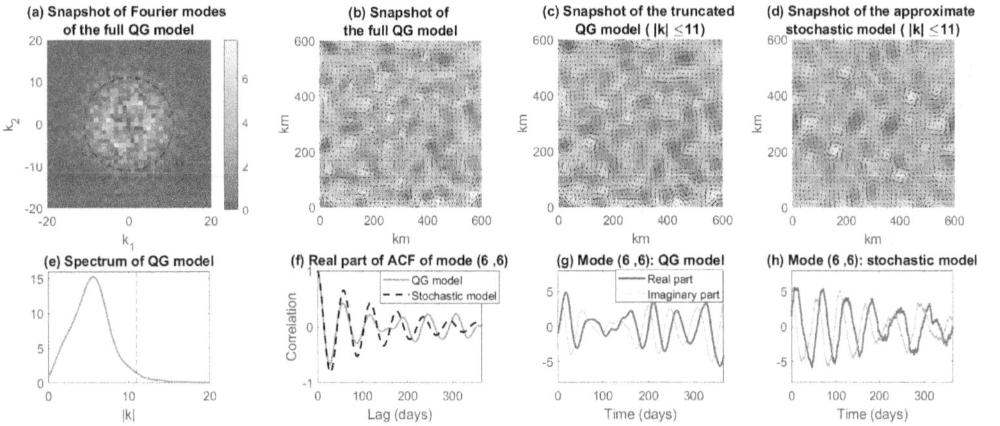

FIGURE 10.3: Comparison between the up-layer stream function simulated from the nonlinear QG model and the stochastic surrogate model. Panel (a): energy spectrum from the QG model. Panel (b): one snapshot of the up-layer stream function from the QG model. Panel (c): stream function at the same time instant as that in Panel (b) by truncating up to modes $|\mathbf{k}| \leq 11$. Panel (d): stream function from the stochastic model using modes up to $|\mathbf{k}| \leq 11$. Note that since the two simulations correspond to different realizations, pointwise agreement is not expected. Nevertheless, the two flow fields are qualitatively similar. Panel (e): spectrum of the QG model. The vertical line shows $|\mathbf{k}| = 11$. Panel (f): comparison of the ACF of Fourier mode $(6, 6)$ between the two models. Panels (g) and (h): time series of Fourier mode $(6, 6)$.

10.2.2 Linear stochastic models with multiplicative noise

The linear Gaussian model in Section 10.2.1 is a suitable reduced-order model when the long-term statistics of a time series are close to Gaussian and the temporal correlations decay exponentially. However, many real-world time series exhibit strong non-Gaussian features such as skewness, heavy tails, or intermittency, which cannot be captured by such a model.

As discussed in Chapter 6, reproducing non-Gaussian statistics within a stochastic differential equation framework requires either nonlinear drift, multiplicative noise, or a combination of both. While one may attempt to construct nonlinear models directly and identify them via sparse regression techniques (see for example, Section 9.4), such approaches often rely on problem-specific ansatzes and can obscure the physical interpretation of the resulting reduced model. Moreover, regression procedures typically optimize short-term transition likelihoods from t to $t+\Delta t$, which does not guarantee that the stationary PDF of the inferred model exactly matches the observed long-term statistics.

An alternative and complementary way is to retain a linear drift while introducing a state-dependent (multiplicative) noise. This preserves a simple and interpretable model structure while enabling the reproduction of non-Gaussian equilibrium statistics. In one dimension, such a model takes the form

$$du(t) = -\lambda\big(u(t) - m\big)\,\mathrm{d}t + \sigma(u)\,\mathrm{d}W(t), \tag{10.48}$$

where $\lambda > 0$ controls the decorrelation time scale, m is the mean state, and $\sigma(u)$ is a state-dependent noise amplitude.

The key advantage of (10.48) is that, despite its linear drift, it can generate a broad class of prescribed non-Gaussian stationary PDFs, provided the resulting diffusion coefficient remains nonnegative and sufficiently regular through an appropriate choice of $\sigma(u)$. Moreover, exact analytic formulae are available for calibrating the model parameters. In particular, the mean m and the damping rate λ can be estimated directly from the empirical mean and decorrelation time of the observed time series.

To determine the multiplicative noise coefficient $\sigma(u)$, consider the Fokker-Planck equation associated with (10.48),

$$\frac{\partial p(u,t)}{\partial t} = \frac{\partial}{\partial u}(\lambda(u-m)p(u,t)) + \frac{1}{2}\frac{\partial^2}{\partial u^2}(\sigma^2(u)p(u,t)).$$

Assuming the stationary probability density $p(u)$ decays sufficiently fast at infinity so that the probability vanishes, the stationary solution satisfies

$$\lambda(u-m)p(u) + \frac{1}{2}\frac{\mathrm{d}}{\mathrm{d}u}(\sigma^2(u)p(u)) = 0.$$

Integrating once yields

$$\sigma^2(u) = \frac{2}{p(u)}\Big(-\lambda\,\Phi(u)\Big), \qquad \Phi(u) = \int_{-\infty}^{u}(y-m)p(y)\,\mathrm{d}y. \tag{10.49}$$

Therefore, given a target stationary distribution $p(u)$ (such that the mean value m is also known) and a prescribed λ (estimated, for example, from the decorrelation), the multiplicative noise $\sigma(u)$ can be determined explicitly through (10.49). This provides a systematic way to build a linear stochastic model that exactly reproduces prescribed equilibrium statistics.

Closed-form expressions for $\sigma(u)$ are available for many classical distributions. See [23] for a comprehensive collection.

Example 10.10 (Linear stochastic model with a uniform equilibrium distribution). Assume the target stationary distribution is the uniform distribution on (b,c),

$$p(x) = \frac{1}{c-b}, \qquad b < x < c. \tag{10.50}$$

Our goal is to determine m and $\sigma(x)$ in the one-dimensional linear stochastic model with multiplicative noise

$$\mathrm{d}x(t) = -\lambda\big(x(t)-m\big)\,\mathrm{d}t + \sigma(x)\,\mathrm{d}W(t), \tag{10.51}$$

so that the stationary density equals (10.50).

First, for the uniform distribution on (b,c), the mean value m is given by

$$m = \mathbb{E}[x] = \frac{b+c}{2}. \tag{10.52}$$

To determine the noise coefficient $\sigma(x)$, we use (10.49). Because the support of p is (b,c), we have $p(y)=0$ outside (b,c), so the integral reduces to

$$\begin{aligned}\int_{-\infty}^{x}(y-m)p(y)\,\mathrm{d}y &= \int_{b}^{x}(y-m)\frac{1}{c-b}\,\mathrm{d}y, \qquad b<x<c\\ &= \frac{1}{c-b}\left[\frac{1}{2}(x-m)^2 - \frac{(c-b)^2}{8}\right].\end{aligned} \tag{10.53}$$

Substituting into (10.49) and using $p(x) = 1/(c-b)$ gives

$$\begin{aligned}\sigma^2(x) &= -2\lambda(c-b)\cdot\frac{1}{c-b}\left[\frac{1}{2}(x-m)^2 - \frac{(c-b)^2}{8}\right] = -\lambda\left[(x-m)^2 - \frac{(c-b)^2}{4}\right]\\ &= \lambda\left[\left(\frac{c-b}{2}\right)^2 - \left(x - \frac{b+c}{2}\right)^2\right] = \lambda(x-b)(c-x).\end{aligned} \tag{10.54}$$

Plugging (10.52) and (10.54) into (10.51) gives the calibrated SDE

$$\mathrm{d}x(t) = -\lambda\left(x(t) - \frac{b+c}{2}\right)\mathrm{d}t + \sqrt{\lambda(x(t)-b)(c-x(t))}\,\mathrm{d}W(t), \qquad b < x(t) < c, \quad (10.55)$$

which has the uniform stationary density on (b, c).

It is seen from (10.55) that the noise amplitude $\sigma(x)$ vanishes at the boundaries $x = b$ and $x = c$, preventing the diffusion from pushing the trajectory outside (b, c), while it is largest near the center, promoting exploration of the interior. The linear drift relaxes toward the mean $(b + c)/2$ with decorrelation time $1/\lambda$. □

Example 10.11 (Linear stochastic model with a Gamma equilibrium distribution). Assume the target stationary distribution is a Gamma distribution (see also Figure 2.2):

$$p(u) = \frac{1}{\beta^{\gamma+1}\Gamma(\gamma+1)}u^{\gamma}e^{-u/\beta}, \qquad u > 0,\ \gamma > -1,\ \beta > 0. \quad (10.56)$$

The corresponding linear stochastic differential equation with multiplicative noise is

$$\mathrm{d}u(t) = -\lambda\big(u(t) - \beta(\gamma+1)\big)\,\mathrm{d}t + \sqrt{2\lambda\beta u(t)}\,\mathrm{d}W(t), \quad (10.57)$$

whose unique stationary density is exactly the Gamma distribution (10.56).

The drift term in (10.57) is linear and relaxes the state $u(t)$ toward the mean value $\beta(\gamma + 1)$ at a rate λ, implying an exponential decorrelation time scale $1/\lambda$. This term alone would generate a Gaussian equilibrium distribution. The non-Gaussianity arises entirely from the multiplicative noise. The noise amplitude $\sqrt{2\lambda\beta u}$ increases with the state magnitude, meaning that larger values of u experience stronger stochastic forcing. This state-dependent amplification promotes intermittent bursts and enhances the probability of extreme events, which leads to the characteristic right-skewed, heavy-tailed shape of the Gamma distribution. At the same time, the noise amplitude vanishes as $u \to 0$, preventing the process from crossing into negative values and ensuring that the support of the stationary distribution remains on $(0, \infty)$.

Gamma distributions are widely used to characterize variables with intermittent and extreme-event behavior, such as precipitation intensity and convective activity. An application of (10.57) to modeling convective variability in the equatorial region can be found in [63]. □

10.3 Stochastic parameterizations

10.3.1 The basic idea

A different yet powerful approach to constructing reduced-order models with rich dynamics is through stochastic parameterization. Such models naturally arise when the feedback from unresolved or small-scale processes must be represented in an effective manner. From this perspective, stochastic parameterization can be regarded as a stochastic closure of standard reduced-order models.

Rather than introducing increasingly complicated deterministic nonlinear terms into the governing equations, stochastic parameterization augments the resolved system with a small number of auxiliary stochastic processes that interact with the resolved variables. These additional processes preserve a relatively simple model structure, while their stochastic nature enables the system to reproduce important features of complex dynamics, including intermittency, regime transitions, and non-Gaussian variability.

The fundamental idea behind stochastic parameterization is to approximate the net influence of unresolved dynamics on resolved variables using simple random processes instead of explicitly resolving the underlying deterministic mechanisms. In this sense, stochastic parameterization can be viewed as a form of "lifting": the dimension of the system is increased modestly by introducing a few stochastic degrees of freedom, yet the resulting model is capable of representing a much broader range of statistical and dynamical behaviors. Importantly, the objective of a stochastically parameterized model is not to recover the exact trajectories of the unresolved processes, but rather to capture their statistical feedback on the resolved variables. Consequently, these models are designed to reproduce the evolution of probability distributions, correlation structures, and extreme event statistics, rather than pointwise accuracy.

Stochastic parameterization is particularly well suited for turbulent systems, where unresolved scales play a critical role in shaping large-scale variability. It has become a central concept in practical modeling across many disciplines [33, 78]. For example, in large-scale geophysical fluid dynamics, direct numerical simulation of all relevant small-scale processes is computationally infeasible. When the primary interest lies in large-scale flow variables, the effects of subgrid-scale motions can be represented through stochastic parameterizations that feed back statistically into the resolved dynamics. This approach often leads to dramatic reductions in computational cost while still capturing essential variability, intermittency, and extremes.

10.3.2 Simple stochastic parameterized systems

The goal here is to introduce the guiding ideas for developing simple yet effective stochastic parameterizations. We begin with the complex OU process, which can be viewed as the governing equation for a single spectral mode in a spatiotemporal dynamical system, as discussed in Section 10.2.1. The complex OU process is written as [111]

$$\frac{du}{dt} = (-\hat{\gamma} + i\hat{\omega})\, u + F(t) + \hat{b} + \sigma_u \dot{W}_u, \tag{10.58}$$

where $\hat{\gamma}$, $\hat{\omega}$, and $\hat{b}$ are constants. The OU process is linear and Gaussian, and hence its long-term statistics are fully characterized by its mean and covariance.

To construct more realistic parameterizations, we can modify the OU process in several ways to produce non-Gaussian behavior. As shown in Chapter 6, introducing a cubic nonlinearity or incorporating multiplicative noise can both generate non-Gaussian statistics. In Chapter 6, it was also demonstrated that allowing the damping coefficient γ to vary in time and modeling its evolution with a simple two-state Markov jump process naturally produces intermittent time series whose PDFs are strongly non-Gaussian. This two-state Markov jump process thus represents one of the simplest stochastic parameterizations for γ that can effectively capture key turbulent features such as intermittency and bursts.

Below, we present a systematic stochastic parameterized system that can be applied to many practical scenarios. The system starts with the complex OU process for u and then replaces the three constant parameters $\hat{\gamma}$, $\hat{\omega}$, and $\hat{b}$ by three stochastic equations. The three auxiliary processes are independent OU processes. Such stochastic parameterizations allow the overall system to remain in a relatively simple form. The OU processes for γ and ω are real since these two parameters are real, while a complex OU process is used for b. The stochastic processes of γ, ω, and b are called stochastic damping, stochastic phase, and stochastic forcing. The stochastic parameterized system reads [114]:

$$
\begin{aligned}
\frac{\mathrm{d}u}{\mathrm{d}t} &= \Big((-\gamma + i\omega)u + F(t) + b\Big) + \sigma_u \dot{W}_u, \\
\frac{\mathrm{d}\gamma}{\mathrm{d}t} &= -d_\gamma(\gamma - \hat{\gamma}) + \sigma_\gamma \dot{W}_\gamma, \\
\frac{\mathrm{d}\omega}{\mathrm{d}t} &= -d_\omega(\omega - \hat{\omega}) + \sigma_\omega \dot{W}_\omega, \\
\frac{\mathrm{d}b}{\mathrm{d}t} &= -d_b(b - \hat{b}) + \sigma_b \dot{W}_b.
\end{aligned}
\tag{10.59}
$$

Each OU process has a prescribed mean $\hat{\gamma}, \hat{\omega}, \hat{b}$, relaxation rates $d_\gamma, d_\omega, d_b > 0$, and noise strengths σ_γ, σ_ω, $\sigma_b > 0$. Although these processes are individually linear and Gaussian, the stochastic damping and stochastic phase couple nonlinearly with the state variable u through multiplicative interactions. As a result, the combined system produces rich nonlinear and non-Gaussian statistics in u. The term $F(t)$ denotes a deterministic external forcing, and σ_u is the noise amplitude associated with the complex Wiener process W_u. Among the nine parameters $\{d_\gamma, d_\omega, d_b, \hat{\gamma}, \hat{\omega}, \hat{b}, \sigma_\gamma, \sigma_\omega, \sigma_b\}$, only $\hat{b}$ is complex while the rest are real constants.

The role of the stochastic damping is to generate intermittency and extreme bursts. These are the mechanisms that lead to non-Gaussian PDFs with heavy tails. The stochastic phase triggers time-varying oscillation frequencies, which can characterize slow-fast structures. The stochastic forcing creates local means that drift away from the long-term average, mimicking nontrivial external modulation or transient forcing.

Example 10.12 (Generating intermittent time series using the stochastic parameterized system). Consider the following parameters in the stochastic parameterized system (10.59):

$$
\begin{gathered}
\sigma_u = 1, \quad d_\gamma = 1, \quad \hat{\gamma} = 1, \quad \sigma_\gamma = 1.2, \\
d_\omega = 0.5, \quad \hat{\omega} = 5, \quad \sigma_\omega = 1.5, \quad d_b = 0.25, \quad \hat{b} = 0, \quad \sigma_b = 25.
\end{gathered}
\tag{10.60}
$$

The time series of the four variables are shown in Figure 10.4. The parameters in (10.60) are chosen such that the stochastic damping γ changes sign, alternating between damping ($\gamma > 0$) and anti-damping ($\gamma \leq 0$) phases. As a result, intermittent bursts are produced in the time series of u. These bursts give rise to extreme events and lead to a strongly non-Gaussian probability distribution for u. On the other hand, the stochastic phase ω introduces slow-fast variability, where large ω corresponds to rapid oscillations, whereas small ω results in slower dynamics. Finally, the stochastic forcing b intermittently drives the system away from its long-term mean state, which further enriches the variability of u. These stochastic components significantly alter the dynamics relative to the standard complex OU process, enabling the model to reproduce intermittent, non-Gaussian, and multiscale behavior with only a few simple parameterizations. □

10.3.3 Analytic tractability and practical advantages in data assimilation

One of the remarkable properties of the stochastic parameterized model in (10.59) is that, despite its intrinsic nonlinearity, the time evolution of its statistical moments can be solved in closed analytic form [114, 49, 125]. This makes the system extremely efficient as a forecast model for data assimilation, where the forecast step can be computed analytically without

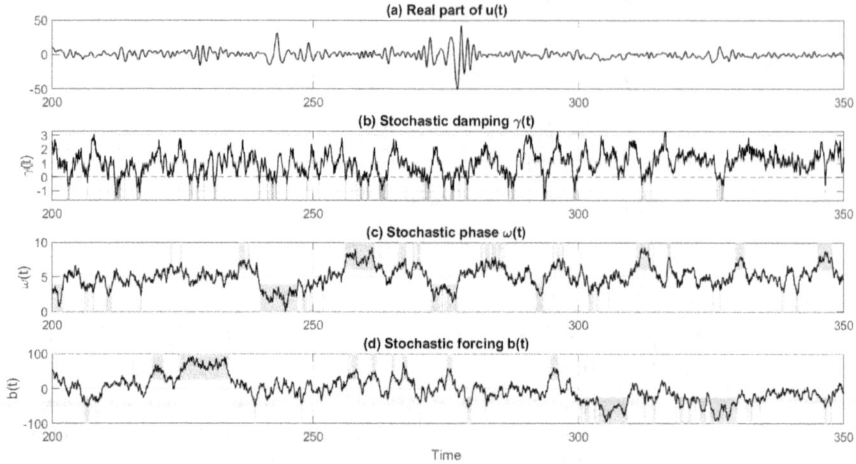

FIGURE 10.4: Simulation of the stochastic parameterized model in (10.59). Panel (a): Real part of the complex state variable $u(t)$ showing intermittent bursts. Panel (b): Stochastic damping $\gamma(t)$, where shading marks anti-damping phases that trigger large fluctuations in u. Panel (c): Stochastic frequency $\omega(t)$; shaded regions indicate fast and slow oscillations. Panel (d): Stochastic forcing $b(t)$; shaded bands represent strong positive and negative impulses that modulate the system's mean state.

resorting to Monte Carlo sampling, thereby eliminating ensemble noise and improving accuracy.

Note that, in many applications, only the nonlinear evolution of the mean and covariance is advanced in time, while the posterior is approximated as Gaussian to maintain computational tractability. Nevertheless, this fundamentally distinguishes it from the classical extended Kalman filter, which relies on local linearization at every assimilation step.

Example 10.13 (Using stochastic parameterized equations as forecast models in data assimilation). Consider two simplified scenarios. In the first case, only the damping coefficient evolves stochastically in time, while in the second, only the oscillation frequency is stochastic. The systems that generate the true signals are given by

$$\begin{aligned} \frac{\mathrm{d}u}{\mathrm{d}t} &= (-\gamma + i\hat{\omega})u + f_u + \sigma_u \dot{W}_u, \\ \gamma &: \text{ a two-state Markov jump process,} \end{aligned} \tag{10.61}$$

and

$$\begin{aligned} \frac{\mathrm{d}u}{\mathrm{d}t} &= (-\hat{\gamma} + i\omega)u + f_u + \sigma_u \dot{W}_u, \\ \omega &: \text{ a two-state Markov jump process,} \end{aligned} \tag{10.62}$$

respectively. In (10.61), the hidden variable $\gamma(t)$ switches between $\gamma_+ = 2.5$ and $\gamma_- = -0.05$ with symmetric transition rates $\mu = \nu = 0.2$. In (10.62), the hidden variable $\omega(t)$ switches between $\omega_+ = 5$ and $\omega_- = 0.2$ with transition rates $\mu = \nu = 0.05$. The other parameters are $\sigma_u = 0.4$, $\hat{\omega} = 2$, and $f_u = 0$ for the stochastic damping case, and $\sigma_u = 0.4$, $\hat{\gamma} = 0.25$, and $f_u = 5$ for the stochastic phase case. In both experiments, the observational interval is $\Delta t^{\text{obs}} = 0.5$, the observational noise level is $r^o = 0.5$, and only the time series of u is observed. The blue solid curve and black dots in Figure 10.5 show the real part of the true signal and noisy observations, respectively.

In the filtering experiments, the stochastic parameterized equations are employed as the forecast models that evolve the first two moments of the coupled state variables (u, γ) or (u, ω) during the forecast step:

$$\begin{aligned}\frac{\mathrm{d}u}{\mathrm{d}t} &= (-\gamma + i\hat{\omega})u + f_u + \sigma_u \dot{W}_u, \\ \frac{\mathrm{d}\gamma}{\mathrm{d}t} &= -d_\gamma(\gamma - \hat{\gamma}) + \sigma_\gamma \dot{W}_\gamma,\end{aligned} \tag{10.63}$$

and

$$\begin{aligned}\frac{\mathrm{d}u}{\mathrm{d}t} &= (-\hat{\gamma} + i\omega)u + f_u + \sigma_u \dot{W}_u, \\ \frac{\mathrm{d}\omega}{\mathrm{d}t} &= -d_\omega(\omega - \hat{\omega}) + \sigma_\omega \dot{W}_\omega.\end{aligned} \tag{10.64}$$

The parameters in the u equation are assumed to be known, while those in the auxiliary stochastic parameterizations are calibrated by matching the mean, variance, and decorrelation time of the underlying two-state Markov process. For comparison, a filter based on the standard complex OU process,

$$\frac{\mathrm{d}u}{\mathrm{d}t} = (-\hat{\gamma} + i\hat{\omega})u + f_u + \sigma_u \dot{W}_u, \tag{10.65}$$

is also implemented as a reference forecast model.

Figure 10.5 illustrates the significant differences between using the stochastic parameterized forecast model and the OU process in data assimilation. The OU model employs constant coefficients that describe only the mean behavior of the hidden processes. This simplification underestimates the system's variability and causes the filter to over-rely on the model forecast, leading to large systematic errors during intermittent transitions. In contrast, the stochastic parameterized forecast models (10.63)–(10.64) successfully capture the time-varying variability of the hidden driving mechanisms. As a result, the posterior estimates not only reproduce the statistical variability of the intermittent signals but also roughly track the regime shifts of the true two-state Markov processes, yielding more physically consistent and accurate filtering performance.

□

Example 10.14 (Linear model with multiplicative noise versus simple stochastic parameterized model). Since both the linear model with multiplicative noise (10.48) and the simple stochastic parameterized model (10.59) are capable of generating non-Gaussian statistics, it is useful to clarify the situations in which each modeling strategy is most appropriate.

The linear model with multiplicative noise (10.48) can be constructed to reproduce a prescribed long-term stationary distribution. Therefore, if the primary goal is to generate samples from a targeted equilibrium distribution, the linear model with multiplicative noise is a preferable choice. In contrast, the simple stochastic parameterized model (10.59) allows closed analytic formulae for the time evolution of low-order moments. As a result, when the goal is short-term statistical forecasting, such as in data assimilation applications, the model (10.59) is typically more suitable. Beyond filtering and predicting intermittent signals in idealized systems [185, 181, 66], the stochastic parameterizations in (10.59) have proven effective in more complex settings such as stochastic dynamical super-resolution [48] and efficient reduced filters for the Navier-Stokes equations [49]. In these contexts, the stochastic

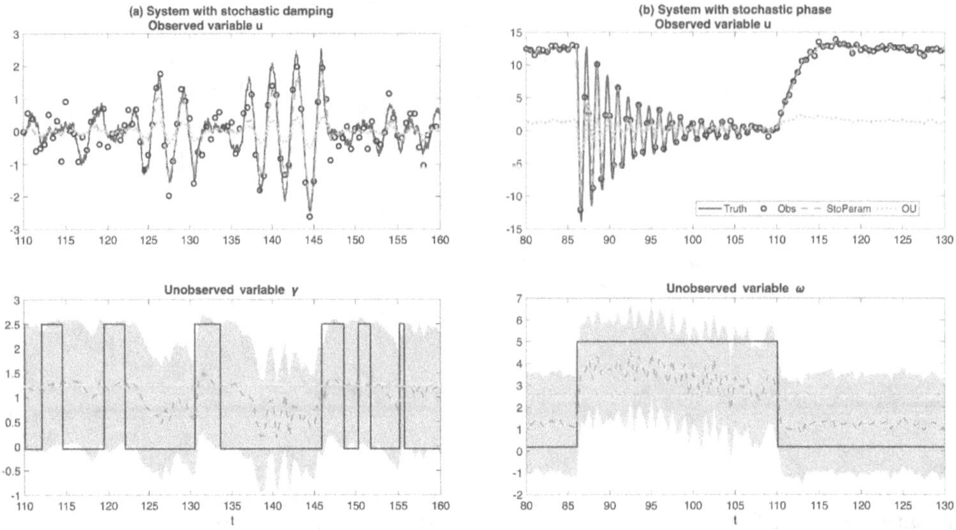

FIGURE 10.5: Filtering of intermittent time series using stochastic parameterized forecast models compared with the complex OU process. The left column corresponds to the case with a stochastically varying damping coefficient $\gamma(t)$ generated by a two-state Markov jump process, and the right column shows the case with a stochastically varying frequency $\omega(t)$. In each case, the solid line denotes the true signal, black circles indicate noisy observations, dashed lines show the posterior mean from the filter using the stochastic parameterized forecast model, and dotted lines represent the filter based on the OU process. The shaded regions indicate one standard deviation of posterior uncertainty for the estimated hidden variables. The stochastic parameterized approach better captures the intermittent transitions and underlying variability of the hidden processes.

parameterized system in (10.59) significantly outperforms simpler OU processes, where all parameterized variables γ, ω, and b are treated as constants rather than as stochastic processes.

It is also important to emphasize that (10.59) represents only one of the simplest examples of stochastic parameterization. In general, there is substantial freedom in constructing stochastic parameterization equations. While closed analytic formulae can be derived for one-dimensional linear models with multiplicative noise, obtaining such analytic results becomes significantly more challenging as the system dimension increases. □

10.4 Analytically solvable nonlinear stochastic models

10.4.1 Revisit: Conditional Gaussian nonlinear stochastic models

The simple stochastic parameterizations discussed above provide an efficient way to enrich linear systems and reproduce certain nonlinear features such as intermittency and non-Gaussian variability. However, these models also have some inherent limitations. First, their primary focus is on capturing the observed behavior of the resolved variable u, while the equations for the parameterized (or unresolved) variables are modeled using very simple structures, such as the OU processes. If the goal is to represent the dynamics of these unresolved components themselves, the OU formulation is often too simple to reproduce their detailed evolution and higher-order statistics. Second, the simple stochastic parameterized systems generally fail to satisfy key physical constraints, such as energy conservation in the quadratic nonlinear interactions as in most fluid and geophysical systems. These properties are essential not only for maintaining physical interpretability but also for ensuring model robustness and preventing nonphysical energy growth and enhancing long-term numerical stability.

Structural simplicity is desirable to ensure that stochastic models remain analytically tractable and computationally feasible in high-dimensional settings. Yet, this simplicity must be balanced against the need for physical realism and statistical accuracy. Although most general nonlinear stochastic systems are not analytically solvable, the conditional Gaussian nonlinear system (CGNS) introduced in Section 7.4.3 effectively maintains this balance. It contains a broad class of nonlinear stochastic dynamics, while many other systems can be reformulated or approximated within the CGNS framework with only mild structural modifications. Let us change the notation of describing CGNS to make it more suitable for a modeling framework [66, 62]:

$$\frac{\mathrm{d}\mathbf{u}_{\mathrm{I}}}{\mathrm{d}t} = \mathbf{A}_0(\mathbf{u}_{\mathrm{I}}, t) + \mathbf{A}_1(\mathbf{u}_{\mathrm{I}}, t)\mathbf{u}_{\mathrm{II}} + \boldsymbol{\Sigma}_{\mathrm{I}}(\mathbf{u}_{\mathrm{I}}, t)\dot{\mathbf{W}}_{\mathrm{I}}, \tag{10.66a}$$

$$\frac{\mathrm{d}\mathbf{u}_{\mathrm{II}}}{\mathrm{d}t} = \mathbf{a}_0(\mathbf{u}_{\mathrm{I}}, t) + \mathbf{a}_1(\mathbf{u}_{\mathrm{I}}, t)\mathbf{u}_{\mathrm{II}} + \boldsymbol{\Sigma}_{\mathrm{II}}(\mathbf{u}_{\mathrm{I}}, t)\dot{\mathbf{W}}_{\mathrm{II}}, \tag{10.66b}$$

where $\mathbf{u}_{\mathrm{I}}$ and $\mathbf{u}_{\mathrm{II}}$ are the resolved/observed and unresolved/unobserved variables (corresponding to $\mathbf{y}$ and $\mathbf{x}$ in (7.55)–(7.56) in Section 7.4.3), respectively. A key distinction of the CGNS framework is that it allows the unresolved variables $\mathbf{u}_{\mathrm{II}}$ to evolve according to a nonlinear conditional process, rather than being restricted to an independent OU formulation. In (10.66b), the coefficients $\mathbf{a}_0$, $\mathbf{a}_1$, and $\boldsymbol{\Sigma}_{\mathrm{II}}$ may depend on the resolved variables $\mathbf{u}_{\mathrm{I}}$ in a nonlinear and time-dependent way. This structure greatly broadens the class of dynamics that the CGNS can represent. In particular, it allows the system to capture nonlinear feedback loops between $\mathbf{u}_{\mathrm{I}}$ and $\mathbf{u}_{\mathrm{II}}$, a feature that is absent in the simpler stochastic parameterizations. Through this coupling, unresolved-scale variability can dynamically influence the evolution of the resolved-scale state, and vice versa, in a statistically consistent way. Moreover, the additional degrees of freedom in (10.66) make it possible to impose important physical constraints, such as energy conservation in the quadratic nonlinear terms. These constraints not only preserve the interpretability of the reduced model but also prevent nonphysical growth and enhance numerical stability.

As mentioned in Section 10.4.3, constructing reduced-order or approximate models within the CGNS framework in (10.66) can significantly enhance end-to-end data assimilation performance.

Example 10.15 (Systems belonging to the CGNS framework). A wide range of classical systems naturally possess the CGNS structure, including numerous nonlinear stochastic models, stochastically coupled reaction-diffusion models in neuroscience and ecology, and many large-scale models in turbulence, fluids, and geophysical flows [66]. Notably, when augmented with small noise, classical models like Lorenz 63, Lorenz 84, and certain two-layer Lorenz 96 variants [65] fall within this framework. For instance, the noisy Lorenz 63 model from Chapter 4 belongs to CGNS when either $\mathbf{u}_{\mathrm{I}} = x, \mathbf{u}_{\mathrm{II}} = (y, z)^{\mathrm{T}}$ or $\mathbf{u}_{\mathrm{I}} = (y, z)^{\mathrm{T}}, \mathbf{u}_{\mathrm{II}} = x$.

The stochastically coupled FitzHugh-Nagumo (FHN) model, a prototype for excitable systems describing neuronal spiking dynamics [169], also exhibits CGNS structure. This model and its stochastic variants have been extensively studied in contexts like stochastic resonance [262, 170, 174, 282] and as noisy excitable media [204, 131, 140, 144]. The lattice version with periodic boundary conditions ($u_0 = u_N$, $u_1 = u_{N+1}$) reads:

$$\begin{aligned}\epsilon\, \mathrm{d}u_i &= \left(d_u(u_{i+1} + u_{i-1} - 2u_i) + u_i - \frac{1}{3}u_i^3 - v_i\right) \mathrm{d}t + \sqrt{\epsilon}\delta_1\, \mathrm{d}W_{u_i}, \\ \mathrm{d}v_i &= \left(u_i + a\right) \mathrm{d}t + \delta_2\, \mathrm{d}W_{v_i}, \qquad i = 1, \ldots, N,\end{aligned} \tag{10.67}$$

where $d_u(u_{i+1} + u_{i-1} - 2u_i)$ represents diffusion discretization. The system becomes CGNS when $\mathbf{u_I} = \{u_i\}$ and $\mathbf{u_{II}} = \{v_i\}$.

Many complex PDE systems can also be easily reformulated into the CGNS framework through appropriate manipulations. This significantly expands the use of the CGNS. The Boussinesq equation serves as an important example. It can be derived from the Navier-Stokes equations under the Boussinesq approximation that density variations only affect buoyancy forces [246, 233]:

$$\begin{aligned}&\frac{\partial \mathbf{u}}{\partial t} + \mathbf{u} \cdot \nabla \mathbf{u} = -\frac{1}{\rho_0}\nabla p + \nu\nabla^2\mathbf{u} - g\alpha T + \mathbf{F_u}, \qquad \nabla \cdot \mathbf{u} = 0, \\ &\frac{\partial T}{\partial t} + \mathbf{u} \cdot \nabla T = \kappa\nabla^2 T + F_T.\end{aligned} \tag{10.68}$$

Here T denotes temperature, $\mathbf{u}$ velocity fields, ρ_0 reference density, p pressure, and $\mathbf{F_u}$, F_T external forcing (possibly stochastic). The system finds applications in Rayleigh-Bénard convection [56, 273] and stratified geophysical flows [183]. Crucially, conditioned on the velocity field $\mathbf{u}$, the temperature equation is linear in T. With the inclusion of stochastic forcing, this conditional linearity places the coupled system within the CGNS framework.□

10.4.2 Energy-conserving quadratic nonlinearity

Energy-conserving quadratic nonlinearity is a fundamental requirement in the construction of reliable reduced-order models for turbulent and geophysical systems. Without such a physics-based constraint, reduced-order models, especially those obtained through regression or truncation, may exhibit pathological behavior, including numerical blow-up or spurious energy growth. Enforcing energy conservation at the nonlinear level provides an essential stabilizing mechanism and preserves the qualitative structure of the underlying dynamics. This is motivated by many nonlinear PDE systems with quadratic nonlinearities in fluids and geophysics.

Example 10.16 (Energy-conserving quadratic nonlinearities in a simple PDE). Energy-conserving quadratic nonlinearities naturally arise in fluid and geophysical flows, most notably through advection. Physically, advection represents the transport of a quantity by the flow itself: fluid parcels carry their energy as they move, but no energy is created or destroyed in the absence of external forcing or dissipation. Under suitable boundary conditions, advection therefore redistributes energy in space without changing its total amount.

To illustrate this property, consider the one-dimensional inviscid advection equation on $x \in [a, b]$,

$$\frac{\partial u}{\partial t} + u\frac{\partial u}{\partial x} = 0, \tag{10.69}$$

with periodic boundary condition $u(a) = u(b)$. Define the energy

$$E = \int_a^b \frac{u^2}{2}\, \mathrm{d}x, \tag{10.70}$$

which is motivated by kinetic energy. Multiplying (10.69) by u and integrating over the domain yields

$$2\frac{\mathrm{d}E}{\mathrm{d}t} + \int_a^b u^2 \frac{\partial u}{\partial x}\,\mathrm{d}x = 0. \tag{10.71}$$

Using the identity $\frac{\partial}{\partial x}\left(\frac{u^3}{3}\right) = u^2 u_x$ and integration by parts, we obtain

$$\int_a^b u^2 \frac{\partial u}{\partial x}\,\mathrm{d}x = \frac{u^3}{3}\bigg|_{x=a}^{x=b} = 0, \tag{10.72}$$

where the last equality follows from the periodic boundary condition. Mathematically, the nonlinear advection term can be written as a spatial divergence of an energy flux, so its contribution to the energy budget reduces to a boundary term. With periodic boundaries, this flux neither enters nor leaves the domain, leading to exact energy conservation.

Therefore, when advection is the only nonlinearity present, the total energy is exactly conserved. This simple example highlights the physical origin and mathematical structure of the energy-conserving quadratic nonlinearities enforced in (10.73). □

For a multivariate dynamical system, which can also be regarded as the discretization of a PDE, the general form of nonlinear stochastic models with energy-conserving quadratic nonlinearity can be written in the form [186, 124]

$$\begin{aligned} \mathrm{d}\mathbf{u} &= \big[(\mathbf{L}+\mathbf{D})\mathbf{u} + \mathbf{B}(\mathbf{u},\mathbf{u}) + \mathbf{F}(t)\big]\,\mathrm{d}t + \boldsymbol{\Sigma}(\mathbf{u})\,\mathrm{d}\mathbf{W}, \\ \text{with} \qquad & \mathbf{u}\cdot\mathbf{B}(\mathbf{u},\mathbf{u}) = 0, \end{aligned} \tag{10.73}$$

where $\mathbf{u} \in \mathbb{R}^n$ is the state vector. The linear operator $\mathbf{L}+\mathbf{D}$ represents dispersion and damping: $\mathbf{L}$ is skew-symmetric and accounts for dispersive effects, while $\mathbf{D}$ is a symmetric negative-definite operator modeling dissipative processes such as surface drag, viscosity, or radiative damping. The term $\mathbf{F}(t)$ denotes deterministic forcing, $\boldsymbol{\Sigma}$ is an $n \times m$ noise coefficient matrix, and $\mathbf{W}$ is an m-dimensional Wiener process.

The quadratic term $\mathbf{B}(\mathbf{u},\mathbf{u})$ encodes the nonlinear interactions. The key physics constraint, $\mathbf{u}\cdot\mathbf{B}(\mathbf{u},\mathbf{u}) = 0$ ensures that the quadratic nonlinearity does not contribute to the evolution of the total energy. This condition is the finite-dimensional analogue of the energy conservation property in continuous systems, where multiplying the nonlinear term by the state variable and integrating over space yields zero. Here, the summation over components in $\mathbf{u}\cdot\mathbf{B}(\mathbf{u},\mathbf{u})$ plays the role of the spatial integral.

It is important to note that energy injection and dissipation still occur through the linear and stochastic terms. Deterministic forcing and stochastic excitation inject energy into the system, while linear damping removes it. These competing effects typically reach a statistical balance, which, together with the energy-conserving quadratic nonlinearity, leads to a statistically steady state when $\mathbf{F}$ is constant, or to a statistical attractor when $\mathbf{F}(t)$ is time-periodic. A major advantage of such physics-constrained stochastic models is that they simultaneously avoid finite-time blow-up and retain physical interpretability, in contrast to many ad hoc regression-based models [188].

Example 10.17 (The role of the energy-conserving nonlinearity). Let us compare the performance of the following two models. The first model (denoted as `StoParam`) is a stochastic parameterized system where an OU process characterizes the stochastic damping:

$$\begin{aligned} \frac{\mathrm{d}u}{\mathrm{d}t} &= -\gamma u + \sigma_u \dot{W}_u, \\ \frac{\mathrm{d}\gamma}{\mathrm{d}t} &= -d_\gamma(\gamma - \hat{\gamma}) + \sigma_\gamma \dot{W}_\gamma. \end{aligned} \tag{10.74}$$

The second model belongs to the CGNS framework (10.66) and includes an additional feedback term, u^2, in the γ equation. This term represents the influence of the resolved variable u on the stochastic damping, introducing a physics-constrained energy-conserving nonlinearity:

$$\begin{aligned}\frac{\mathrm{d}u}{\mathrm{d}t} &= -\gamma u + \sigma_u \dot{W}_u,\\ \frac{\mathrm{d}\gamma}{\mathrm{d}t} &= -d_\gamma(\gamma - \hat{\gamma}) + u^2 + \sigma_\gamma \dot{W}_\gamma.\end{aligned} \tag{10.75}$$

This model is referred to as PhyConst for short. The energy-conserving property of the quadratic coupling can be seen by multiplying the first equation by u and the second by γ, and then adding the resulting identities. Defining the (formal) total energy $E = \frac{1}{2}(u^2 + \gamma^2)$, the left-hand side becomes $\mathrm{d}E/\,\mathrm{d}t$, while the quadratic terms on the right-hand side cancel exactly: $(-\gamma u)\,u = -\gamma u^2$ in the first equation and $(u^2)\,\gamma = +\gamma u^2$ in the second.

With the parameters $\sigma_u = 0.25$, $d_\gamma = 0.5$, and $\sigma_\gamma = 1$, Panels (a)–(b) of Figure 10.6 show the variance and kurtosis of u as functions of the mean damping parameter $\hat{\gamma}$. When $\hat{\gamma}$ is large and positive, the strong damping $-\gamma$ stabilizes the process for u and suppressing extreme events. In this regime, the StoParam model remains stable and the kurtosis of u is close to 3, indicating nearly Gaussian statistics. However, as $\hat{\gamma}$ decreases, the dynamics change dramatically. Both the variance and kurtosis increase rapidly, and the system becomes unstable when $\hat{\gamma}$ falls below 1.5. The corresponding time series (Panel (c)) shows intermittent bursts of unrealistically large amplitudes, eventually leading to numerical blow-up.

In contrast, the PhyConst model in (10.75) remains statistically stable across the same parameter range. The feedback term u^2 acts as a self-regulating mechanism: whenever u grows excessively, the feedback term u^2 increases the value of γ, introducing stronger damping that prevents a further growth of u. Panel (e) illustrates this effect for $\hat{\gamma} = 1.5$, where the time series exhibits realistic intermittent events (approximately three to five standard deviations from the mean) without numerical instability. Even when $\hat{\gamma}$ is negative (Panel (d)), the PhyConst system avoids blow-up because the energy-conserving feedback maintains a bounded total energy in the nonlinear terms. In contrast, in the StoParam model, the absence of this feedback violates the energy-conservation property. When γ becomes negative, the nonlinear term $-\gamma u^2$ continuously injects energy into the system, leading to unbounded growth.

As a final remark, it is worth emphasizing that although the simple stochastic parameterized model based on the OU process in (10.74) may exhibit pathological behavior in long-term simulations due to the absence of energy-conserving nonlinear interactions, it remains highly useful for data assimilation applications when only short-range forecasts are required. Different models are suited to different purposes, and their limitations should be interpreted in the context of their intended use.

□

10.4.3 Systematic model development

From an approximation perspective, CGNS can be employed in two primary ways. First, many complex dynamical systems in geophysics and engineering can be naturally approximated or idealized as CGNS. For instance, turbulent systems can often be cast in this framework by applying stochastic mode reduction strategies [187] to approximate the small-scale modes. The resulting model is both nonlinear and computationally tractable [199, 70, 67]. Second, data-driven techniques, such as the sparse model identification technique shown in Section 9.4, can be used to learn a CGNS approximation from data.

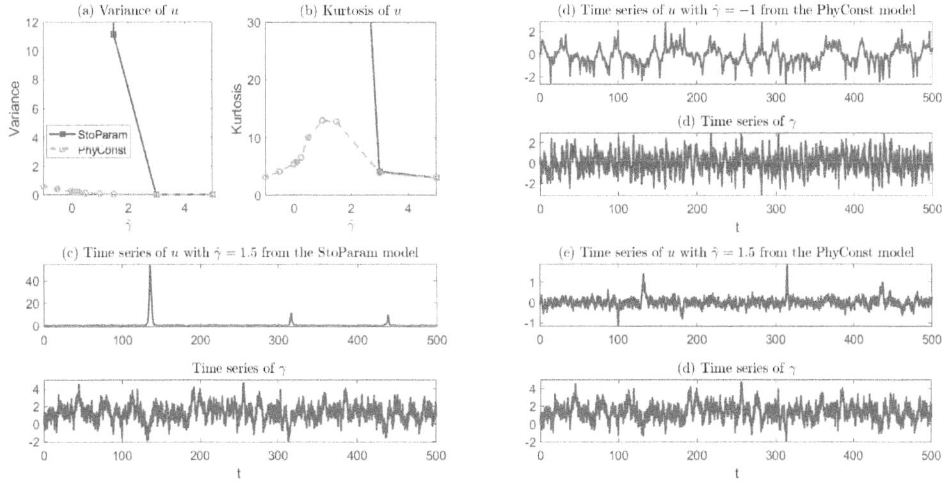

FIGURE 10.6: Comparison between the standard stochastic parameterized model (`StoParam`) in (10.74) and the CGNS model with a physics constraint (`PhyConst`) in (10.75). All parameters are fixed except for $\hat{\gamma}$. Panels (a)–(b): variance and kurtosis of u as functions of $\hat{\gamma}$. Panel (c): time series of `StoParam` with $\hat{\gamma} = 1.5$. Panels (d)–(e): time series of `PhyConst` with $\hat{\gamma} = -1$ and $\hat{\gamma} = 1.5$, respectively.

Example 10.18 (Embedding a multiscale quadratic system into the CGNS class via stochastic closure). Consider a deterministic multiscale system with quadratic nonlinearity,

$$\mathrm{d}\mathbf{u} = \Big[\mathbf{L}\mathbf{u} + \mathbf{B}(\mathbf{u}, \mathbf{u}) + \mathbf{F}\Big]\, dt, \tag{10.76}$$

where $\mathbf{L}\mathbf{u}$ contains all the linear terms, $\mathbf{B}(\mathbf{u}, \mathbf{u})$ represents the quadratic nonlinear terms, and $\mathbf{F}$ is a forcing term. Decompose $\mathbf{u} = (\mathbf{u}_{\mathbf{I}}, \mathbf{u}_{\mathbf{II}})^{\mathtt{T}}$ into resolved variables $\mathbf{u}_{\mathbf{I}}$ and unresolved variables $\mathbf{u}_{\mathbf{II}}$. Then

$$\begin{aligned}
\mathrm{d}\mathbf{u}_{\mathbf{I}} &= \Big[\mathbf{L}_{11}\mathbf{u}_{\mathbf{I}} + \mathbf{L}_{12}\mathbf{u}_{\mathbf{II}} + \mathbf{B}^1_{11}(\mathbf{u}_{\mathbf{I}}, \mathbf{u}_{\mathbf{I}}) + \mathbf{B}^1_{12}(\mathbf{u}_{\mathbf{I}}, \mathbf{u}_{\mathbf{II}}) + \mathbf{B}^1_{22}(\mathbf{u}_{\mathbf{II}}, \mathbf{u}_{\mathbf{II}}) + \mathbf{F}_1\Big]\, dt,\\
\mathrm{d}\mathbf{u}_{\mathbf{II}} &= \Big[\mathbf{L}_{22}\mathbf{u}_{\mathbf{II}} + \mathbf{L}_{21}\mathbf{u}_{\mathbf{I}} + \mathbf{B}^2_{11}(\mathbf{u}_{\mathbf{I}}, \mathbf{u}_{\mathbf{I}}) + \mathbf{B}^2_{12}(\mathbf{u}_{\mathbf{I}}, \mathbf{u}_{\mathbf{II}}) + \mathbf{B}^2_{22}(\mathbf{u}_{\mathbf{II}}, \mathbf{u}_{\mathbf{II}}) + \mathbf{F}_2\Big]\, dt.
\end{aligned} \tag{10.77}$$

A key obstacle to the CGNS structure is that the fast-scale self-interaction terms $\mathbf{B}^1_{22}(\mathbf{u}_{\mathbf{II}}, \mathbf{u}_{\mathbf{II}})$ and $\mathbf{B}^2_{22}(\mathbf{u}_{\mathbf{II}}, \mathbf{u}_{\mathbf{II}})$ are nonlinear in $\mathbf{u}_{\mathbf{II}}$. A common stochastic-closure approximation replaces these quadratic self-interactions by linear damping and additive noise [187]:

$$\mathbf{B}^1_{22}(\mathbf{u}_{\mathbf{II}}, \mathbf{u}_{\mathbf{II}})\, dt \approx -\frac{\mathbf{\Gamma}_1}{\epsilon}\mathbf{u}_{\mathbf{II}}\, dt + \frac{\mathbf{\Sigma}_{\mathbf{I}}}{\sqrt{\epsilon}}\, \mathrm{d}\mathbf{W}_{\mathbf{I}}, \qquad \mathbf{B}^2_{22}(\mathbf{u}_{\mathbf{II}}, \mathbf{u}_{\mathbf{II}})\, dt \approx -\frac{\mathbf{\Gamma}_2}{\epsilon}\mathbf{u}_{\mathbf{II}}\, dt + \frac{\mathbf{\Sigma}_{\mathbf{II}}}{\sqrt{\epsilon}}\, \mathrm{d}\mathbf{W}_{\mathbf{II}}, \tag{10.78}$$

where $\epsilon \ll 1$ quantifies the time-scale separation (and $\mathbf{W}_{\mathbf{I}}, \mathbf{W}_{\mathbf{II}}$ may be taken independent, or correlated if desired). The resulting stochastic system becomes:

$$\begin{aligned}
\mathrm{d}\mathbf{u}_{\mathbf{I}} &= \Big[\mathbf{L}_{11}\mathbf{u}_{\mathbf{I}} + \Big(\mathbf{L}_{12} - \frac{\mathbf{\Gamma}_1}{\epsilon}\Big)\mathbf{u}_{\mathbf{II}} + \mathbf{B}^1_{11}(\mathbf{u}_{\mathbf{I}}, \mathbf{u}_{\mathbf{I}}) + \mathbf{B}^1_{12}(\mathbf{u}_{\mathbf{I}}, \mathbf{u}_{\mathbf{II}}) + \mathbf{F}_1\Big]\, dt + \frac{\mathbf{\Sigma}_{\mathbf{I}}}{\sqrt{\epsilon}}\, \mathrm{d}\mathbf{W}_{\mathbf{I}},\\
\mathrm{d}\mathbf{u}_{\mathbf{II}} &= \Big[\Big(\mathbf{L}_{22} - \frac{\mathbf{\Gamma}_2}{\epsilon}\Big)\mathbf{u}_{\mathbf{II}} + \mathbf{L}_{21}\mathbf{u}_{\mathbf{I}} + \mathbf{B}^2_{11}(\mathbf{u}_{\mathbf{I}}, \mathbf{u}_{\mathbf{I}}) + \mathbf{B}^2_{12}(\mathbf{u}_{\mathbf{I}}, \mathbf{u}_{\mathbf{II}}) + \mathbf{F}_2\Big]\, dt + \frac{\mathbf{\Sigma}_{\mathbf{II}}}{\sqrt{\epsilon}}\, \mathrm{d}\mathbf{W}_{\mathbf{II}}.
\end{aligned} \tag{10.79}$$

Conditioned on $\mathbf{u}_{\mathbf{I}}$, the dynamics of $\mathbf{u}_{\mathbf{II}}$ are linear with additive noise, and hence the system falls into the CGNS framework. □

10.5 Koopman theory: Linear perspectives on nonlinear dynamics

Most of the dynamical systems in science and engineering problems involve strong nonlinearities. Fundamentally different from a direct linearization of these equations, which often leads to significant errors and completely changes the underlying physics and mechanisms, Koopman theory provides a simple yet powerful alternative. It represents nonlinear dynamics as a linear evolution in a higher-dimensional space of measurable features. Koopman theory bridges classical dynamical systems and linear operator theory. It has also become the foundation for modern data-driven modeling techniques such as the dynamic mode decomposition (DMD) and neural operator learning.

10.5.1 Motivation

A linear system is easy to analyze because every variable changes in proportion to its current value. For example, if $x_{k+1} = 0.5x_k$, then every step just multiplies the current value by 0.5. We can easily predict that $x_k = 0.5^k x_0$, and plotting it gives a smooth exponential decay.

However, nonlinear systems are different. Consider the following simple scalar equation as an example:

$$x_{k+1} = x_k^2. \tag{10.80}$$

The change depends on x_k in a more complicated way. If the amplitude of the initial condition $|x_0| < 1$, then the solution at later time instants quickly goes to zero. In contrast, if $|x_0| > 1$, the solution will blow up to infinity. It seems there is no single simple rule.

Koopman's idea is that maybe we are looking at the wrong quantity. If we measure or "observe" the system in a clever way, the evolution might again look linear.

Example 10.19 (A nonlinear system that looks linear after a change of variable).
Consider again the nonlinear map in (10.80). Let us define a new variable $y_k = \log(x_k)$. Assume $x_k > 0$ so that the logarithm is well defined. Then

$$y_{k+1} = \log(x_{k+1}) = \log\left(x_k^2\right) = 2\log(x_k) = 2y_k, \tag{10.81}$$

which simply becomes a linear system in terms of the y-variable. Therefore, although the original rule in (10.80) was nonlinear in x, the evolution in y is a simple doubling every step. This gives an example of "hidden linearity." □

10.5.2 Koopman operator

The function $y_k = \log(x_k)$ in Example 10.19 is called a measurable feature (or observable in the language of Koopman theory). It represents a new way of describing the system. Instead of tracking the raw state variable x_k, we monitor a function of it. Choosing appropriate measurable features is crucial, because they determine whether the underlying nonlinear dynamics can be represented as a linear evolution (possibly in this higher-dimensional space).

The key idea of Koopman theory is that, even though the dynamics of x_k may be nonlinear, the evolution of certain functions of x_k can still be linear [155, 52]. Formally, suppose the system evolves as

$$x_{k+1} = F(x_k), \tag{10.82}$$

where F may be nonlinear. For any function $g(x)$ that we can measure or compute, its evolution follows

$$g(x_{k+1}) = g(F(x_k)). \tag{10.83}$$

This defines a linear operator $\mathcal{K}$, called the Koopman operator, such that

$$(\mathcal{K}g)(x_k) = g(F(x_k)). \tag{10.84}$$

Although F may be nonlinear, the Koopman operator $\mathcal{K}$ acts linearly on the space of all measurable functions g. This linear perspective provides a powerful new way to analyze nonlinear systems by lifting them into a space where linear methods apply.

Example 10.20 (Revealing hidden linearity by lifting a 2D nonlinear system into 3D). Consider the following nonlinear system

$$\begin{aligned} x_{k+1} &= a\,x_k, \\ y_{k+1} &= b\,y_k + c\,x_k^2, \end{aligned} \tag{10.85}$$

with constants $a, b, c \in \mathbb{R}$. By introducing the observable (measurable feature) vector

$$\mathbf{g}(x, y) = \begin{bmatrix} g_1 \\ g_2 \\ g_3 \end{bmatrix} = \begin{bmatrix} x \\ y \\ x^2 \end{bmatrix}, \tag{10.86}$$

we have

$$\begin{aligned} g_1(x_{k+1}, y_{k+1}) &= x_{k+1} = a\,x_k = a\,g_1(x_k, y_k), \\ g_2(x_{k+1}, y_{k+1}) &= y_{k+1} = b\,y_k + c\,x_k^2 = b\,g_2(x_k, y_k) + c\,g_3(x_k, y_k), \\ g_3(x_{k+1}, y_{k+1}) &= x_{k+1}^2 = (a\,x_k)^2 = a^2\,x_k^2 = a^2\,g_3(x_k, y_k). \end{aligned} \tag{10.87}$$

Therefore, in the lifted coordinates $\mathbf{g}$, the evolution is exactly linear:

$$\begin{bmatrix} g_1 \\ g_2 \\ g_3 \end{bmatrix}_{k+1} = \underbrace{\begin{bmatrix} a & 0 & 0 \\ 0 & b & c \\ 0 & 0 & a^2 \end{bmatrix}}_{\text{Koopman matrix on the chosen features}} \begin{bmatrix} g_1 \\ g_2 \\ g_3 \end{bmatrix}_k . \tag{10.88}$$

Hence $\{x,\, y,\, x^2\}$ is a finite, closed set of observables for (10.85), giving an exact finite-dimensional Koopman representation.

The geometric interpretation of the result is as follows. In the (x, y) plane, the trajectory of (10.85) bends because the y-update receives a nonlinear forcing x^2. Although both x and y decay, their relation is not linear. When we lift the system to (x, y, z) with $z = x^2$, the motion lies on the paraboloid surface $z = x^2$. On this manifold, the evolution of the lifted variables (x, y, z) is exactly linear as in (10.88). Thus, while the trajectory appears curved in the original coordinates, it follows a linear rule in the lifted space. Figure 10.7 shows the trajectory in the original (x, y) plane (Panel (b)) and on the paraboloid $z = x^2$ (the nonlinear manifold), illustrating how Koopman lifting transforms nonlinear dynamics into linear evolution on a curved manifold.

□

Choosing the right observables is crucial for revealing the hidden linear structure of a nonlinear system. Different choices can either simplify the dynamics dramatically or make them unnecessarily complicated.

Example 10.21 (Choosing the right observables). Let us recall the simple scalar equation in (10.80). If we choose only one observable $g_1(x) = x$, representing the original state variable, then

$$g_1(x_{k+1}) = x_{k+1} = x_k^2 = (g_1(x_k))^2. \tag{10.89}$$

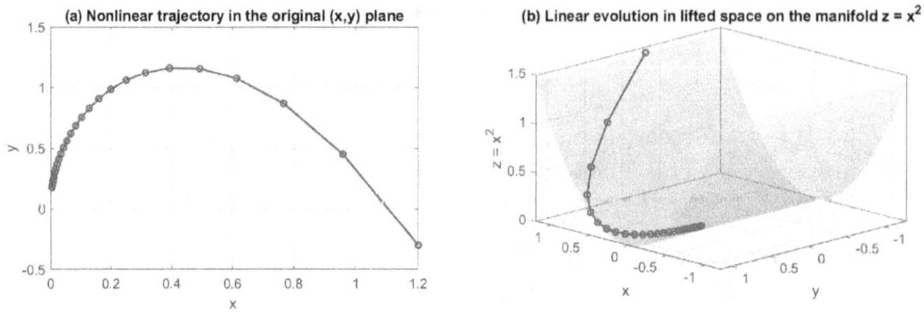

FIGURE 10.7: Panel (a): Trajectory of the nonlinear discrete system in (10.85) in the original (x, y) plane, showing a curved path caused by the quadratic coupling. Panel (b): The same trajectory after Koopman lifting to the augmented coordinates (x, y, z) with $z = x^2$. The points evolve linearly under a constant matrix on the nonlinear manifold $z = x^2$. Although the motion appears curved on the surface, each update step follows an exact linear rule in the lifted variables, illustrating how an appropriate choice of observable can convert nonlinear dynamics into linear evolution.

This expression cannot be written as a constant times $g_1(x_k)$, because the squaring operation introduces a new function, x^2, that lies outside our current set of observables. In other words, the space spanned by $\{x\}$ is not closed under the system dynamics.

To make it closed, we can enlarge our set of observables to include x^2, namely, $g_1(x) = x$ and $g_2(x) = x^2$. Therefore,

$$\begin{bmatrix} g_1(x_{k+1}) \\ g_2(x_{k+1}) \end{bmatrix} = \begin{bmatrix} x_{k+1} \\ x_{k+1}^2 \end{bmatrix} = \begin{bmatrix} x_k^2 \\ x_k^4 \end{bmatrix} = \begin{bmatrix} g_2(x_k) \\ g_2(x_k)^2 \end{bmatrix}. \tag{10.90}$$

Now the first equation in (10.90) becomes linear, but the second introduces yet another higher-order term, x^4, which again lies outside our current set $\{x, x^2\}$. As the iteration continues, we would generate $x^4, x^8, \ldots,$ and so on. Therefore, no finite collection of polynomial observables can make this system perfectly closed. For this example, the logarithmic observable used in Example 10.19 perfectly closes the system in one step, illustrating how the right observable can completely transform our understanding of the dynamics. □

In most practical nonlinear systems, the Koopman representation is not perfectly closed. Therefore, we often truncate the infinite-dimensional sequence and work with a finite set of observables, which leads to an approximate linear system. This is the principle behind data-driven Koopman and DMD methods. The accuracy of this approximation depends critically on how we choose the observables.

10.6 Dynamic mode decomposition (DMD) and its link to Koopman theory

Koopman theory shows that even nonlinear systems can be represented as linear evolutions in a higher-dimensional space of observables. However, in practice, we often do not know the true governing equations or the appropriate observables. DMD is a data-driven method that approximates this linear evolution directly from measurements, without explicitly performing a lifting step [264, 238]. It can therefore be viewed as an approximation of the Koopman operator acting on the original state variables.

10.6.1 Basic idea of the DMD

Assume we have snapshots of the system state $\mathbf{x}_k \in \mathbb{R}^n$ at discrete times $\mathbf{x}_1, \mathbf{x}_2, \ldots, \mathbf{x}_m$. We arrange these snapshots into two data matrices:

$$X_1 = \begin{bmatrix} \mathbf{x}_1 & \mathbf{x}_2 & \cdots & \mathbf{x}_{m-1} \end{bmatrix}, \quad X_2 = \begin{bmatrix} \mathbf{x}_2 & \mathbf{x}_3 & \cdots & \mathbf{x}_m \end{bmatrix}. \tag{10.91}$$

We then find a best-fit linear map A such that

$$X_2 \approx A X_1, \tag{10.92}$$

where the matrix A is given by the least-squares solution $A = X_2 X_1^\dagger$ with $X_1^\dagger$ denoting the Moore-Penrose pseudoinverse of X_1. The solution provides a finite-dimensional linear operator that best advances the data in time. The eigenvalues and eigenvectors of A provide the DMD modes and DMD eigenvalues, which approximate the dominant Koopman modes and their temporal dynamics.

Note that this basic DMD uses the original state vector $\mathbf{x}_k$ as the observable. Thus, it only approximates the Koopman operator acting on the identity observable $g(\mathbf{x}) = \mathbf{x}$. If the true dynamics involves nonlinearities (or the true Koopman dynamics require nonlinear observables), standard DMD will capture only part of the true dynamics.

DMD provides a practical first step toward Koopman analysis when only limited measurements are available. While standard DMD captures the linear behavior in the measured variables, it cannot fully represent nonlinear interactions that live outside this measurement space. In contrast, extended DMD (EDMD) generalizes standard DMD by including nonlinear features, effectively performing the Koopman lifting step using data and discovering the hidden linearity of nonlinear systems.

Example 10.22 (DMD versus Koopman lifting).
Consider the nonlinear system

$$\begin{aligned} x_{k+1} &= 0.9\, x_k, \\ y_{k+1} &= 0.8\, y_k + 0.2\, x_k^2. \end{aligned} \tag{10.93}$$

The true Koopman lifting (as shown earlier) introduces the observable $z = x^2$, giving an exact 3D linear system. Let us now compare what happens when we apply DMD directly to the observed pairs (x_k, y_k).

If we collect data from (10.93) and compute

$$A_{\text{DMD}} = X_2 X_1^\dagger, \tag{10.94}$$

we obtain an approximate linear model such as

$$A_{\text{DMD}} \approx \begin{bmatrix} 0.9 & 0 \\ 0.18 & 0.8 \end{bmatrix}. \tag{10.95}$$

This DMD matrix captures the leading linear trend (decay in both variables) and even infers an apparent linear coupling between x and y. However, it cannot perfectly reproduce the nonlinear effect of x^2, because this observable was not included in the state vector.

To better approximate the Koopman operator, we can explicitly lift the data by defining a vector of observables (or features)

$$\mathbf{g}(\mathbf{x}) = \begin{bmatrix} x \\ y \\ x^2 \end{bmatrix}. \tag{10.96}$$

We then build new snapshot matrices in this lifted space and compute

$$A_{\text{EDMD}} = G_2 G_1^{\dagger}, \quad G_1 = \begin{bmatrix} \mathbf{g}(\mathbf{x}_1) & \cdots & \mathbf{g}(\mathbf{x}_{m-1}) \end{bmatrix}, \quad G_2 = \begin{bmatrix} \mathbf{g}(\mathbf{x}_2) & \cdots & \mathbf{g}(\mathbf{x}_m) \end{bmatrix}. \tag{10.97}$$

With this lifted feature vector, the resulting matrix A_{EDMD} exactly matches the Koopman matrix derived analytically earlier:

$$A_{\text{EDMD}} = \begin{bmatrix} 0.9 & 0 & 0 \\ 0 & 0.8 & 0.2 \\ 0 & 0 & 0.9^2 \end{bmatrix}. \tag{10.98}$$

□

10.6.2 The difference between DMD and PCA

PCA identifies spatial patterns that explain the largest variance in a dataset, but it does not explicitly account for how the system evolves over time. It is a purely static method. Each snapshot is treated as an independent sample, and temporal ordering plays no role. As a result, the principal components describe dominant directions of variability in space, rather than dynamical mechanisms governing transitions between states. In contrast, DMD is explicitly designed to capture dynamics. Instead of analyzing snapshots individually, DMD focuses on the change between consecutive snapshots by seeking a best-fit linear operator that advances the system from one time step to the next. In this sense, DMD models the evolution rule of the data rather than its static distribution. The resulting DMD modes are therefore associated with specific growth or decay rates and oscillation frequencies, providing a direct link to the underlying temporal dynamics. While PCA summarizes how the data are distributed in state space, DMD reveals how the data evolve through state space. This distinction makes DMD particularly powerful for identifying physically meaningful dynamical features such as oscillations, traveling waves, and transient decay modes that are invisible to variance-based methods alone.

Despite these differences, PCA and DMD are closely related. Both are linear, data-driven techniques that extract low-dimensional structure from high-dimensional datasets, and both rely on linear algebraic decompositions. In practice, DMD almost always employs a PCA (or singular value decomposition) step to reduce dimensionality and filter noise before estimating the dynamical operator. From this perspective, PCA provides an optimal static subspace, while DMD augments this subspace with dynamical information by learning how states move within it.

Example 10.23 (DMD versus PCA on an oscillating signal). Consider a simple system composed of two spatial sensors measuring a time-dependent process:

$$\mathbf{x}(t) = \begin{bmatrix} \sin(\omega t) \\ \cos(\omega t) \end{bmatrix}. \tag{10.99}$$

We collect m snapshots at equal time intervals Δt to form the data matrix

$$X = \begin{bmatrix} \mathbf{x}(t_1) & \mathbf{x}(t_2) & \cdots & \mathbf{x}(t_m) \end{bmatrix}. \tag{10.100}$$

If we apply PCA to X, the two dominant principal components correspond to the directions of maximum and second-maximum variance in the oscillatory data. Since the data trace a circular trajectory centered at the origin, these components form two orthogonal directions that together span the plane of motion. Their orientation depends on the mean and sampling phase, so they are not necessarily aligned with the coordinate axes. PCA therefore reveals the spatial subspace where the motion occurs, but it does not explicitly model temporal evolution or identify the governing evolution operator, such as the oscillation frequency or phase rotation between the two components.

In contrast, by applying DMD, we form the matrices

$$X_1 = \begin{bmatrix}\mathbf{x}(t_1) & \cdots & \mathbf{x}(t_{m-1})\end{bmatrix}, \quad X_2 = \begin{bmatrix}\mathbf{x}(t_2) & \cdots & \mathbf{x}(t_m)\end{bmatrix}, \tag{10.101}$$

and compute $A = X_2 X_1^{\dagger}$. For this example, the exact DMD operator is a rotation matrix:

$$A = \begin{bmatrix}\cos(\omega\Delta t) & -\sin(\omega\Delta t)\\ \sin(\omega\Delta t) & \cos(\omega\Delta t)\end{bmatrix}. \tag{10.102}$$

Its eigendecomposition gives

$$A\phi_{\pm} = e^{\pm i\omega\Delta t}\,\phi_{\pm}, \qquad \phi_{\pm} = \frac{1}{\sqrt{2}}\begin{bmatrix}1\\ \mp i\end{bmatrix}. \tag{10.103}$$

These complex-conjugate eigenpairs correspond to oscillations at frequency ω. Thus, the DMD identifies two rotating modes that capture the circular motion of (x_1, x_2) through time, which is a clear dynamical signature invisible to PCA.

This example illustrates that DMD modes can have a direct physical interpretation, such as oscillation frequency or decay rate, whereas PCA modes are purely statistical.

□

10.7 Neural differential equations from the generalized Koopman theory

Recall the CGNS in (10.66). As mentioned above, there exist several analytic methods for specifying the coefficients $\mathbf{A}_0$, $\mathbf{A}_1$, $\mathbf{a}_0$, $\mathbf{a}_1$, $\boldsymbol{\Sigma}_{\text{I}}$, and $\boldsymbol{\Sigma}_{\text{II}}$ when constructing reduced-order models. These parametric choices are often guided by physical insight or asymptotic analysis, but the model identification process can be challenging when the underlying interactions are strongly nonlinear, high-dimensional, or the true dynamics are beyond the prior knowledge (e.g., the function library in sparse identification methods) used to determine the system.

It is therefore worth emphasizing that these functions can also be approximated using neural networks. In this case, the resulting framework naturally becomes a neural conditional Gaussian nonlinear system (neural CGNS), which belongs to the broader class of neural differential equations. By leveraging the universal approximation capability of neural networks, the neural CGNS significantly expands the expressiveness of traditional parametric closures, allowing the model to capture complex interactions that are difficult to prescribe analytically. Meanwhile, the CGNS structure imposes strong statistical and physical constraints, such as conditional Gaussianity and linearity in the latent variables, that improve interpretability, stability, and data efficiency compared with unconstrained neural ordinary differential equations.

Different from standard neural ordinary differential equations that operate directly in the observational space, the second component of the neural CGNS is formulated in a latent space. This design is motivated by the fact that the original observational variables may not admit a conditional Gaussian structure, even though such a structure can naturally emerge after a suitable nonlinear transformation. By introducing latent variables, the neural CGNS provides a compact representation in which the nonlinear dependencies between resolved and unresolved components can be well approximated by conditionally linear dynamics with additive noise.

A key advantage of the CGNS formulation is that it possesses closed analytical expressions for the data assimilation update. This property allows the incorporation of the analytically solvable data-assimilation loss into neural network training. By doing so, the training process simultaneously optimizes both predictive accuracy and assimilation consistency. This tight coupling between neural modeling and analytical assimilation brings two major benefits. First, it enables efficient and accurate state estimation and forecasting, even in the presence of partially observed data. Second, it mitigates overfitting, as an overfitted model yields unstable, overly sensitive inverse solutions when reconstructing hidden states. In contrast, accurate recovery of unobserved variables depends on correctly capturing the dynamical coupling structure, so good assimilation performance naturally reflects the correct causal and physical relationships within the system. Essentially, this neural CGNS leads to a stochastic digital twin system for forecasting, data assimilation, and uncertainty quantification.

Conceptually, the neural CGNS provides a highly expressive approximation framework for a broad class of nonlinear stochastic systems while retaining conditional linearity. This differs fundamentally from standard Koopman methods, which approximate the original nonlinear system with a high-dimensional linear model. In the neural CGNS, the nonlinear conditional structure can be designed to preserve intrinsic energy exchange and other physical invariants through appropriate architectural constraints, providing a more realistic and dynamically consistent approximation. The unresolved physical variables $\mathbf{u_{II}}$ can be embedded into an augmented latent variable $\mathbf{v} = \psi(\mathbf{u_{II}})$, governed by a high-dimensional nonlinear equation that remains conditionally linear. This representation can be viewed as a generalized Koopman theory in a nonlinear latent space. Unlike classical Koopman theory, which seeks a linear evolution in an infinite-dimensional observable space, the neural CGNS introduces a nonlinear latent transformation ψ such that the lifted dynamics remain conditionally linear, thereby combining nonlinear feature learning with analytic tractability. Nevertheless, by choosing an appropriate transformation $\psi(\cdot)$, for example, via an autoencoder or another dimensionality reduction technique, the system can be projected into a minimal latent space while retaining key dynamics. Because of this nonlinear compression, the latent neural differential system can capture comparable levels of explainable variance as high-dimensional linear Koopman models, often with a substantially lower effective dimension. The resulting neural differential equations take the form

$$\frac{\mathrm{d}\mathbf{u_I}}{\mathrm{d}t} = \mathbf{A_0}(\mathbf{u_I}, t) + \mathbf{A_1}(\mathbf{u_I}, t)\mathbf{v} + \boldsymbol{\Sigma}_\mathbf{I}(\mathbf{u_I}, t)\dot{\mathbf{W}}_\mathbf{I}, \tag{10.104a}$$

$$\frac{\mathrm{d}\mathbf{v}}{\mathrm{d}t} = \mathbf{a_0}(\mathbf{u_I}, t) + \mathbf{a_1}(\mathbf{u_I}, t)\mathbf{v} + \boldsymbol{\Sigma}_\mathbf{v}(\mathbf{u_I}, t)\dot{\mathbf{W}}_\mathbf{v}. \tag{10.104b}$$

Notably, although the conditional distribution of $\mathbf{v}$ given a realization of $\mathbf{u_I}$ is Gaussian, the nonlinear mapping ψ ensures that the conditional distribution of the original unresolved variables $\mathbf{u_{II}}$ given $\mathbf{u_I}$ remains non-Gaussian, thus capturing essential nonlinear and non-Gaussian dynamics.

Such a neural CGNS framework has been successfully applied to estimate the states of the Navier-Stokes equations and other turbulent systems from sparse observations. When the governing equations are decomposed into observed large-scale variables $\mathbf{u_I}$ and unobserved small-scale variables $\mathbf{u_{II}}$, the resulting system does not exactly satisfy the CGNS form. Nevertheless, the neural CGNS serves as an effective approximation, capturing the essential conditional dependencies between the resolved and unresolved components. In practice, the neural CGNS achieves forecast accuracy comparable to that of state-of-the-art neural network predictors, while its data assimilation performance even slightly exceeds that of a finely tuned EnKF with localization. Moreover, the analytical tractability of the CGNS framework leads to remarkable computational efficiency, which allows state estimation that is hundreds of times faster than the EnKF. Some examples can be found in [59, 58].

10.8 Code listings for figures and examples

- `Chapter_10_Galerkin_ROMs.m`: generates Figure 10.1 in Example 10.6, which compares a full three-dimensional system, the Galerkin reduced-order model without closure, and the reduced-order model with closure.
- `Chapter_10_ROM_L63.m`: generates Figure 10.2 in Example 10.7, which uses the Lorenz 63 model to demonstrate that the Galerkin reduced-order model may perform poorly for chaotic systems.
- `Chapter_10_Stochastic_Parameterized_Model.m`: generates Figure 10.4 in Example 10.12, which presents a representative simulation of the stochastic parameterized model.
- `Chapter_10_Stochastic_Parameterized_Model.m`: generates Figure 10.5 in Example 10.13, which demonstrates how stochastic parameterized forecast models improve data assimilation performance compared with forecast models based on complex Ornstein-Uhlenbeck processes.
- `Chapter_10_Comparison_Dyad_Model.m`: generates Figure 10.6 in Example 10.17, which compares a standard stochastic parameterized model without physical constraints with one that incorporates physical constraints.
- `Chapter_10_Koopman_Nonlinear_Discrete_System.m`: generates Figure 10.7 in Example 10.20, which reveals hidden linearity by lifting a two-dimensional nonlinear system into three dimensions.

Bibliography

[1] Hervé Abdi and Lynne J Williams. Principal component analysis. *Wiley interdisciplinary reviews: computational statistics*, 2(4):433–459, 2010.

[2] Milton Abramowitz and Irene A Stegun. *Handbook of mathematical functions: with formulas, graphs, and mathematical tables*, volume 55. Courier Corporation, 1965.

[3] János Aczél, Bruno Forte, and Che Tat Ng. Why the Shannon and Hartley entropies are 'natural'. *Advances in applied probability*, 6(1):131–146, 1974.

[4] Robert A Adams and John JF Fournier. *Sobolev spaces*, volume 140. Elsevier, 2003.

[5] Ronald J Adrian, Kenneth T Christensen, and Z-C Liu. Analysis and interpretation of instantaneous turbulent velocity fields. *Experiments in fluids*, 29(3):275–290, 2000.

[6] Mohiuddin Ahmed, Raihan Seraj, and Syed Mohammed Shamsul Islam. The k-means algorithm: A comprehensive survey and performance evaluation. *Electronics*, 9(8):1295, 2020.

[7] Shady E Ahmed, Suraj Pawar, Omer San, Adil Rasheed, Traian Iliescu, and Bernd R Noack. On closures for reduced order models—A spectrum of first-principle to machine-learned avenues. *Physics of Fluids*, 33(9), 2021.

[8] Ken Aho, DeWayne Derryberry, and Teri Peterson. Model selection for ecologists: the worldviews of AIC and BIC. *Ecology*, 95(3):631–636, 2014.

[9] Hirotugu Akaike. A new look at the statistical model identification. *IEEE transactions on automatic control*, 19(6):716–723, 2003.

[10] Brian DO Anderson and John B Moore. *Optimal filtering*. Courier Corporation, 2005.

[11] Jeffrey L Anderson. An ensemble adjustment Kalman filter for data assimilation. *Monthly weather review*, 129(12):2884–2903, 2001.

[12] Jeffrey L Anderson. Exploring the need for localization in ensemble data assimilation using a hierarchical ensemble filter. *Physica D: Nonlinear Phenomena*, 230(1-2):99–111, 2007.

[13] Jeffrey L Anderson. Localization and sampling error correction in ensemble Kalman filter data assimilation. *Monthly Weather Review*, 140(7):2359–2371, 2012.

[14] Jeffrey L Anderson and Stephen L Anderson. A Monte Carlo implementation of the nonlinear filtering problem to produce ensemble assimilations and forecasts. *Monthly weather review*, 127(12):2741–2758, 1999.

[15] Marios Andreou and Nan Chen. A martingale-free introduction to conditional Gaussian nonlinear systems. *Entropy*, 27(1):2, 2024.

[16] Marios Andreou and Nan Chen. Bridging prediction and attribution: Identifying forward and backward causal influence ranges using assimilative causal inference. *arXiv preprint arXiv:2510.21889*, 2025.

[17] Marios Andreou, Nan Chen, and Erik Bollt. Assimilative causal inference. *Nature Communications*, 2026.

[18] Marios Andreou, Nan Chen, and Yingda Li. An adaptive online smoother with closed-form solutions and information-theoretic lag selection for conditional Gaussian nonlinear systems. *arXiv preprint arXiv:2411.05870*, 2024.

[19] Christophe Andrieu and Johannes Thoms. A tutorial on adaptive MCMC. *Statistics and computing*, 18(4):343–373, 2008.

[20] Rossella Arcucci, Jiangcheng Zhu, Shuang Hu, and Yi-Ke Guo. Deep data assimilation: integrating deep learning with data assimilation. *Applied Sciences*, 11(3):1114, 2021.

[21] Robert B Ash. *Information theory.* Courier Corporation, 2012.

[22] Peter Ashwin, Sebastian Wieczorek, Renato Vitolo, and Peter Cox. Tipping points in open systems: bifurcation, noise-induced and rate-dependent examples in the climate system. *Philosophical Transactions of the Royal Society A: Mathematical, Physical and Engineering Sciences*, 370(1962):1166–1184, 2012.

[23] TA Averina and SS Artemiev. Numerical solution of systems of stochastic differential equations. *Russian Journal of Numerical Analysis and Mathematical Modelling*, 3(4):267–286, 1988.

[24] Eviatar Bach and Michael Ghil. A multi-model ensemble Kalman filter for data assimilation and forecasting. *Journal of Advances in Modeling Earth Systems*, 15(1):e2022MS003123, 2023.

[25] Marco Baldovin, Fabio Cecconi, and Angelo Vulpiani. Understanding causation via correlations and linear response theory. *Physical Review Research*, 2(4):043436, 2020.

[26] Jørgen Bang-Jensen and Gregory Z Gutin. *Digraphs: theory, algorithms and applications.* Springer Science & Business Media, 2008.

[27] Matthew F Barone, Irina Kalashnikova, Daniel J Segalman, and Heidi K Thornquist. Stable Galerkin reduced order models for linearized compressible flow. *Journal of Computational Physics*, 228(6):1932–1946, 2009.

[28] Francis Bashforth and John Couch Adams. *An attempt to test the theories of capillary action: by comparing the theoretical and measured forms of drops of fluid.* University Press, 1883.

[29] Pouria Behnoudfar, Charlotte Moser, Marc Bocquet, Sibo Cheng, and Nan Chen. Bridging idealized and operational models: an explainable AI framework for Earth system emulators. *npj Climate and Atmospheric Science*, 2026.

[30] Kay Bergemann and Sebastian Reich. An ensemble Kalman-Bucy filter for continuous data assimilation. *Meteorologische Zeitschrift*, 21(3):213, 2012.

[31] Adam Berger, Stephen A Della Pietra, and Vincent J Della Pietra. A maximum entropy approach to natural language processing. *Computational linguistics*, 22(1):39–71, 1996.

[32] Joseph Berkson. Bayes' theorem. *The Annals of Mathematical Statistics*, 1(1):42–56, 1930.

[33] Judith Berner, Ulrich Achatz, Lauriane Batte, Lisa Bengtsson, Alvaro de la Cámara, Hannah M Christensen, Matteo Colangeli, Danielle RB Coleman, Daan Crommelin, Stamen I Dolaptchiev, et al. Stochastic parameterization: Toward a new view of weather and climate models. *Bulletin of the American Meteorological Society*, 98(3):565–588, 2017.

[34] Michael Betancourt. A conceptual introduction to Hamiltonian Monte Carlo. *arXiv preprint arXiv:1701.02434*, 2017.

[35] James C Bezdek. *Pattern recognition with fuzzy objective function algorithms*. Springer Science & Business Media, 2013.

[36] Craig H Bishop, Brian J Etherton, and Sharanya J Majumdar. Adaptive sampling with the ensemble transform Kalman filter. Part I: Theoretical aspects. *Monthly weather review*, 129(3):420–436, 2001.

[37] Stefano Boccaletti, Jürgen Kurths, Grigory Osipov, DL Valladares, and CS Zhou. The synchronization of chaotic systems. *Physics reports*, 366(1-2):1–101, 2002.

[38] Marc Bocquet, Julien Brajard, Alberto Carrassi, and Laurent Bertino. Bayesian inference of chaotic dynamics by merging data assimilation, machine learning and expectation-maximization. *arXiv preprint arXiv:2001.06270*, 2020.

[39] Marc Bocquet, Alban Farchi, and Quentin Malartic. Online learning of both state and dynamics using ensemble Kalman filters. *arXiv preprint arXiv:2006.03859*, 2020.

[40] Marc Bocquet and Lin Wu. Bayesian design of control space for optimal assimilation of observations. Part II: Asymptotic solutions. *Quarterly Journal of the Royal Meteorological Society*, 137(658):1357–1368, 2011.

[41] Diogo Bolster, Robert E Hershberger, and Russell J Donnelly. Dynamic similarity, the dimensionless science. *Physics Today*, 64(9):42–47, 2011.

[42] Massimo Bonavita and Patrick Laloyaux. Machine learning for model error inference and correction. *Journal of Advances in Modeling Earth Systems*, 12(12):e2020MS002232, 2020.

[43] Pierre Boudier, Anthony Fillion, Serge Gratton, and Selime Gürol. Dan–an optimal data assimilation framework based on machine learning recurrent networks. *arXiv preprint arXiv:2010.09694*, 2020.

[44] Pierre Boudier, Anthony Fillion, Serge Gratton, Selime Gürol, and Sixin Zhang. Data assimilation networks. *Journal of Advances in Modeling Earth Systems*, 15(4):e2022MS003353, 2023.

[45] Stephen P Boyd and Lieven Vandenberghe. *Convex optimization*. Cambridge university press, 2004.

[46] Julien Brajard, Alberto Carrassi, Marc Bocquet, and Laurent Bertino. Combining data assimilation and machine learning to emulate a dynamical model from sparse and noisy observations: A case study with the Lorenz 96 model. *Journal of Computational Science*, 44:101171, 2020.

[47] Michal Branicki, Nan Chen, and Andrew J Majda. Non-Gaussian test models for prediction and state estimation with model errors. *Chinese Annals of Mathematics, Series B*, 34(1):29–64, 2013.

[48] Michal Branicki and Andrew J Majda. Dynamic stochastic superresolution of sparsely observed turbulent systems. *Journal of Computational Physics*, 241:333–363, 2013.

[49] Michal Branicki, Andrew J Majda, and Kody JH Law. Accuracy of some approximate Gaussian filters for the Navier–Stokes equation in the presence of model error. *Multiscale Modeling & Simulation*, 16(4):1756–1794, 2018.

[50] Fred Brauer, Carlos Castillo-Chavez, and Carlos Castillo-Chavez. *Mathematical models in population biology and epidemiology*, volume 2. Springer, 2012.

[51] Jason Brownlee. A gentle introduction to the rectified linear unit (ReLU). *Machine learning mastery*, 6, 2019.

[52] Steven L Brunton, Marko Budišić, Eurika Kaiser, and J Nathan Kutz. Modern Koopman theory for dynamical systems. *arXiv preprint arXiv:2102.12086*, 2021.

[53] Caterina Buizza, César Quilodrán Casas, Philip Nadler, Julian Mack, Stefano Marrone, Zainab Titus, Clémence Le Cornec, Evelyn Heylen, Tolga Dur, Luis Baca Ruiz, et al. Data learning: Integrating data assimilation and machine learning. *Journal of Computational Science*, 58:101525, 2022.

[54] William F Campbell, Craig H Bishop, and Daniel Hodyss. Vertical covariance localization for satellite radiances in ensemble Kalman filters. *Monthly Weather Review*, 138(1):282–290, 2010.

[55] Marek Capiński and Ekkehard Kopp. *The Black–Scholes Model.* Cambridge University Press, 2012.

[56] Bernard Castaing, Gemunu Gunaratne, François Heslot, Leo Kadanoff, Albert Libchaber, Stefan Thomae, Xiao-Zhong Wu, Stéphane Zaleski, and Gianluigi Zanetti. Scaling of hard thermal turbulence in Rayleigh-Bénard convection. *Journal of Fluid Mechanics*, 204:1–30, 1989.

[57] Ashesh Chattopadhyay, Ebrahim Nabizadeh, Eviatar Bach, and Pedram Hassanzadeh. Deep learning-enhanced ensemble-based data assimilation for high-dimensional nonlinear dynamical systems. *Journal of Computational Physics*, 477:111918, 2023.

[58] Chuanqi Chen, Nan Chen, Yinling Zhang, and Jin-Long Wu. CGKN: A deep learning framework for modeling complex dynamical systems and efficient data assimilation. *Journal of Computational Physics*, 532:113950, 2025.

[59] Chuanqi Chen, Zhongrui Wang, Nan Chen, and Jin-Long Wu. Modeling partially observed nonlinear dynamical systems and efficient data assimilation via discrete-time conditional Gaussian Koopman network. *Computer Methods in Applied Mechanics and Engineering*, 445:118189, 2025.

[60] Nan Chen. Learning nonlinear turbulent dynamics from partial observations via analytically solvable conditional statistics. *Journal of Computational Physics*, 418:109635, 2020.

[61] Nan Chen. *Stochastic Methods for Modeling and Predicting Complex Dynamical Systems.* Springer, 2023.

[62] Nan Chen, Yingda Li, and Honghu Liu. Conditional Gaussian nonlinear system: A fast preconditioner and a cheap surrogate model for complex nonlinear systems. *Chaos: An Interdisciplinary Journal of Nonlinear Science*, 32(5), 2022.

[63] Nan Chen and Andrew J Majda. Filtering nonlinear turbulent dynamical systems through conditional Gaussian statistics. *Monthly Weather Review*, 144(12):4885–4917, 2016.

[64] Nan Chen and Andrew J Majda. Model error in filtering random compressible flows utilizing noisy Lagrangian tracers. *Monthly Weather Review*, 144(11):4037–4061, 2016.

[65] Nan Chen and Andrew J Majda. Beating the curse of dimension with accurate statistics for the Fokker–Planck equation in complex turbulent systems. *Proceedings of the National Academy of Sciences*, 114(49):12864–12869, 2017.

[66] Nan Chen and Andrew J Majda. Conditional Gaussian systems for multiscale nonlinear stochastic systems: Prediction, state estimation and uncertainty quantification. *Entropy*, 20(7):509, 2018.

[67] Nan Chen and Andrew J Majda. Predicting observed and hidden extreme events in complex nonlinear dynamical systems with partial observations and short training time series. *Chaos: An Interdisciplinary Journal of Nonlinear Science*, 30(3), 2020.

[68] Nan Chen, Andrew J Majda, and Xin T Tong. Information barriers for noisy Lagrangian tracers in filtering random incompressible flows. *Nonlinearity*, 27(9):2133, 2014.

[69] Nan Chen, Andrew J Majda, and Xin T Tong. Noisy Lagrangian tracers for filtering random rotating compressible flows. *Journal of Nonlinear Science*, 25(3):451–488, 2015.

[70] Nan Chen, Changhong Mou, Leslie M Smith, and Yeyu Zhang. A stochastic precipitating quasi-geostrophic model. *Physics of Fluids*, 36(11), 2024.

[71] Nan Chen, Stephen Wiggins, and Marios Andreou. Taming uncertainty in a complex world: The rise of uncertainty quantification—a tutorial for beginners. *Notices of the American Mathematical Society*, 72(3), 2025.

[72] Nan Chen and Yinling Zhang. A causality-based learning approach for discovering the underlying dynamics of complex systems from partial observations with stochastic parameterization. *Physica D: Nonlinear Phenomena*, 449:133743, 2023.

[73] Yen-Chi Chen. A tutorial on kernel density estimation and recent advances. *Biostatistics & Epidemiology*, 1(1):161–187, 2017.

[74] Sibo Cheng and Mingming Qiu. Observation error covariance specification in dynamical systems for data assimilation using recurrent neural networks. *Neural Computing and Applications*, 34(16):13149–13167, 2022.

[75] Sibo Cheng, César Quilodrán-Casas, Said Ouala, Alban Farchi, Che Liu, Pierre Tandeo, Ronan Fablet, Didier Lucor, Bertrand Iooss, Julien Brajard, et al. Machine learning with data assimilation and uncertainty quantification for dynamical systems: a review. *IEEE/CAA Journal of Automatica Sinica*, 10(6):1361–1387, 2023.

[76] Siddhartha Chib and Edward Greenberg. Understanding the metropolis-hastings algorithm. *The american statistician*, 49(4):327–335, 1995.

[77] S-N Chow and Jack K Hale. *Methods of bifurcation theory*, volume 251. Springer Science & Business Media, 2012.

[78] Hannah M Christensen, Judith Berner, Danielle RB Coleman, and TN Palmer. Stochastic parameterization and El Niño–southern oscillation. *Journal of Climate*, 30(1):17–38, 2017.

[79] Thomas M Cover, Joy A Thomas, et al. Entropy, relative entropy and mutual information. *Elements of information theory*, 2(1):12–13, 1991.

[80] Jeffrey Covington, Nan Chen, and Monica M Wilhelmus. Bridging gaps in the climate observation network: A physics-based nonlinear dynamical interpolation of Lagrangian ice floe measurements via data-driven stochastic models. *Journal of Advances in Modeling Earth Systems*, 14(9):e2022MS003218, 2022.

[81] John Crank and Phyllis Nicolson. A practical method for numerical evaluation of solutions of partial differential equations of the heat-conduction type. In *Mathematical proceedings of the Cambridge philosophical society*, volume 43, pages 50–67. Cambridge University Press, 1947.

[82] John David Crawford. Introduction to bifurcation theory. *Reviews of modern physics*, 63(4):991, 1991.

[83] Daan T Crommelin, JD Opsteegh, and Ferdinand Verhulst. A mechanism for atmospheric regime behavior. *Journal of the atmospheric sciences*, 61(12):1406–1419, 2004.

[84] DT Crommelin and Andrew J Majda. Strategies for model reduction: Comparing different optimal bases. *Journal of the Atmospheric Sciences*, 61(17):2206–2217, 2004.

[85] Imre Csiszár. I-divergence geometry of probability distributions and minimization problems. *The annals of probability*, pages 146–158, 1975.

[86] Mengyao Cui et al. Introduction to the k-means clustering algorithm based on the elbow method. *Accounting, Auditing and Finance*, 1(1):5–8, 2020.

[87] Haskell B Curry. The method of steepest descent for non-linear minimization problems. *Quarterly of Applied Mathematics*, 2(3):258–261, 1944.

[88] Pierre Del Moral. Nonlinear filtering: Interacting particle resolution. *Comptes Rendus de l'Académie des Sciences-Series I-Mathematics*, 325(6):653–658, 1997.

[89] Robert P Dobrow. *Introduction to stochastic processes with R*. John Wiley & Sons, 2016.

[90] David L Donoho et al. High-dimensional data analysis: The curses and blessings of dimensionality. *AMS math challenges lecture*, 1(2000):32, 2000.

[91] Philip G Drazin. *Nonlinear systems*, volume 10. Cambridge University Press, 1992.

[92] Georg Duffing. *Erzwungene Schwingungen bei veränderlicher Eigenfrequenz und ihre technische Bedeutung*, volume 41-42. Vieweg, 1918.

[93] Joseph C Dunn. A fuzzy relative of the ISODATA process and its use in detecting compact well-separated clusters. *Journal of Cybernetics*, 3:32–57, 1973.

[94] David Eberly. Stability analysis for systems of differential equations. *Geometric Tools, LLC*, pages 1–15, 2008.

[95] Bradley Efron. Bayes' theorem in the 21st century. *Science*, 340(6137):1177–1178, 2013.

[96] Lawrence C Evans. *Partial differential equations*, volume 19. American mathematical society, 2022.

[97] Geir Evensen. *Data assimilation: the ensemble Kalman filter.* Springer, 2009.

[98] Geir Evensen and Peter Jan Van Leeuwen. An ensemble Kalman smoother for nonlinear dynamics. *Monthly Weather Review*, 128(6):1852–1867, 2000.

[99] Geir Evensen, Femke C Vossepoel, and Peter Jan Van Leeuwen. *Data assimilation fundamentals: A unified formulation of the state and parameter estimation problem.* Springer Nature, 2022.

[100] Alban Farchi, Marc Bocquet, Patrick Laloyaux, Massimo Bonavita, and Quentin Malartic. A comparison of combined data assimilation and machine learning methods for offline and online model error correction. *Journal of Computational Science*, 55:101468, 2021.

[101] Alban Farchi, Marcin Chrust, Marc Bocquet, Patrick Laloyaux, and Massimo Bonavita. Online model error correction with neural networks in the incremental 4D-Var framework. *Journal of Advances in Modeling Earth Systems*, 15(9):e2022MS003474, 2023.

[102] Alban Farchi, Patrick Laloyaux, Massimo Bonavita, and Marc Bocquet. Using machine learning to correct model error in data assimilation and forecast applications. *Quarterly Journal of the Royal Meteorological Society*, 147(739):3067–3084, 2021.

[103] Elana J Fertig, Brian R Hunt, Edward Ott, and Istvan Szunyogh. Assimilating nonlocal observations with a local ensemble Kalman filter. *Tellus A: Dynamic Meteorology and Oceanography*, 59(5):719–730, 2007.

[104] Evelyn Fix. *Discriminatory analysis: nonparametric discrimination, consistency properties*, volume 1. USAF school of Aviation Medicine, 1985.

[105] Adriaan Daniël Fokker. Die mittlere energie rotierender elektrischer dipole im strahlungsfeld. *Annalen der Physik*, 348(5):810–820, 1914.

[106] Edward W Forgy. Cluster analysis of multivariate data: efficiency versus interpretability of classifications. *Biometrics*, 21:768–769, 1965.

[107] David A Freedman. *Statistical models: theory and practice.* Cambridge university press, 2009.

[108] David John Gagne, Hannah M Christensen, Aneesh C Subramanian, and Adam H Monahan. Machine learning for stochastic parameterization: Generative adversarial networks in the lorenz'96 model. *Journal of Advances in Modeling Earth Systems*, 12(3):e2019MS001896, 2020.

[109] Xiang Gao, Emilio Gallicchio, and Adrian E Roitberg. The generalized Boltzmann distribution is the only distribution in which the Gibbs-Shannon entropy equals the thermodynamic entropy. *The Journal of Chemical Physics*, 151(3), 2019.

[110] Crispin Gardiner. *Stochastic methods*, volume 4. Springer Berlin Heidelberg, 2009.

[111] Crispin W Gardiner et al. *Handbook of stochastic methods*, volume 3. Springer Berlin, 2004.

[112] Carl Friedrich Gauss. Theoria motus corporum coelestum. *Werke*, 1809.

[113] Stuart Geman and Donald Geman. Stochastic relaxation, Gibbs distributions, and the Bayesian restoration of images. *IEEE Transactions on pattern analysis and machine intelligence*, 6:721–741, 1984.

[114] Boris Gershgorin, John Harlim, and Andrew J Majda. Test models for improving filtering with model errors through stochastic parameter estimation. *Journal of Computational Physics*, 229(1):1–31, 2010.

[115] Robin C Gilbert, Michael B Richman, Theodore B Trafalis, and Lance M Leslie. Machine learning methods for data assimilation. *Computational Intelligence in Architecturing Complex Engineering Systems*, pages 105–112, 2010.

[116] Georg A Gottwald and Sebastian Reich. Combining machine learning and data assimilation to forecast dynamical systems from noisy partial observations. *Chaos: An Interdisciplinary Journal of Nonlinear Science*, 31(10):101103, 2021.

[117] Georg A Gottwald and Sebastian Reich. Supervised learning from noisy observations: Combining machine-learning techniques with data assimilation. *Physica D: Nonlinear Phenomena*, 423:132911, 2021.

[118] Clive WJ Granger. Investigating causal relations by econometric models and cross-spectral methods. *Econometrica: journal of the Econometric Society*, pages 424–438, 1969.

[119] Robert M Gray. *Entropy and information theory*. Springer Science & Business Media, 2011.

[120] Ian Grooms. Analog ensemble data assimilation and a method for constructing analogs with variational autoencoders. *Quarterly Journal of the Royal Meteorological Society*, 147(734):139–149, 2021.

[121] John Guckenheimer and Philip Holmes. *Nonlinear oscillations, dynamical systems, and bifurcations of vector fields*, volume 42. Springer Science & Business Media, 2013.

[122] Debbie L Hahs-Vaughn and Richard G Lomax. *Statistical concepts-a second course*. Routledge, 2020.

[123] Jun Han and Claudio Moraga. The influence of the sigmoid function parameters on the speed of backpropagation learning. In *International workshop on artificial neural networks*, pages 195–201. Springer, 1995.

[124] John Harlim, Adam Mahdi, and Andrew J Majda. An ensemble kalman filter for statistical estimation of physics constrained nonlinear regression models. *Journal of Computational Physics*, 257:782–812, 2014.

[125] John Harlim and Andrew J Majda. Test models for filtering with superparameterization. *Multiscale Modeling & Simulation*, 11(1):282–308, 2013.

[126] Philip Hartman. *Ordinary differential equations*. SIAM, 2002.

[127] Trevor Hastie, Robert Tibshirani, Jerome Friedman, et al. The elements of statistical learning, 2009.

[128] W Keith Hastings. Monte Carlo sampling methods using Markov chains and their applications. *Biometrika*, 57(1), 1970.

[129] Douglas M Hawkins. The problem of overfitting. *Journal of chemical information and computer sciences*, 44(1):1–12, 2004.

[130] Leslie Hayduk, Greta Cummings, Rainer Stratkotter, Melanie Nimmo, Kostyantyn Grygoryev, Donna Dosman, Michael Gillespie, Hannah Pazderka-Robinson, and Kwame Boadu. Pearl's D-separation: One more step into causal thinking. *Structural Equation Modeling*, 10(2):289–311, 2003.

[131] H Hempel, Lutz Schimansky-Geier, and Jordi García-Ojalvo. Noise-sustained pulsating patterns and global oscillations in subexcitable media. *Physical Review Letters*, 82(18):3713, 1999.

[132] Miguel A Hernán and James M Robins. Causal inference, 2010.

[133] Nicholas J Higham. *Accuracy and stability of numerical algorithms*. SIAM, 2002.

[134] Saddam Hijazi, Giovanni Stabile, Andrea Mola, and Gianluigi Rozza. Data-driven POD-galerkin reduced order model for turbulent flows. *Journal of Computational Physics*, 416:109513, 2020.

[135] Morris W Hirsch, Stephen Smale, and Robert L Devaney. *Differential equations, dynamical systems, and an introduction to chaos*. Academic press, 2013.

[136] Katerina Hlaváčková-Schindler, Milan Paluš, Martin Vejmelka, and Joydeep Bhattacharya. Causality detection based on information-theoretic approaches in time series analysis. *Physics Reports*, 441(1):1–46, 2007.

[137] Mark H Holmes. *Introduction to perturbation methods*, volume 20. Springer Science & Business Media, 2012.

[138] Frank Hoppensteadt. Predator-prey model. *Scholarpedia*, 1(10):1563, 2006.

[139] Peter L Houtekamer and Fuqing Zhang. Review of the ensemble Kalman filter for atmospheric data assimilation. *Monthly Weather Review*, 144(12):4489–4532, 2016.

[140] Bambi Hu and Changsong Zhou. Phase synchronization in coupled nonidentical excitable systems and array-enhanced coherence resonance. *Physical Review E*, 61(2):R1001, 2000.

[141] Brian R Hunt, Eric J Kostelich, and Istvan Szunyogh. Efficient data assimilation for spatiotemporal chaos: A local ensemble transform Kalman filter. *Physica D: Nonlinear Phenomena*, 230(1-2):112–126, 2007.

[142] Edwin T Jaynes. Information theory and statistical mechanics. *Physical review*, 106(4):620, 1957.

[143] Simon J Julier and Jeffrey K Uhlmann. Unscented filtering and nonlinear estimation. *Proceedings of the IEEE*, 92(3):401–422, 2004.

[144] Peter Jung, Ann Cornell-Bell, Kathleen Shaver Madden, and Frank Moss. Noise-induced spiral waves in astrocyte syncytia show evidence of self-organized criticality. *Journal of neurophysiology*, 79(2):1098–1101, 1998.

[145] Rudolph E Kalman and Richard S Bucy. New results in linear filtering and prediction theory. *Journal of Fluids Engineering*, 83:95–108, 1961.

[146] Eugenia Kalnay. *Atmospheric modeling, data assimilation and predictability.* Cambridge University Press, 2003.

[147] Bhargav Kanagal and Amol Deshpande. Online filtering, smoothing and probabilistic modeling of streaming data. In *2008 IEEE 24th International Conference on Data Engineering*, pages 1160–1169. IEEE, 2008.

[148] Wei Kang and Liang Xu. Optimal placement of mobile sensors for data assimilations. *Tellus A: Dynamic Meteorology and Oceanography*, 64(1):17133, 2012.

[149] Pileun Kim, Jonathan Rogers, Jie Sun, and Erik Bollt. Causation entropy identifies sparsity structure for parameter estimation of dynamic systems. *Journal of Computational and Nonlinear Dynamics*, 12(1):011008, 2017.

[150] Scott Kirkpatrick, C Daniel Gelatt Jr, and Mario P Vecchi. Optimization by simulated annealing. *science*, 220(4598):671–680, 1983.

[151] Peter E Kloeden. Stochastic differential equations. In *International Encyclopedia of Statistical Science*, pages 1520–1521. Springer, 2011.

[152] Peter Eris Kloeden, Eckhard Platen, and Henri Schurz. *Numerical solution of SDE through computer experiments.* Springer Science & Business Media, 2012.

[153] Trupti M Kodinariya, Prashant R Makwana, et al. Review on determining number of cluster in k-means clustering. *International Journal*, 1(6):90–95, 2013.

[154] Armin Köhl and Detlef Stammer. Optimal observations for variational data assimilation. *Journal of Physical Oceanography*, 34(3):529–542, 2004.

[155] Bernard O Koopman. Hamiltonian systems and transformation in Hilbert space. *Proceedings of the National Academy of Sciences*, 17(5):315–318, 1931.

[156] Boris Kramer, Benjamin Peherstorfer, and Karen E Willcox. Learning nonlinear reduced models from data with operator inference. *Annual Review of Fluid Mechanics*, 56(1):521–548, 2024.

[157] Boris Kramer and Karen E Willcox. Nonlinear model order reduction via lifting transformations and proper orthogonal decomposition. *AIAA journal*, 57(6):2297–2307, 2019.

[158] Paruchuri R Krishnaiah. *A Hand Book of Statistics*, volume 1. Motilal Banarsidass Publishe, 1980.

[159] Dirk P Kroese, Tim Brereton, Thomas Taimre, and Zdravko I Botev. Why the Monte Carlo method is so important today. *Wiley Interdisciplinary Reviews: Computational Statistics*, 6(6):386–392, 2014.

[160] Solomon Kullback. *Information theory and statistics.* Courier Corporation, 1997.

[161] Solomon Kullback and Richard A Leibler. On information and sufficiency. *The annals of mathematical statistics*, 22(1):79–86, 1951.

[162] Wilhelm Kutta. *Beitrag zur näherungsweisen Integration totaler Differentialgleichungen.* Teubner, 1901.

[163] Nikolay Kuznetsov and Volker Reitmann. *Attractor dimension estimates for dynamical systems: theory and computation.* Springer, 2020.

[164] Yuri A Kuznetsov. Saddle-node bifurcation. *Scholarpedia*, 1(10):1859, 2006.

[165] Kody Law, Andrew Stuart, and Kostas Zygalakis. Data assimilation. *Cham, Switzerland: Springer*, 214:52, 2015.

[166] Timothy M Lenton. Tipping points in the climate system. *Weather, 76 (10)*, 2021.

[167] Timothy M Lenton, Johan Rockström, Owen Gaffney, Stefan Rahmstorf, Katherine Richardson, Will Steffen, and Hans Joachim Schellnhuber. Climate tipping points—too risky to bet against. *Nature*, 575(7784):592–595, 2019.

[168] X San Liang and Richard Kleeman. Information transfer between dynamical system components. *Physical Review Letters*, 95(24):244101, 2005.

[169] Benjamin Lindner, Jordi Garcıa-Ojalvo, Alexander Neiman, and Lutz Schimansky-Geier. Effects of noise in excitable systems. *Physics Reports*, 392(6):321–424, 2004.

[170] Benjamin Lindner and Lutz Schimansky-Geier. Coherence and stochastic resonance in a two-state system. *Physical Review E*, 61(6):6103, 2000.

[171] Robert S Liptser and Albert N Shiryaev. *Statistics of random processes II: Applications*, volume 6. Springer Science & Business Media, 2013.

[172] Jun S Liu and Rong Chen. Sequential Monte Carlo methods for dynamic systems. *Journal of the American Statistical Association*, 93(443):1032–1044, 1998.

[173] Katherine Liu, Kyel Ok, William Vega-Brown, and Nicholas Roy. Deep inference for covariance estimation: Learning Gaussian noise models for state estimation. In *2018 IEEE International Conference on Robotics and Automation (ICRA)*, pages 1436–1443. IEEE, 2018.

[174] André Longtin. Stochastic resonance in neuron models. *Journal of statistical physics*, 70(1):309–327, 1993.

[175] Edward N Lorenz. Deterministic nonperiodic flow 1. In *Universality in Chaos, 2nd edition*, pages 367–378. Routledge, 2017.

[176] Edward N Lorenz and K Haman. The essence of chaos. *Pure and Applied Geophysics*, 147(3):598–599, 1996.

[177] Valerio Lucarini. Revising and extending the linear response theory for statistical mechanical systems: Evaluating observables as predictors and predictands. *Journal of Statistical Physics*, 173(6):1698–1721, 2018.

[178] David JC MacKay et al. Introduction to Gaussian processes. *NATO ASI series F computer and systems sciences*, 168:133–166, 1998.

[179] Andrew Majda. *Introduction to PDEs and Waves for the Atmosphere and Ocean*, volume 9. American Mathematical Soc., 2003.

[180] Andrew Majda and Xiaoming Wang. *Nonlinear dynamics and statistical theories for basic geophysical flows.* Cambridge University Press, 2006.

[181] Andrew J Majda and Nan Chen. Model error, information barriers, state estimation and prediction in complex multiscale systems. *Entropy*, 20(9):644, 2018.

[182] Andrew J Majda, Christian Franzke, and Daan Crommelin. Normal forms for reduced stochastic climate models. *Proceedings of the National Academy of Sciences*, 106(10):3649–3653, 2009.

[183] Andrew J Majda and Marcus J Grote. Model dynamics and vertical collapse in decaying strongly stratified flows. *Physics of fluids*, 9(10):2932–2940, 1997.

[184] Andrew J Majda and John Harlim. Information flow between subspaces of complex dynamical systems. *Proceedings of the National Academy of Sciences*, 104(23):9558–9563, 2007.

[185] Andrew J Majda and John Harlim. *Filtering complex turbulent systems.* Cambridge University Press, 2012.

[186] Andrew J Majda and John Harlim. Physics constrained nonlinear regression models for time series. *Nonlinearity*, 26(1):201, 2012.

[187] Andrew J Majda, Ilya Timofeyev, and Eric Vanden Eijnden. A mathematical framework for stochastic climate models. *Communications on Pure and Applied Mathematics: A Journal Issued by the Courant Institute of Mathematical Sciences*, 54(8):891–974, 2001.

[188] Andrew J Majda and Yuan Yuan. Fundamental limitations of ad hoc linear and quadratic multi-level regression models for physical systems. *Discrete & Continuous Dynamical Systems-Series B*, 17(4), 2012.

[189] Quentin Malartic, Alban Farchi, and Marc Bocquet. State, global, and local parameter estimation using local ensemble Kalman filters: Applications to online machine learning of chaotic dynamics. *Quarterly Journal of the Royal Meteorological Society*, 148(746):2167–2193, 2022.

[190] Georgy E Manucharyan, Lia Siegelman, and Patrice Klein. A deep learning approach to spatiotemporal sea surface height interpolation and estimation of deep currents in geostrophic ocean turbulence. *Journal of Advances in Modeling Earth Systems*, 13(1):e2019MS001965, 2021.

[191] Romit Maulik, Vishwas Rao, Jiali Wang, Gianmarco Mengaldo, Emil Constantinescu, Bethany Lusch, Prasanna Balaprakash, Ian Foster, and Rao Kotamarthi. AIEADA 1.0: Efficient high-dimensional variational data assimilation with machine-learned reduced-order models. *Geoscientific Model Development Discussions*, 2022:1–20, 2022.

[192] Dulani Meedeniya. *Deep learning: A beginners' guide.* Chapman and Hall/CRC, 2023.

[193] Nicholas Metropolis, Arianna W Rosenbluth, Marshall N Rosenbluth, Augusta H Teller, and Edward Teller. Equation of state calculations by fast computing machines. *The Journal of Chemical Physics*, 21(6):1087–1092, 1953.

[194] Nicholas Metropolis and Stanislaw Ulam. The Monte Carlo method. *Journal of the American Statistical Association*, 44(247):335–341, 1949.

[195] Sean P Meyn and Richard L Tweedie. *Markov chains and stochastic stability.* Springer Science & Business Media, 2012.

[196] Grigori Noyhovich Mil'shtein. A method of second-order accuracy integration of stochastic differential equations. *Theory of Probability & Its Applications*, 23(2):396–401, 1979.

[197] Todd K Moon. The expectation-maximization algorithm. *IEEE Signal processing magazine*, 13(6):47–60, 1996.

[198] Charlotte Moser, Nan Chen, and Stephen Wiggins. A mathematical framework for quantifying nonlinear uncertainty propagation in eddy identification criteria. *Physica D: Nonlinear Phenomena*, page 135220, 2026.

[199] Changhong Mou, Nan Chen, and Traian Iliescu. An efficient data-driven multiscale stochastic reduced order modeling framework for complex systems. *Journal of Computational Physics*, 493:112450, 2023.

[200] Changhong Mou, Leslie M Smith, and Nan Chen. Combining stochastic parameterized reduced-order models with machine learning for data assimilation and uncertainty quantification with partial observations. *Journal of Advances in Modeling Earth Systems*, 15(10):e2022MS003597, 2023.

[201] Peter Müller. *The equations of oceanic motions*. Cambridge University Press, 2006.

[202] JD Murray. Laplace's method for integrals. In *Asymptotic Analysis*, pages 19–39. Springer, 1984.

[203] Alireza Namdari and Zhaojun Li. A review of entropy measures for uncertainty quantification of stochastic processes. *Advances in Mechanical Engineering*, 11(6):1687814019857350, 2019.

[204] Alexander Neiman, Lutz Schimansky-Geier, Ann Cornell-Bell, and Frank Moss. Noise-enhanced phase synchronization in excitable media. *Physical Review Letters*, 83(23):4896, 1999.

[205] Michael A Nielsen. *Neural networks and deep learning*, volume 25. Determination Press San Francisco, CA, USA, 2015.

[206] Akira Okubo. Horizontal dispersion of floatable particles in the vicinity of velocity singularities such as convergences. In *Deep sea research and oceanographic abstracts*, volume 17, pages 445–454. Elsevier, 1970.

[207] Samuel E Otto and Clarence W Rowley. Linearly recurrent autoencoder networks for learning dynamics. *SIAM Journal on Applied Dynamical Systems*, 18(1):558–593, 2019.

[208] Said Ouala, Ronan Fablet, Cédric Herzet, Bertrand Chapron, Ananda Pascual, Fabrice Collard, and Lucile Gaultier. Neural network based Kalman filters for the spatio-temporal interpolation of satellite-derived sea surface temperature. *Remote Sensing*, 10(12):1864, 2018.

[209] Athanasios Papoulis. *Random variables and stochastic processes*. McGraw Hill, 1965.

[210] Athanasios Papoulis. *Probability and statistics*. Prentice-Hall, Inc., 1990.

[211] Kun Il Park, M Park, and James. *Fundamentals of probability and stochastic processes with applications to communications*. Springer, 2018.

[212] Suraj Pawar, Shady E Ahmed, Omer San, Adil Rasheed, and Ionel M Navon. Long short-term memory embedded nudging schemes for nonlinear data assimilation of geophysical flows. *Physics of Fluids*, 32(7):076606, 2020.

[213] Judea Pearl. Causal inference. *Causality: objectives and assessment*, pages 39–58, 2010.

[214] Stephen G Penny, Timothy A Smith, T-C Chen, Jason A Platt, H-Y Lin, Michael Goodliff, and Henry DI Abarbanel. Integrating recurrent neural networks with data assimilation for scalable data-driven state estimation. *Journal of Advances in Modeling Earth Systems*, 14(3):e2021MS002843, 2022.

[215] Mark A Pinsky. *Introduction to Fourier analysis and wavelets*, volume 102. American Mathematical Society, 2023.

[216] VM Planck. Über einen satz der statistischen dynamik und seine erweiterung in der quantentheorie. *Sitzungberichte der*, 1917.

[217] Henri Poincaré. Sur l'équilibre d'une masse fluide animée d'un mouvement de rotation. *Bulletin astronomique, Observatoire de Paris*, 2(1):109–118, 1885.

[218] Henri Poincaré. *Les méthodes nouvelles de la mécanique céleste*, volume 2. Gauthier-Villars et fils, imprimeurs-libraires, 1893.

[219] Sk M Rahman, Shady E Ahmed, and Omer San. A dynamic closure modeling framework for model order reduction of geophysical flows. *Physics of Fluids*, 31(4), 2019.

[220] María-Luisa Rapún and José M Vega. Reduced order models based on local POD plus Galerkin projection. *Journal of Computational Physics*, 229(8):3046–3063, 2010.

[221] Tariq Rashid. *Make your own neural network*. CreateSpace Independent Publishing Platform, 2016.

[222] Stephan Rasp, Michael S Pritchard, and Pierre Gentine. Deep learning to represent subgrid processes in climate models. *Proceedings of the National Academy of Sciences*, 115(39):9684–9689, 2018.

[223] Herbert E Rauch, F Tung, and Charlotte T Striebel. Maximum likelihood estimates of linear dynamic systems. *AIAA journal*, 3(8):1445–1450, 1965.

[224] Guy Revach, Nir Shlezinger, Xiaoyong Ni, Adria Lopez Escoriza, Ruud JG Van Sloun, and Yonina C Eldar. Kalmannet: Neural network aided Kalman filtering for partially known dynamics. *IEEE Transactions on Signal Processing*, 70:1532–1547, 2022.

[225] Douglas Reynolds. Gaussian mixture models. In *Encyclopedia of biometrics*, pages 827–832. Springer, 2015.

[226] Osborne Reynolds. Iv. on the dynamical theory of incompressible viscous fluids and the determination of the criterion. *Philosophical transactions of the royal society of london.(a.)*, 186:123–164, 1895.

[227] Gareth O Roberts and Jeffrey S Rosenthal. Examples of adaptive MCMC. *Journal of computational and graphical statistics*, 18(2):349–367, 2009.

[228] Sheldon M Ross. Variations on Brownian motion. *Introduction to Probability Models*, pages 612–614, 2014.

[229] Gianluigi Rozza, Giovanni Stabile, and Francesco Ballarin. *Advanced reduced order methods and applications in computational fluid dynamics*. SIAM, 2022.

[230] Reuven Y Rubinstein and Dirk P Kroese. *Simulation and the Monte Carlo method.* John Wiley & Sons, 2016.

[231] Yvonne Ruckstuhl, Tijana Janjić, and Stephan Rasp. Training a convolutional neural network to conserve mass in data assimilation. *Nonlinear Processes in Geophysics*, 28(1):111–119, 2021.

[232] Carl Runge. Über die numerische auflösung von differentialgleichungen. *Mathematische Annalen*, 46(2):167–178, 1895.

[233] Rick Salmon. *Lectures on geophysical fluid dynamics.* Oxford University Press, 1998.

[234] Omer San and Traian Iliescu. A stabilized proper orthogonal decomposition reduced-order model for large scale quasigeostrophic ocean circulation. *Advances in Computational Mathematics*, 41(5):1289–1319, 2015.

[235] Omer San and Romit Maulik. Neural network closures for nonlinear model order reduction. *Advances in Computational Mathematics*, 44(6):1717–1750, 2018.

[236] Daniel Sanz-Alonso, Andrew Stuart, and Armeen Taeb. *Inverse problems and data assimilation*, volume 107. Cambridge University Press, 2023.

[237] Simo Särkkä and Lennart Svensson. *Bayesian filtering and smoothing*, volume 17. Cambridge University Press, 2023.

[238] Peter J Schmid. Dynamic mode decomposition and its variants. *Annual Review of Fluid Mechanics*, 54(1):225–254, 2022.

[239] Thomas Schreiber. Measuring information transfer. *Physical review letters*, 85(2):461, 2000.

[240] Eric Schulz, Maarten Speekenbrink, and Andreas Krause. A tutorial on Gaussian process regression: Modelling, exploring, and exploiting functions. *Journal of Mathematical Psychology*, 85:1–16, 2018.

[241] Larry Schumaker. *Spline functions: basic theory.* Cambridge University Press, 2007.

[242] Teddy Seidenfeld. Entropy and uncertainty. *Philosophy of Science*, 53(4):467–491, 1986.

[243] Lawrence F Shampine and Skip Thompson. Stiff systems. *Scholarpedia*, 2(3):2855, 2007.

[244] Claude E Shannon. A mathematical theory of communication. *The Bell system technical journal*, 27(3):379–423, 1948.

[245] Steven S Skiena. *The algorithm design manual*, volume 2. Springer, 2008.

[246] Leslie M Smith and Fabian Waleffe. Generation of slow large scales in forced rotating stratified turbulence. *Journal of Fluid Mechanics*, 451:145–168, 2002.

[247] Michael Steinbach, Levent Ertöz, and Vipin Kumar. The challenges of clustering high dimensional data. In *New directions in statistical physics: econophysics, bioinformatics, and pattern recognition*, pages 273–309. Springer, 2004.

[248] Walter A Strauss. *Partial differential equations: An introduction.* John Wiley & Sons, 2007.

[249] Steven H Strogatz. *Nonlinear dynamics and chaos: with applications to physics, biology, chemistry, and engineering (studies in nonlinearity)*, volume 1. Westview press, 2001.

[250] Steven H Strogatz. *Nonlinear dynamics and chaos: with applications to physics, biology, chemistry, and engineering.* Chapman and Hall/CRC, 2024.

[251] Student. The probable error of a mean. *Biometrika*, pages 1–25, 1908.

[252] Endre Süli and David F Mayers. *An introduction to numerical analysis.* Cambridge university press, 2003.

[253] Jie Sun, Dane Taylor, and Erik M Bollt. Causal network inference by optimal causation entropy. *SIAM Journal on Applied Dynamical Systems*, 14(1):73–106, 2015.

[254] Naoya Takeishi, Yoshinobu Kawahara, and Takehisa Yairi. Learning Koopman invariant subspaces for dynamic mode decomposition. *Advances in Neural Information Processing Systems*, 30, 2017.

[255] Nico M Temme. *Asymptotic methods for integrals*, volume 6. World Scientific, 2014.

[256] Stephen J Thomas, JP Hacker, and JL Anderson. A robust formulation of the ensemble Kalman filter. *Quarterly Journal of the Royal Meteorological Society: A journal of the atmospheric sciences, applied meteorology and physical oceanography*, 135(639):507–521, 2009.

[257] J Michael T Thompson and Jan Sieber. Climate tipping as a noisy bifurcation: a predictive technique. *IMA Journal of Applied Mathematics*, 76(1):27–46, 2011.

[258] Luke Tierney. Markov chains for exploring posterior distributions. *the Annals of Statistics*, pages 1701–1728, 1994.

[259] Michalis Titsias and Petros Dellaportas. Gradient-based adaptive markov chain monte carlo. *Advances in neural information processing systems*, 32, 2019.

[260] Juliana Tolles and William J Meurer. Logistic regression: relating patient characteristics to outcomes. *Jama*, 316(5):533–534, 2016.

[261] Georgi P Tolstov. *Fourier series.* Courier Corporation, 2012.

[262] Herbert Treutlein and Klaus Schulten. Noise induced limit cycles of the Bonhoeffer-Van der Pol model of neural pulses. *Berichte der Bunsengesellschaft für physikalische Chemie*, 89(6):710–718, 1985.

[263] Tadashi TSUYUKI and Ryosuke TAMURA. Nonlinear data assimilation by deep learning embedded in an ensemble Kalman filter. *Journal of the Meteorological Society of Japan. Ser. II*, 2022.

[264] Jonathan H Tu. *Dynamic mode decomposition: Theory and applications.* PhD thesis, Princeton University, 2013.

[265] Geoffrey K Vallis. *Atmospheric and oceanic fluid dynamics.* Cambridge University Press, 2017.

[266] William Vega-Brown, Abraham Bachrach, Adam Bry, Jonathan Kelly, and Nicholas Roy. Cello: A fast algorithm for covariance estimation. In *2013 IEEE International Conference on Robotics and Automation*, pages 3160–3167. IEEE, 2013.

[267] Ferdinand Verhulst. Differential equations and dynamical systems. In *Henri Poincaré: Impatient Genius*, pages 109–177. Springer, 2012.

[268] Jean-Philippe Vert, Koji Tsuda, and Bernhard Schölkopf. A primer on kernel methods. In *Kernel methods in computational biology*, pages 35–70. MIT Press, 2004.

[269] Matti Vihola. Robust adaptive Metropolis algorithm with coerced acceptance rate. *Statistics and Computing*, 22(5):997–1008, 2012.

[270] Cornelis Boudewijn Vreugdenhil. *Numerical methods for shallow-water flow*, volume 13. Springer Science & Business Media, 2013.

[271] Angelo Vulpiani. *Chaos: from simple models to complex systems*, volume 17. World Scientific, 2010.

[272] Wolfgang Walter. *Ordinary differential equations*, volume 182. Springer Science & Business Media, 2013.

[273] Xiaoming Wang. Infinite Prandtl number limit of Rayleigh-Bénard convection. *Communications on Pure and Applied Mathematics: A Journal Issued by the Courant Institute of Mathematical Sciences*, 57(10):1265–1282, 2004.

[274] Yasi Wang, Hongxun Yao, and Sicheng Zhao. Auto-encoder based dimensionality reduction. *Neurocomputing*, 184:232–242, 2016.

[275] Wolfgang Richard Wasow. *on boundary layer problems in the theory of ordinary differential equation.* New York University, 1942.

[276] Daniel M Watkins, Angela C Bliss, Jennifer K Hutchings, and Monica M Wilhelmus. Evidence of abrupt transitions between sea ice dynamical regimes in the East Greenland marginal ice zone. *Geophysical Research Letters*, 50(15):e2023GL103558, 2023.

[277] Claus Weiland. *Mechanics of Flow Similarities.* Springer, 2020.

[278] John Weiss. The dynamics of enstrophy transfer in two-dimensional hydrodynamics. *Physica D: Nonlinear Phenomena*, 48(2-3):273–294, 1991.

[279] Jeremy West, Dan Ventura, and Sean Warnick. Spring research presentation: A theoretical foundation for inductive transfer. *Brigham Young University, College of Physical and Mathematical Sciences*, 1(08), 2007.

[280] Jeffrey S Whitaker and Thomas M Hamill. Ensemble data assimilation without perturbed observations. *Monthly weather review*, 130(7):1913–1924, 2002.

[281] Paula A Whitlock and MH Kalos. *Monte Carlo Methods*, volume 1. Wiley Hoboken, NJ, USA, 1986.

[282] Kurt Wiesenfeld, David Pierson, Eleni Pantazelou, Chris Dames, and Frank Moss. Stochastic resonance on a circle. *Physical Review Letters*, 72(14):2125, 1994.

[283] Stephen Wiggins. *Introduction to applied nonlinear dynamical systems and chaos.* Springer, 2003.

[284] Alexander Wikner, Jaideep Pathak, Brian R Hunt, Istvan Szunyogh, Michelle Girvan, and Edward Ott. Using data assimilation to train a hybrid forecast system that combines machine-learning and knowledge-based components. *Chaos: An Interdisciplinary Journal of Nonlinear Science*, 31(5), 2021.

[285] Christopher Williams and Carl Rasmussen. Gaussian processes for regression. *Advances in neural information processing systems*, 8, 1995.

[286] Ernst Wit, Edwin van den Heuvel, and Jan-Willem Romeijn. 'all models are wrong...': an introduction to model uncertainty. *Statistica Neerlandica*, 66(3):217–236, 2012.

[287] Weng Kee Wong. Comparing robust properties of A, D, E and G-optimal designs. *Computational statistics & data analysis*, 18(4):441–448, 1994.

[288] Qin Xu. Measuring information content from observations for data assimilation: Relative entropy versus Shannon entropy difference. *Tellus A: Dynamic Meteorology and Oceanography*, 59(2):198–209, 2007.

[289] Liang Xue and Dongxiao Zhang. A multimodel data assimilation framework via the ensemble Kalman filter. *Water Resources Research*, 50(5):4197–4219, 2014.

[290] L Minah Yang and Ian Grooms. Machine learning techniques to construct patched analog ensembles for data assimilation. *Journal of Computational Physics*, 443:110532, 2021.

[291] Miin-Shen Yang, Chien-Yo Lai, and Chih-Ying Lin. A robust EM clustering algorithm for gaussian mixture models. *Pattern Recognition*, 45(11):3950–3961, 2012.

[292] Shouto Yonekura and Alexandros Beskos. Online smoothing for diffusion processes observed with noise. *Journal of Computational and Graphical Statistics*, 31(4):1344–1360, 2022.

[293] Zijian Zheng and Geoffrey I Webb. Lazy learning of Bayesian rules. *Machine learning*, 41(1):53–84, 2000.

Index

For Product Safety Concerns and Information please contact our EU
representative GPSR@taylorandfrancis.com
Taylor & Francis Verlag GmbH, Kaufingerstraße 24, 80331 München, Germany

www.ingramcontent.com/pod-product-compliance
Lightning Source LLC
LaVergne TN
LVHW061801280826
846697LV00031B/317